# Structure & Function of the Human Body

# Ruth L. Memmler, MD

Professor Emeritus
Life Sciences
formerly Coordinator
Health, Life Sciences, Nursing
East Los Angeles College
Los Angeles, California

•

# Barbara Janson Cohen, BA, MSEd.

Assistant Professor
Delaware County Community College
Media, Pennsylvania

•

# Dena Lin Wood, RN, MS

Staff Nurse
Verdugo Hills Visiting Nurse Association
Glendale, California

**Illustrated by Janice A. Schwegler, MS, CMI**
**and**
**Anthony Ravielli**

# Structure & Function of the Human Body

*Sixth Edition*

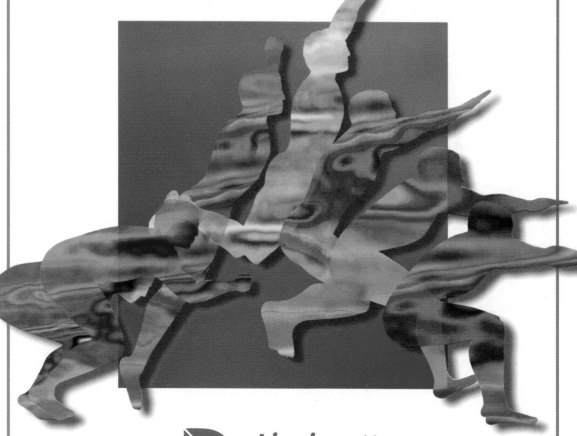

 **Lippincott**

*Philadelphia • New York*

*Acquisitions Editor:* Andrew Allen
*Editorial Assistant:* Laura Dover
*Project Editor:* Barbara Ryalls
*Production Manager/Managing Editor:* Helen Ewan
*Production Coordinator:* Nannette Winski
*Design Coordinator:* Doug Smock
*Indexer:* Alberta Morrison

6th Edition

Library of Congress Cataloging-in-Publication Data

Memmler, Ruth Lundeen.
    Structure and function of the human body / Ruth L. Memmler, Barbara Janson Cohen,
Dena Lin Wood; illustrated by Janice A. Schwegler and Anthony Ravielli. — 6th ed.
      p.    cm.
    Includes bibliographical references and index.
    ISBN 0-397-55172-X
    1. Human physiology.   2. Human anatomy.   3. Body, Human.   I. Cohen, Barbara J.
II. Wood, Dena Lin.   III. Title.
    [DNLM:  1. Anatomy.   2. Physiology.   QS 4 M533s 1996]
    QP36.M54   1996
    612—dc20
    DNLM/DLC                               95-38675
    for Library of Congress                      CIP

The material contained in this volume was submitted as previously unpublished material, except in the instances in which credit has been given to the source from which some of the illustrative material was derived.

Any procedure or practice described in this book should be applied by the health-care practitioner under appropriate supervision in accordance with professional standards of care used with regard to the unique circumstances that apply in each practice situation. Care has been taken to confirm the accuracy of information presented and to describe generally accepted practices. However, the authors, editors, and publisher cannot accept any responsibility for errors or omissions or for any consequences from application of the information in this book and make no warranty, express or implied, with respect to the contents of the book.

The authors and publisher have exerted every effort to ensure that drug selection and dosage set forth in this text are in accordance with current recommendations and practice at the time of publication. However, in view of ongoing research, changes in government regulations, and the constant flow of information relating to drug therapy and drug reactions, the reader is urged to check the package insert for each drug for any change in indications and dosage and for added warnings and precautions. This is particularly important when the recommended agent is a new or infrequently employed drug.

Materials appearing in this book prepared by individuals as part of their official duties as U.S. Government employees are not covered by the above-mentioned copyright.

9  8  7  6  5  4  3

# Preface

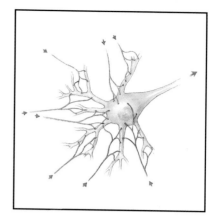

**Structure and Function of the Human Body, Sixth Edition**, traces the organization of the body from the single cell to the coordinated whole. A major theme is the interaction of all body systems for the maintenance of a stable internal state, a condition termed *homeostasis*. The concepts of human anatomy (structure) and physiology (function) are presented in a manner simple enough for beginning students.

## ▶ Organization

1. Chapters with a common theme are grouped into units. These include units on basic body structure, the systems involved in movement and support, coordination and control, circulation, obtaining and using energy, and reproduction.
2. In this edition, the previous chapter on reproduction has been divided into two chapters. One describes the male and female reproductive systems and the second combines information on development and birth with a discussion of heredity. This balances the size of the chapters and gives greater flexibility in planning the curriculum.

## ▶ Aids to Learning

1. Each chapter begins with Behavioral Objectives that state the learning goals for the chapter.
2. Selected Key Terms are listed at the start of each chapter.
3. New terms are emphasized in the text with ***boldface italic*** type. Each new term is followed by a phonetic pronunciation.
4. Abundant illustrations are included to clarify and expand on the text. In many chapters information is summarized in charts to organize the material and aid in review.
5. Each chapter is followed by a Summary outline for review.
6. Questions for Study and Review based on the text appear at the end of each chapter.
7. There is a complete Glossary with pronunciations.
8. Appendices summarize information from the text and give additional related material.

9. A section on Biologic Terminology shows how scientific words are built from prefixes, roots, and suffixes. Learning these word parts will help the student remember terms and understand the meaning of new terms.

## New with This Edition

1. More information on physiology (body functions) has been included. This appears in chapters on the muscular system, nervous system, urinary system, and others. Descriptions of changes that occur in the body systems with age also have been added.
2. The number of illustrations has been increased considerably. In addition, many of the original illustrations have been redrawn for greater clarity. Photographs have been added, including photographs of tissues seen under the microscope.
3. Each chapter includes a special interest box which expands on the text or adds related information. Each box is signified with a logo:

## Supplements

### 1. Study Guide
The study guide that accompanies **Structure and Function of the Human Body** has been revised and expanded. New illustrations have been included for labeling. New forms of questions have been added, including short essay questions.

### 2. Instructor's Manual
The Instructor's Manual includes:
—A test file with questions similar to those in the text and workbook. This test file will also be provided in computerized form as described below.
—Answers to the Workbook exercises
—Suggestions for teaching and for additional studies

### 3. Computerized Test Bank
With this edition, a Computerized Test Bank is available which allows instructors to select, modify, or add questions as required.

### 4. Color Transparencies
A set of color transparencies of key illustrations in the text is available for classroom instruction.

## Guide to Pronunciation

The stressed syllable in each word is shown with capital letters. The vowel pronunciations are as follows:

Any vowel at the end of a syllable is given as a long sound:

a as in say
e as in be
i as in nice
o as in go
u as in true

A vowel followed by a consonant and the letter e (as in *rate*) also is given a long pronunciation.

Any vowel followed by a consonant receives a short pronunciation:

a as in absent
e as in end
i as in bin
o as in not
u as in up

In some cases, the letter h is added to a syllable to make a short vowel sound.

# Acknowledgments

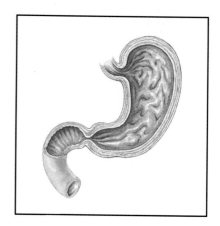

The authors are grateful to all those who helped in the preparation of this sixth edition of *Structure and Function of the Human Body*. Major changes have been made in the art program. We thank Janice Schwegler, the artist for this edition, for her talent, creativity, and patience. For supervising the art program and designing the book we owe many thanks to Susan Blaker. We also wish to thank Dana Morse Bittus for giving generously of her time and expertise in preparing the photomicrographs.

We are indebted to those who reviewed the text, especially Margaret M. Nowack, who evaluated the entire manuscript. Above all, we thank the many instructors nationwide who have sent reviews in response to user surveys. We value their opinions and use their suggestions in striving to improve this text.

At Lippincott-Raven Publishers we thank our editor, Andrew Allen, who sees each edition through from start to finish. We extend our appreciation for the great effort, support, and attention to detail shown by the Editorial Assistant, Laura Dover. Finally, we thank the Project Editor for this edition, Barbara Ryalls.

# Contents

# Unit V
## Energy: Supply and Use . . . . . . . . . . . . . . . . . 213

# UNIT VI
## Perpetuating Life . . . . . . . . . . . . . . . . . 271

# Special Interest Boxes

Topics of interest are highlighted in boxes within each chapter.

# UNIT 1

# The Body as a Whole

This unit presents the basic levels of organization within the human body. Included are brief descriptions of the systems, which are made of groups of organs. Organs are made of tissues, which in turn are made of the smallest units, called *cells*. A short survey of chemistry, which deals with the composition of all matter and is important for the understanding of human physiology, is incorporated in this unit. The final chapter in the unit illustrates some of these basic principles with a detailed study of the skin and its associated structures, which together make up the integumentary system.

# CHAPTER 1

# Introduction to the Human Body

## Selected Key Terms

The following terms are defined in the Glossary:

anabolism

anatomy

ATP

catabolism

cell

gram

homeostasis

liter

metabolism

meter

negative feedback

organ

physiology

system

tissue

## Behavioral Objectives

After careful study of this chapter, you should be able to:

- Define the terms *anatomy* and *physiology*
- Describe the organization of the body from cells to the whole organism
- List 10 body systems and give the general function of each
- Define *metabolism* and name the two phases of metabolism
- Differentiate between extracellular and intracellular fluids
- Briefly explain the role of ATP in the body
- Define *homeostasis* and *negative feedback*
- List and define the main directional terms for the body
- List and define the three planes of division of the body
- Name the subdivisions of the dorsal and ventral cavities
- Name the basic metric units of length, weight, and volume
- Name and locate the nine regions of the abdomen
- Define the metric prefixes *kilo-, centi-, milli-,* and *micro-*

Memmler, RL, Cohen, BJ, Wood, DL. *STRUCTURE AND FUNCTION OF THE HUMAN BODY*, 6/e,
© 1996 Lippincott-Raven Publishers

# ▶ Studies of the Human Body

Everyone is interested from the earliest age in the body and how it works. Picture an infant carefully examining its fingers and toes and exploring their uses. The scientific term for the study of body structure is *anatomy* (ah-NAT-o-me). Part of this word means "to cut," because one of the many ways to learn about the human body is to cut it apart, or *dissect* it. *Physiology* (fiz-e-OL-o-je) is the term for the study of how the body functions, and the two sciences are closely related. It is hoped that from these studies you will gain an appreciation for the design and balance of the human body and for living organisms in general.

All living things are organized from very simple levels to more complex levels. Living matter begins with simple chemicals. These chemicals are formed into the complex substances that make living *cells*—the basic units of all life. Specialized groups of cells form *tissues,* and tissues may function together as *organs.* Organs functioning together for the same general purpose make up organ *systems,* which maintain the body.

# ▶ Body Systems

The body systems have been listed as 9, 10, or 11 in number, depending on how much detail one wishes to include. Here is one such list:

1. The *skeletal system.* The basic framework of the body is a system of over 200 bones and the joints between them, collectively known as the *skeleton.*
2. The *muscular system.* Body movements result from the action of the skeletal muscles, which are attached to the bones. Other types of muscles are present in the walls of body organs such as the intestine and the heart.
3. The *circulatory system.* The heart and blood vessels make up the system that pumps blood to all the body tissues, bringing with it nutrients, oxygen, and other substances and carrying away waste materials. Lymphatic vessels play an important auxiliary role in circulation.
4. The *digestive system.* This system comprises all the organs that are involved with taking in food and converting it into substances that body cells can use. Examples of these organs are

the mouth, esophagus, stomach, intestine, liver, and pancreas.

5. The *respiratory system.* This system includes the lungs and the passages leading to the lungs. The purpose of this system is to take in air and conduct it to the areas designed for gas exchange. Oxygen passes from the air into the blood and is carried to all tissues by the circulatory system. In like manner, carbon dioxide, as a gaseous waste product, is taken from the tissues to be expelled from the body.
6. The *integumentary* (in-teg-u-MEN-tar-e) *system.* The word *integument* (in-TEG-u-ment) means skin. The skin with its appendages is considered to be a separate body system. The structures associated with the skin include the hair, the nails, and the sweat and oil glands.
7. The *urinary system.* This system is also called the *excretory system.* Its main components are the kidneys, the ureters, the bladder, and the urethra. Its chief purpose is to rid the body of waste products and excess water. (Note that other waste products are removed by the digestive and respiratory systems and by the skin).
8. The *nervous system.* The brain, the spinal cord, and the nerves make up this very complex system by which the body is controlled and coordinated. The organs of special sense (such as the eyes, ears, taste buds, and organs of smell), together with the receptors for touch and other senses, receive stimuli from the outside world. These stimuli are converted into impulses that are transmitted to the brain. The brain directs the body's responses to these messages and also to messages coming from within the body. Such higher functions as memory and reasoning also occur in the brain.
9. The *endocrine* (EN-do-krin) *system.* The scattered organs known as *endocrine glands* are grouped together because they share a similar function. All produce special substances called *hormones,* which regulate such body activities as growth, food utilization within the cells, and reproduction. Examples of endocrine glands are the thyroid and the pituitary glands.
10. The *reproductive system.* This system includes the external sex organs and all related internal structures that are concerned with the production of offspring.

## ▶ Body Processes

All the life-sustaining reactions that go on within the body systems together make up *metabolism* (meh-TAB-o-lizm). Metabolism can be divided into two types of activities:

1. In *catabolism* (kah-TAB-o-lizm), complex substances, including the nutrients from digested foods, are broken down into simpler compounds with the release of energy.

2. In *anabolism* (ah-NAB-o-lizm), simple compounds are used to manufacture complex materials needed for growth, function, and repair of tissues.

The energy obtained from the breakdown of nutrients is used to form a compound often described

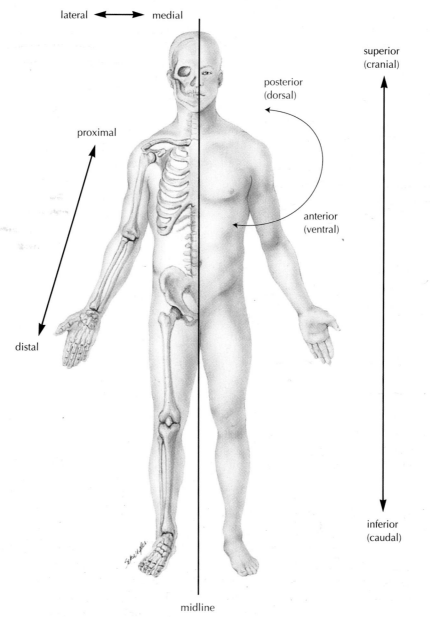

**Figure 1-1**

Directional terms.

as the "energy currency" of the cell. It has the long name of *adenosine triphosphate* (ah-DEN-o-sene tri-FOS-fate), but is commonly abbreviated *ATP.* Chapter 18 has more information on metabolism and ATP.

It is important to recognize that our bodies are composed of large amounts of fluids. Certain kinds of fluids bathe the cells, carry nutrient substances to and from the cells, and transport the nutrients into and out of the cells. This group of fluids is called *extracellular fluid* because it includes all body fluids outside the cells. A second type of fluid, *intracellular fluid*, is contained within the cells. Extracellular and intracellular fluids account for approximately 60% of an adult's weight.

### Homeostasis

The purpose of metabolism is to maintain a state of balance within the organism, an important characteristic of all living things. Such conditions as body temperature, the composition of body fluids, heart rate, respiration rate, and blood pressure must be kept within set limits to maintain health. This steady state within the organism is called *homeostasis* (ho-me-o-STA-sis), which literally means "staying (stasis) the same (homeo)."

Homeostasis is maintained by the feedback of information. Sensors throughout the body monitor internal conditions and bring them back to normal when they shift, much as a thermostat regulates the temperature of a house to a set level. Since each change, up or down, must be reversed to restore the norm, this mechanism is described as *negative feedback.* The main controllers in this self-regulation are the nervous and endocrine systems. You will find many examples of negative feedback throughout this book.

## ❚ Directions in the Body

Because it would be awkward and inaccurate to speak of bandaging the "southwest part" of the chest, a number of terms are used to designate position and directions in the body. They refer to the body in the *anatomic position*—upright with face front, arms at the sides with palms forward, and feet parallel. The main directional terms are as follows (Fig. 1-1):

1. *Superior* is a term meaning above, or in a higher position. Its opposite, *inferior,* means below, or lower. The heart, for example, is superior to the intestine.

2. *Ventral* and *anterior* mean the same thing in humans: located toward the belly surface or front of the body. Their corresponding opposites, *dorsal* and *posterior,* refer to locations nearer the back.

3. *Cranial* means nearer to the head; *caudal,* nearer to the sacral region of the spinal column (*i.e.,* where the tail is located in lower animals).

4. *Medial* means nearer to an imaginary plane that passes through the midline of the body, dividing it into left and right portions. *Lateral,* its opposite, means farther away from the midline, toward the side.

5. *Proximal* means nearer the origin of a structure; *distal,* farther from that point. For example, the part of your thumb where it joins your hand is its proximal region; the tip of the thumb is its distal region.

## A Map of the Body

The body's surface landmarks provide a good introduction to body structure. These landmarks are useful for orientation in surgery and other clinical procedures. For example, the lower tip of the sternum, the xiphoid process, is used as a reference point in the administration of CPR (cardiopulminary resuscitation).

Feel for some of the other superficial markings to learn how your body is organized. The joint between the mandible and the temporal bone of the skull (the temporomandibular, or TM, joint) can be felt in front of the ear canal as you move your lower jaw up and down. The pulse of the facial artery can be felt in a small depression about 1 inch in front of the angle of the lower jaw. The pulse of the carotid artery can be felt at the side of the neck.

Touch behind your neck for the spine of the 7th cervical vertebra, the prominent vertebra at the base of the neck. At the distal end of the forearm , you can feel the pulse of the radial artery at the lateral surface of the anterior wrist.

In the torso, you can feel the ribs and the lower margin of the rib cage. By placing your hands on your hips, you can feel the crest of the hip bone. When it hurts to sit on a hard chair, you become aware of the lower bone of the pelvic girdle on each side.

Many other features can be seen and felt at the surface to tell us about the structure and the condition of the body.

## Planes of Division

For convenience in visualizing the spatial relations of various body structures to each other, anatomists have divided the body into three imaginary planes, each of which is a cut through the body in a different direction (Fig. 1-2).

1. The **sagittal** (SAJ-ih-tal) **plane.** If you were to cut the body in two from front to back, separating it into right and left portions, the sections you would see would be sagittal sections. A cut exactly down the midline of the body, separating it into equal right and left halves, is a **midsagittal** section.

2. The **frontal plane.** If the cut were made in line with the ears and then down the middle of the body, creating a front and a rear portion, you would see a front (anterior or ventral) section and a rear (posterior or dorsal) section. Another name for this plane is *coronal plane.*

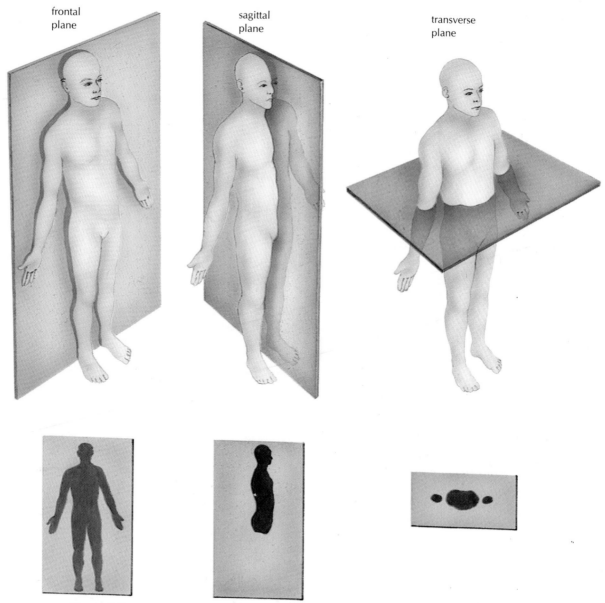

**Figure 1-2**

Planes of division.

3. The ***transverse plane.*** If the cut were made horizontally, across the other two planes, it would divide the body into an upper (superior) part and a lower (inferior) part. There could be many such cross sections, each of which would be on a transverse plane, also called a *horizontal plane.*

## ❱ Body Cavities

The body contains a few large internal spaces, or ***cavities,*** within which various organs are located. There are two groups of cavities: ***dorsal*** and ***ventral*** (Fig. 1-3).

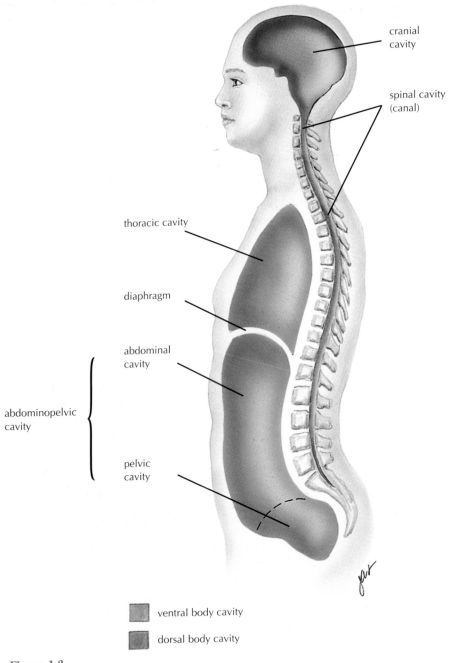

cranial cavity

spinal cavity (canal)

thoracic cavity

diaphragm

abdominal cavity

abdominopelvic cavity

pelvic cavity

⬛ ventral body cavity

▨ dorsal body cavity

**Figure 1-3**

Side view of the body cavities.

### Dorsal Cavity

The dorsal body cavity has two regions: the *cranial cavity,* containing the brain, and the *spinal canal,* enclosing the spinal cord. These two areas form one continuous space.

### Ventral Cavity

The ventral cavity is much larger than the dorsal cavity. Unlike the dorsal cavity, it is not continuous, but is separated by the *diaphragm* (DI-ah-fram), a muscle used in breathing. The *thoracic* (tho-RAS-ik) *cavity* is located above the diaphragm. Its contents include the heart, the lungs, and the large blood vessels that join the heart. The heart is contained in the pericardial sac; the lungs are in the pleural sacs (Fig. 1-4). The *abdominopelvic* (ab-dom-ih-no-PEL-vik) *cavity* is located below the diaphragm. This space is subdivided into two regions. The uppermost *abdominal* cavity contains the stomach, most of the intestine, the kidneys, the liver, the gallbladder, the pancreas, and the spleen. The lower portion, set off by an imaginary line across the top of the hip bones, is the *pelvic cavity.* This cavity contains the urinary bladder, the rectum, and the internal parts of the reproductive system.

### REGIONS OF THE ABDOMEN

Because the abdomen is so large, it is helpful to divide it for reference into nine regions (see Fig. 1-4). The three central regions are the *epigastric* (ep-ih-GAS-trik) *region,* located just below the breastbone; the *umbilical* (um-BIL-ih-kal) *region* around the umbilicus (um-BIL-ih-kus), commonly called the *navel;* and the *hypogastric* (hi-po-GAS-trik) *region,* the lowest of all the midline regions. At each side are the right and left *hypochondriac* (hi-po-KON-dre-ak) *regions,* just below the ribs; then the right and left *lumbar regions;* and finally the right and left *iliac,* or *inguinal* (IN-gwih-nal), *regions.* A simpler but less precise division into four quadrants is sometimes used. These are the right upper quadrant (RUQ), left upper quadrant (LUQ), right lower quadrant (RLQ), and left lower quadrant (LLQ) (Fig. 1-5).

**Figure 1-4**

Front view of the thoracic cavity and the nine regions of the abdomen.

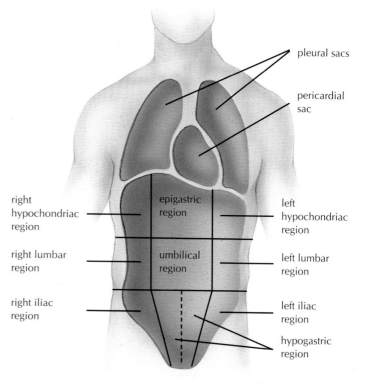

pleural sacs

pericardial sac

right hypochondriac region

epigastric region

left hypochondriac region

right lumbar region

umbilical region

left lumbar region

right iliac region

left iliac region

hypogastric region

# ❱ **The Metric System**

Now that we have set the stage for further study of the body and its structure and processes, a look at the metric system is in order, since this is the system used for all scientific measurements. To use the metric system easily and correctly may require a bit of effort, as is often the case with any new idea. Actually, you already know something about the metric system because our monetary system is similar to it in that both are decimal systems. One hundred cents equal one dollar; one hundred centimeters equal one meter. The decimal system is based on multiples of the number 10, which are indicated by prefixes:

kilo = 1000
centi = 1/100
milli = 1/1000
micro = 1/1,000,000

    The basic unit of length in the metric system is the *meter*. Thus, 1 kilometer is equal to 1000

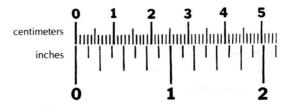

**Figure 1-6**

Comparison of centimeters and inches.

meters. A centimeter is 1/100 of a meter; stated another way, there are 100 centimeters in 1 meter. A changeover to the metric system has been slow in coming to the United States. Often measurements on packages, bottles, and yard goods are now given according to both scales. In this text, equivalents in the more familiar units of inches, feet, and so forth are included along with the metric units for comparison. There are 2.5 centimeters (cm) or 25 millimeters (mm) in 1 inch, as shown in Figure 1-6. Some equivalents that may help you to appreciate the size of various body parts are as follows:

1 mm = 0.04 inch, or 1 inch = 25 mm
1 cm = 0.4 inch, or 1 inch = 2.5 cm
1 m = 3.3 feet, or 1 foot = 30 cm

    The same prefixes used for linear measurements are used for weights and volumes. The *gram* is the standard unit of weight. Thirty grams are approximately equal to 1 ounce, and 1 kilogram to 2.2 pounds. Drug dosages are usually stated in grams or milligrams. One thousand milligrams equal 1 gram; a 500-milligram (mg) dose would be the equivalent of 0.5 gram (g), and 250 mg are equal to 0.25 g.

    Liquids are measured in units of volume. The standard metric measurement for volume is the *liter* (LE-ter). There are 1000 milliliters (mL) in a liter. A liter is slightly greater than a quart, a liter being equal to 1.06 quarts. For smaller quantities, the milliliter is used most of the time. There are 5 mL in a teaspoon and 15 mL in a tablespoon. A fluid ounce contains 30 mL.

    The Celsius (centigrade) temperature scale, now in use by most countries and by scientists in this country, is discussed in Chapter 18. A chart of all the common metric measurements and their equivalents is shown in Appendix 1. A Celsius–Fahrenheit temperature conversion scale appears in Appendix 2.

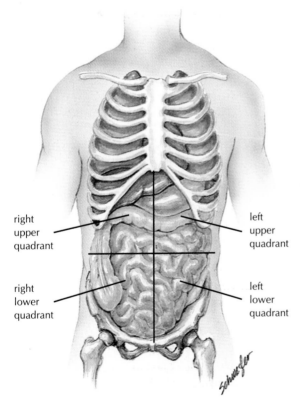

right
upper
quadrant

left
upper
quadrant

right
lower
quadrant

left
lower
quadrant

**Figure 1-5**

Quadrants of the abdomen showing the organs within each quadrant.

## SUMMARY

I. **Studies of the human body**
  A. Anatomy—study of structure
  B. Physiology—study of function

II. **Organization**
  A. Levels—cell, tissue, organ, organ system
  B. Body systems
    1. Skeletal system—support
    2. Muscular system—movement
    3. Circulatory system (includes lymphatic system)—transport
    4. Digestive system—intake and breakdown of food
    5. Respiratory system—intake of oxygen and release of carbon dioxide
    6. Integumentary system—skin and associated structures (appendages).
    7. Urinary system—elimination of waste and water
    8. Nervous system—receipt of stimuli and control of responses
    9. Endocrine system—production of hormones for regulation of growth, metabolism, reproduction
    10. Reproductive system—production of offspring

III. **Body processes**
  A. Metabolism—all the chemical reactions needed to sustain life
    1. Catabolism—breakdown of complex substances into simpler substances
    2. Anabolism—building of body materials
  B. ATP (adenosine triphosphate)—energy compound of cells
  C. Body fluids
    1. Extracellular—outside the cells
    2. Intracellular—inside the cells
  D. Homeostasis—steady state of body conditions; maintained by negative feedback

IV. **Body positions, directions, planes**
  A. Anatomic position—upright, palms forward
  B. Directions
    1. Superior—above or higher; inferior—below or lower
    2. Ventral (anterior)—toward belly or front surface; dorsal (posterior)—nearer to back surface
    3. Cranial—nearer to head; caudal—nearer to sacrum
    4. Medial—toward midline; lateral—toward side
    5. Proximal—nearer to point of origin; distal—farther from point of origin
  C. Planes of division
    1. Sagittal—from front to back, dividing the body into left and right parts
      a. Midsagittal—exactly down the midline
    2. Frontal (coronal)—from left to right, dividing the body into anterior and posterior parts
    3. Transverse—horizontally, dividing the body into superior and inferior parts

V. **Body cavities**
  A. Dorsal—contains cranial and spinal cavities for brain and spinal cord
  B. Ventral
    1. Thoracic—chest cavity containing heart and lungs and divided from abdominal cavity by diaphragm
    2. Abdominopelvic
      a. Abdominal—upper region containing stomach, most of intestine, kidneys, liver, spleen, etc.
      b. Pelvic—lower region containing reproductive organs, urinary bladder, rectum
      c. Abdomen divided for reference into nine regions: epigastric, umbilical, hypogastric, right and left hypochondriac, right and left lumbar, right and left iliac (inguinal)

VI. **Metric system**—based on multiples of 10
  A. Basic units
    1. Meter—length
    2. Liter—volume
    3. Gram—weight
  B. Prefixes—indicate multiples of 10
    1. Kilo—1000 times
    2. Centi—1/100th (0.01)
    3. Milli—1/1000th (0.001)
    4. Micro—1/1,000,000 (0.000001)
  C. Temperature—measured in Celsius (centigrade) scale

## QUESTIONS FOR STUDY AND REVIEW

1. List two types of study of the human body and define each.
2. Define *cell, tissue, organ,* and *system.*
3. List 10 body systems and briefly describe the function of each.
4. Name and define the two phases of *metabolism.*
5. What is ATP?
6. Define the term *homeostasis* and give two examples. How is homeostasis maintained?
7. Stand in the anatomic position.
8. List the opposite term for each of the following body directions: *superior, ventral, anterior, cranial, medial,* and *proximal.* Define each term and its opposite.
9. Describe the location of the elbow in relation to the wrist; in relation to the shoulder.
10. Describe the location of the stomach in relation to the urinary bladder; in relation to the lungs.
11. Describe the location of the ears in relation to the nose.
12. What are the three main body planes? Explain the division each makes.
13. Name the cavities within the dorsal and ventral cavities. Name one organ found in each.
14. Make a rough sketch of the abdomen and label the nine divisions.
15. Name and locate the four quadrants of the abdomen.
16. Why should you learn the metric system? What are its advantages?

# CHAPTER 2

# Chemistry, Matter, and Life

## Selected Key Terms

The following terms are defined in the Glossary:

acid

atom

base

buffer

carbohydrate

compound

electrolyte

electron

element

enzyme

ion

lipid

molecule

neutron

pH

protein

proton

solution

suspension

## Behavioral Objectives

After careful study of this chapter, you should be able to:

- Describe the structure of an atom
- Differentiate between atoms and molecules
- Define the atomic number of an atom
- Differentiate between elements and compounds and give several examples of each
- Define *radioactivity*
- Explain why water is so important to the body
- Define *mixture*; list the three types of mixtures and give two examples of each
- Differentiate between ionic and covalent bonds
- Define the terms *acid, base,* and *salt*
- Explain how the numbers on the pH scale relate to acidity and alkalinity
- Define *buffer* and explain why buffers are important in the body
- List three characteristics of organic compounds
- Name the three main types of organic compounds and the building blocks of each
- Define *enzyme*; describe how enzymes work

Memmler, RL, Cohen, BJ, Wood, DL. *STRUCTURE AND FUNCTION OF THE HUMAN BODY*, 6/e,
© 1996 Lippincott-Raven Publishers

# What Is Chemistry?

Great strides toward an understanding of living organisms, including the human being, have come to us through *chemistry,* the science that deals with the composition of matter. Knowledge of chemistry and chemical changes helps us understand the functioning of the body and its parts. The digestion of food in the intestinal tract, the production of urine by the kidneys, the regularity of breathing—all body processes—are based on chemical principles. Foods, vitamins, and minerals are chemicals, as are the various components of our bodies and the environment. To provide some understanding of the importance of chemistry in the life sciences, this chapter briefly describes *atoms* and *molecules, elements, compounds,* and *mixtures,* which are fundamental forms of matter.

# A Look at Atoms

Atoms are small particles that form the building blocks of matter, the smallest complete units of which all matter is made. To visualize the size of an atom, one can think of placing millions of them on the sharpened end of a pencil and still have room for many more. Everything about us, everything we can see and touch, is made of atoms—plants, the atmosphere, the water in the ocean, the smoke coming out of a chimney.

Despite the fact that the atom is such a tiny particle, it has been carefully studied and has been found to have a definite structure. At the center of the atom is a nucleus, which contains positively charged electric particles called *protons* and non-charged particles called *neutrons.* Outside the nucleus, in regions called *orbitals,* are negatively charged particles called *electrons* (Fig. 2-1).

The protons and electrons always are equal in number in any atom. Collectively, they are responsible for all the atom's characteristics. The neutrons and protons are tightly bound in the nucleus, contributing nearly all the atom's weight. The positively charged protons keep the negatively charged electrons in the orbital area around the nucleus by means of the opposite charges the particles possess. Positively (+) charged protons attract negatively (−) charged electrons. The electrons contribute important chemical characteristics to the atom.

Most atoms have several orbitals of electrons. However, each orbital can hold only two electrons. The orbitals are arranged into energy levels. Energy levels are identified by their distance from the nucleus of the atom. The first energy level, the one closest to the nucleus, is composed of one orbital. The second energy level, the next in distance away from the nucleus, can have four orbitals. Since each orbital contains two electrons, the second energy level has the capacity to hold eight electrons. The electrons farthest away from the nucleus are the particles that give the atom its chemical characteristics.

If the outermost energy level has more than four electrons but fewer than its capacity of eight, the atom normally completes this level by gaining electrons. Such an atom is called a *nonmetal.* The oxygen atom illustrated in Figure 2-1 has six electrons in its second, or outermost, level. When oxygen enters into chemical reactions, its chemical behavior is to gain two electrons. The oxygen atom then has two more electrons than protons. If the outermost shell has fewer than four electrons, the atom normally loses those electrons to attain a complete outer energy level. Such an atom is called a *metal.*

# Elements, Molecules, and Compounds

Atoms are the fundamental units that make up the chemical *elements* from which all matter is made. An element is a substance that cannot be decomposed—that is, changed into something else—by physical means (*e.g.,* by the use of heat, pressure, or electricity). Examples of elements include various gases, such as hydrogen, oxygen, and nitrogen; liquids, such as the mercury used in thermometers and blood pressure instruments; and many solids, such as iron, aluminum, gold, silver, and carbon. Graphite (the so-called "lead" in a pencil), coal, charcoal, and diamonds are examples of the element carbon. The entire universe is made up of about 105 elements.

Elements can be identified by their names, their symbols, or their atomic numbers. Chemists use symbols as a shorthand method to name elements. The atomic number of an element is equal to the number of protons that are present in the nucleus. Because the number of protons is equal to the number of electrons in an atom, the atomic number also

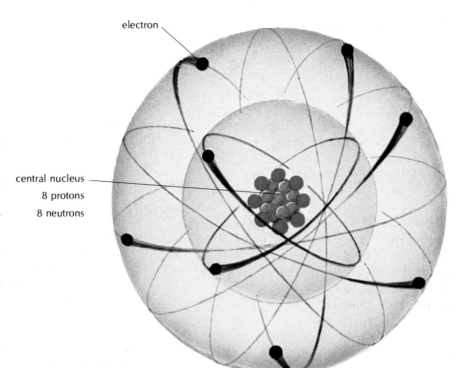

electron

central nucleus
8 protons
8 neutrons

**Figure 2-1**

Representation of the oxygen atom. Eight protons and eight neutrons are tightly bound in the central nucleus, around which the eight electrons revolve.

identifies the number of electrons whirling about the nucleus. Table 2-1 lists some elements found in the human body along with their functions.

Atoms of an element may exist in several forms, called *isotopes.* These forms are alike in their chemical reactions but different in weight (*e.g.,* heavy oxygen and regular oxygen). The greater weight of the heavier isotopes is due to the presence of one or more extra neutrons in the nucleus. Some isotopes are stable and maintain a constant character. Others disintegrate (fall apart) as they give off small particles and rays; these are said to be ***radioactive.*** Radioactive elements may occur naturally, as is the case with such very heavy isotopes as radium and uranium. Others may be produced artificially from nonradioactive elements such as iodine and gold by bombardment (smashing) of the atoms in special machines.

When, on the basis of electron structure, two or more atoms unite, a *molecule* is formed. A molecule can be made of like atoms—the oxygen molecule is made of two identical atoms—but more often it is made of two or more different atoms. For example, a molecule of water ($H_2O$) contains 1 atom of oxygen (O) and 2 atoms of hydrogen (H) (Fig. 2-2). Substances that contain molecules formed by the union of two or more different atoms are called ***compounds.*** Compounds may be made of a few elements

in a simple combination; for example, the gas carbon monoxide (CO) contains 1 atom of carbon (C) and 1 atom of oxygen (O). Alternatively, compounds may be very complex; some proteins, for example, have thousands of atoms.

It is interesting to observe how different a compound is from any of its constituents. For example, a molecule of liquid water is formed from oxygen and hydrogen, both of which are gases. Another example is a crystal sugar, glucose ($C_6H_{12}O_6$). Its constituents include 12 atoms of the gas hydrogen, 6 atoms of the gas oxygen, and 6 atoms of the solid element carbon. In this case, the component gases and the solid carbon do not in any way resemble the glucose.

### More About Water

Water is the most abundant compound in the body. No plant or animal, including the human, can live very long without it. Water is of critical importance in all physiologic processes in body tissues. Water carries substances to and from the cells and makes possible the essential processes of absorption, exchange, secretion, and excretion. What are some of the properties of water that make it such an ideal medium for living cells?

## TABLE 2-1
## Some Common Chemical Elements*

| Name | Symbol | Function |
|---|---|---|
| Oxygen | O | Part of water; needed to metabolize nutrients for energy |
| Carbon | C | Basis of all organic compounds; in carbon dioxide, the  waste gas of metabolism |
| Hydrogen | H | Part of water; participates in energy metabolism, acid–base balance |
| Nitrogen | N | Present in all proteins, ATP (the energy compound), and nucleic acids (DNA and RNA) |
| Calcium | Ca | Builds bones and teeth; needed for muscle contraction, nerve impulse conduction, and blood clotting |
| Phosphorus | P | Active ingredient in the energy-storing compound ATP; builds bones and teeth; in cell membrane and nucleic acids |
| Potassium | K | Nerve impulse conduction; muscle contraction; water balance and acid–base balance |
| Sulfur | S | Part of many proteins |
| Sodium | Na | Water balance; nerve impulse conduction; muscle contraction |
| Chlorine | Cl | Water balance; acid–base balance; in stomach acid |
| Iron | Fe | Part of hemoglobin, the compound that carries oxygen in red blood cells |

*The elements are listed in decreasing order by weight in the body.

1. Water can dissolve many different substances in large amounts. For this reason it is called the *universal solvent*. All the materials needed by the body, such as gases, minerals, and nutrients, dissolve in water to be carried from place to place.
2. Water is very stable as a liquid at ordinary temperatures. Water does not freeze until the temperature drops to 0°C (32°F) and does not boil until the temperature reaches 100°C (212°F).

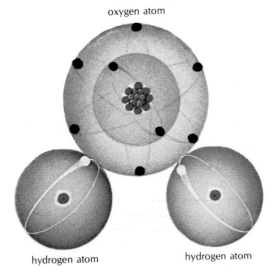

oxygen atom

hydrogen atom          hydrogen atom

Figure 2-2

Molecule of water.

This stability provides a constant environment for body cells. Water can also be used to distribute heat throughout the body and to cool the body by evaporation of sweat from the body surface.
3. Water participates in chemical reactions in the body. It is needed directly in the process of digestion and in many of the metabolic reactions that occur in the cells.

### Mixtures, Solutions, and Suspensions

Not all elements or compounds combine chemically when brought together. The air we breathe every day is a mixture of gases, largely nitrogen, oxygen, and carbon dioxide, along with smaller percentages of other substances. The constituents in the air maintain their identity, although the proportions of each may vary. Blood plasma is also a mixture in which the various components maintain their identity. The many valuable compounds in the plasma remain separate entities with their own properties. Such combinations are called *mixtures*—blends of two or more substances. A mixture, such as salt water, in which the component substances remain evenly distributed is called a *solution*. The dissolving substance, in this case water, is the *solvent;* the substance dissolved, in this case salt, is the *solute.*

In some mixtures, the material distributed in the solvent settles out unless the mixture is constantly shaken; this type of mixture is called a *suspension.* Settling occurs in a suspension because the particles in the mixture are large and heavy. Examples of suspensions are milk of magnesia, india ink, and, in the body, red blood cells suspended in blood plasma.

One other type of mixture is of importance in the body. In a *colloidal suspension,* the particles do not dissolve but remain distributed in the solvent because they are so small. The fluid that fills the cells (cytoplasm) is a colloidal suspension, as is blood plasma.

# ⬤ Chemical Bonds

### Ionic Bonds and Electrolytes

When discussing the structure of the atom, we mentioned the positively charged (+) protons that are located in the nucleus, with the corresponding number of negatively charged (−) electrons found in the surrounding space and neutralizing the protons. If a single electron were removed from the sodium atom, it would leave one proton not neutralized, and the atom would have a positive charge ($Na^+$). This actually happens during a chemical change. Similarly, atoms can gain electrons so that there are more electrons than protons. Chlorine, which has seven electrons in its outermost energy level, tends to gain one electron to fill the level to its capacity. Such an atom of chlorine is negatively charged ($Cl^-$) (Fig. 2-3). An atom with a positive or negative charge is called an *ion.* An ion that is positively charged is a *cation,* whereas a negatively charged ion is an *anion.*

Let us imagine a sodium atom coming in contact with a chlorine atom. The chlorine atom gains an electron from the sodium atom, and the two newly formed ions ($Na^+$ and $Cl^-$), because of their opposite charges, which attract each other, cling together and produce the compound sodium chloride, ordinary table salt. A bond formed by this method of electron transfer is called an *ionic bond* (Fig. 2-4).

In the fluids and cells of the body, ions make it possible for materials to be altered, broken down, and recombined to form new substances. Calcium ions ($Ca^{++}$) are necessary for the clotting of blood, the contraction of muscle, and the health of bone tissue. Bicarbonate ions ($HCO_3^-$) are required for the regulation of acidity and alkalinity of the tissues. The stable condition of the normal organism, homeostasis, is influenced by ions.

Compounds that form ions whenever they are in solution are called *electrolytes* (e-LEK-tro-lites). Elec-

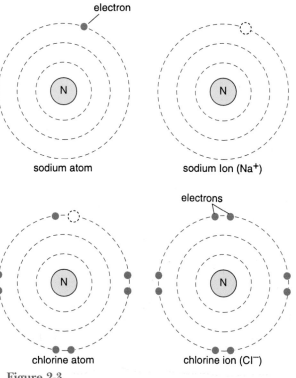

**Figure 2-3**

Formation of $Na^+$ cation and $Cl^-$ anion. Only the electrons in the outermost energy level are shown.

trolytes are responsible for the acidity and the alkalinity of solutions. They also include a variety of mineral salts, such as sodium and potassium chloride. Electrolytes must be present in exactly the right quantities in the fluid within the cell (intracellular) and outside the cell (extracellular), or there will be very damaging effects, preventing the cells in the body from functioning properly.

Because ions are charged particles, electrolyte solutions can conduct an electric current. (In practice, the term *electrolytes* is also used to refer to the ions themselves in body fluids.) Records of electric currents in tissues are valuable indications of the functioning or malfunctioning of tissues and organs. The *electrocardiogram* (e-lek-tro-KAR-de-o-gram) and the *electroencephalogram* (e-lek-tro-en-SEF-ah-lo-gram) are graphic tracings of the electric currents generated by the heart muscle and the brain, respectively (see Chaps. 9 and 13).

### Covalent Bonds

Although many chemical compounds are formed by ionic bonds, a much larger number are

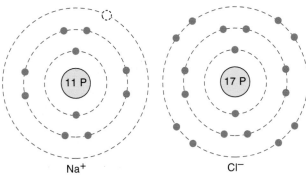

Na$^+$                    Cl$^-$

**Figure 2-4**

A sodium ion with 11 protons in the nucleus and 10 electrons in orbitals is attracted to a chlorine ion with 17 protons in the nucleus and 18 electrons in orbitals to form the compound sodium chloride.

formed by another type of chemical bond. This type of bond involves not the exchange of electrons but a sharing of electrons between the atoms in the molecule. The electrons orbit around both of the atoms in the bond to make both of them stable. The electrons may be equally shared, as in the case of a hydrogen molecule ($H_2$), or they may be held closer to one atom than the other, as in the case of water ($H_2O$), shown in Figure 2-2. These bonds are called *covalent bonds.* Covalently bonded compounds do not conduct an electric current in solution. Carbon, the element that is the basis of organic chemistry, forms covalent bonds. Thus, the compounds that are characteristic of living things are covalently bonded compounds. These bonds may involve the sharing of one, two, or three pairs of electrons between atoms.

## ▶ Acids, Bases, and Buffers

An *acid* is a chemical substance capable of donating a hydrogen ion ($H^+$) to another substance. A common example is hydrochloric acid, the acid found in stomach juices:

| HCl | → | H$^+$ | + | Cl$^-$ |
|---|---|---|---|---|
| hydrochloric acid | | hydrogen ion | | chlorine ion |

A *base* is a chemical substance, usually containing a hydroxide ion ($OH^-$), that can accept a hydrogen ion. Sodium hydroxide, which releases hydroxide ion in solution, is an example of a base:

| NaOH | → | Na$^+$ | + | OH$^-$ |
|---|---|---|---|---|
| sodium hydroxide | | sodium ion | | hydroxide ion |

A reaction between an acid and a base produces a *salt,* such as sodium chloride:

$$HCl + NaOH \rightarrow NaCl + H_2O$$

The greater the concentration of hydrogen ions in a solution, the greater is the acidity of that solution. As the concentration of hydrogen ions becomes less than it is in pure water, the more alkaline (basic) the solution becomes. Acidity is indicated by pH units, which represent the concentration of hydrogen ions in a solution. These units are listed on a scale from 0 to 14, with 0 being the most acidic and 14 being the most basic (Fig. 2-5).

A pH of 7.0 is neutral, having an equal number of hydrogen and hydroxide ions. Each pH unit on the scale represents a 10-fold change in the number of hydrogen and hydroxide ions present. A solution registering 5.0 on the scale has 10 times the number of hydrogen ions as a solution that registers 6.0. A solution registering 9.0 has one tenth the number of hydrogen ions and 10 times the number of hydroxide ions as one registering 8.0. Thus, the lower the pH rating, the greater is the acidity, and the higher the pH, the greater is the alkalinity. Blood is only slightly alkaline, with a pH range of 7.35 to 7.45, or nearly neutral.

A delicate balance exists in the acidity or alkalinity of body fluids. If a person is to remain healthy, these chemical characteristics must remain within narrow limits. The substances responsible for the balanced chemical state are the acids, bases, and buffers. *Buffers* form a chemical system that prevents sharp changes in hydrogen ion concentration and thus maintains a relatively constant pH. Buffers are very important in maintaining the stability of body fluids.

## ▶ The Chemistry of Living Matter

Of the 100 or more elements that have been found in nature, only a relatively small number, mostly

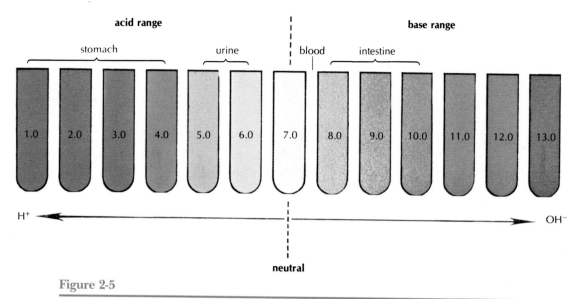

Figure 2-5

The pH scale measures degree of acidity or alkalinity.

the lighter ones, are important components of living cells. Hydrogen, oxygen, carbon, and nitrogen are the elements that make up about 99% of the cells. Calcium, sodium, potassium, phosphorus, sulfur, chlorine, and magnesium are the seven elements that make up most of the remaining 1% of tissue elements. A number of others are present in trace amounts, such as iron, copper, iodine, and fluorine.

The chemical compounds that characterize living things are called *organic compounds.* All of these contain the element *carbon.* Since carbon can combine with a variety of different elements and can even bond to other carbon atoms to form long chains, most organic compounds consist of large, complex molecules. The starch found in potatoes, the fat in the tissue under the skin, and many drugs are examples of organic compounds. These large molecules are often formed from simpler molecules called *building blocks,* which bond together in long chains. The main types of organic compounds are carbohydrates, fats, and proteins. All three contain carbon, hydrogen, and oxygen as their main ingredients.

### Organic Compounds

Carbohydrates are the simple sugars, called *monosaccharides* (mon-o-SAK-ah-rides), or molecules made from simple sugars linked together. Carbohy-

drates in the form of sugars and starches are important sources of energy in the diet. Examples of carbohydrates in the body are the glucose that circulates in the blood as food for the cells and a storage form of glucose called *glycogen* (GLI-ko-jen).

## Trace Elements

Trace elements are required in very small amounts by the body but are absolutely essential for health. Many are used in enzymes and vitamins.

Iodine is needed for the manufacture of thyroid hormones. Selenium helps to prevent damage to cells caused by the chemical process of oxidation. A deficiency of selenium leads to liver and muscular disorders. Growth is impaired if the body lacks chromium. This element works with insulin, the hormone that helps the cells to take up glucose from the blood. Copper is needed for the proper use of iron, and cobalt is a part of vitamin $B_{12}$ needed for blood cell formation. Zinc is used in energy metabolism and for the transport of carbon dioxide, the waste gas of metabolism.

Other elements that are needed in trace amounts include silicon, fluorine, tin, and manganese. More information on trace elements may be discovered as nutrition studies continue.

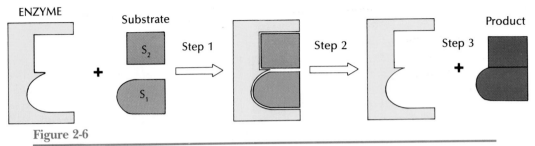

Figure 2-6

Diagram of enzyme action. The enzyme combines with substance 1 ($S_1$) and substance 2 ($S_2$). When a new product is formed, the enzyme is released unchanged.

Fats, or *lipids,* are also stored for energy in the body. In addition, they provide insulation and protection for the body organs. Fats are made from a substance called *glycerol* (glycerine) in combination with fatty acids. A group of complex lipids containing phosphorus, called *phospholipids* (fos-fo-LIP-ids), is important in the body. Among other things, phospholipids make up a major part of the membrane around living cells.

All proteins contain, in addition to carbon, hydrogen, and oxygen, the element *nitrogen.* They may also contain sulfur and phosphorus. Proteins are composed of building blocks called *amino* (ah-ME-no) *acids.* Although there are only about twenty different amino acids found in the body, a vast number of proteins can be made by linking them together in different combinations. Proteins are the structural materials of the body, found in muscle, bone, and connective tissue. They also make up the pigments that give hair, eyes, and skin their color. It is protein that makes each individual physically distinct from others. All three categories of organic compounds, in addition to minerals and vitamins, must be taken in as part of a normal diet. These compounds will be discussed further in Chapters 17 and 18.

### Enzymes

An important group of proteins is the *enzymes.* Enzymes function as *catalysts* in the hundreds of reactions that occur in metabolism. A catalyst is a substance that speeds up the rate of a reaction but is not changed or used up in that reaction. Because they are constantly reused, only very small amounts of enzymes are needed. Each chemical reaction in the body requires a specific enzyme, so enzymes really control the metabolism of the cells. Some of the vitamins and minerals needed in the diet are important because they make up parts of enzymes.

In its action, the shape of the enzyme is important. It must match the shape of the substance or substances with which the enzyme combines in much the same way as a key fits a lock. This so-called "lock and key" mechanism is illustrated in Figure 2-6.

Harsh conditions, such as extremes of temperature or pH, can alter the shape of an enzyme and stop its action. Such an event is always harmful to the cells.

You can usually recognize the names of enzymes because, with few exceptions, they end with the suffix *-ase.* Examples are amylase, lipase, and oxidase. The first part of the name usually refers to the substance acted on or the type of reaction involved.

## SUMMARY

I. **Atoms**—basic units of matter that make up the various elements on earth
   A. Structure
      1. Protons—positively charged particles in the nucleus
      2. Neutrons—neutral particles in the nucleus
      3. Electrons—negatively charged particles in energy levels around the nucleus
   B. Radioactivity
II. **Molecules**—combinations of two or more atoms
   A. Compounds—combinations of different atoms (*e.g.,* water)

III. **Mixtures**—blends of two or more substances
    A. Solution—substance (solute) remains evenly distributed in solvent (*e.g.*, salt in water)
    B. Suspension—material settles out of mixture on standing (*e.g.*, red cells in blood plasma)
    C. Colloidal suspension—particles do not dissolve but remain suspended (*e.g.*, cytoplasm)

IV. **Chemical bonds**
    A. Ionic bonds—atoms exchange electrons to become stable
    B. Covalent bonds—atoms share electrons to become stable

V. **Types of compounds**
    A. Acid—donates hydrogen ions
    B. Base—accepts hydrogen ions
    C. Salt—formed by reaction between acid and base

VI. **pH**
    A. Measure of acidity or alkalinity of a solution
    B. Scale goes from 0 to 14; 7 is neutral
    C. Buffer—maintains constant pH of a solution

VII. **Organic compounds**—all contain carbon
    A. Examples
        1. Carbohydrates
        2. Lipids (*e.g.*, fats)
        3. Proteins (*e.g.*, enzymes–organic catalysts)

## QUESTIONS FOR STUDY AND REVIEW

1. Define *chemistry* and tell something about what is included in this study.
2. Define an atom and describe the basic structure of an atom.
3. Define *molecule, element, compound,* and *mixture.*
4. What is an element and what are some examples of elements?
5. What is meant by *radioactivity*?
6. Why is water so important to life?
7. How does a mixture differ from a compound? Give two examples of each.
8. You dissolve a teaspoon of sugar in a cup of tea. Which is the solute? the solvent?
9. What are ions and how are they related to electrolytes? What are some examples of ions and of what importance are they in the body?
10. Explain how covalent bonds are formed. Give two examples of covalently bonded compounds.
11. What would a pH reading of 6.5 indicate about a solution? 8.5? 7.0?
12. What are organic compounds and what element is found in all of them?
13. What elements are found in the largest amounts in living cells?
14. What are the building blocks of carbohydrates? fats? proteins?
15. What are enzymes? What type of organic compound are they?
16. Why is the shape of an enzyme important to its action?

# CHAPTER 3

# Cells and Their Functions

## Selected Key Terms

The following terms are defined in the Glossary:

**active transport**

**cell membrane**

**centriole**

**chromosome**

**cytoplasm**

**diffusion**

**DNA**

**interphase**

**isotonic**

**micrometer**

**mitochondria**

**mitosis**

**mutation**

**nucleotide**

**nucleus**

**organelle**

**osmosis**

**phagocytosis**

**ribosome**

**RNA**

## Behavioral Objectives

After careful study of this chapter, you should be able to:

- List three types of microscopes used to study cells
- Describe the function and composition of cytoplasm
- Name and describe the main organelles in the cell
- Give the composition, location, and function of DNA in the cell
- Give the composition, location, and function of RNA in the cell
- Explain briefly how cells make proteins
- Name and briefly describe the stages in cell division
- Define six methods by which substances enter and leave cells
- Explain what will happen if cells are placed in solutions with the same or different concentrations than the cell fluids

Memmler, RL, Cohen, BJ, Wood, DL. *STRUCTURE AND FUNCTION OF THE HUMAN BODY*, 6/e, © 1996 Lippincott-Raven Publishers

The cell is the basic unit of all life. It is the simplest structure that shows all the characteristics of life: growth, metabolism, responsiveness, reproduction, and homeostasis. In fact, it is possible for a single cell to live independently of other cells. Examples of such independent cells are microscopic organisms such as bacteria and protozoa. All the activities of the human body, which is composed of millions of cells, result from the activities of individual cells. Any materials produced within the body are produced by cells.

The scientific study of cells began some 350 years ago with the invention by Anton van Leeuwenhoek of the microscope. In time, this single-lens microscope was replaced by the modern compound light microscope, which has two sets of lenses. This is the type in use in most laboratories. A great boon to cell biologists has been the development of the **transmission electron microscope,** which by a combination of magnification and enlargement of the resulting image affords magnification to 1 million times or more (Fig. 3-1). Another type, the **scanning electron microscope,** does not magnify as much ($\times 250,000$) but gives a three-dimensional picture of an object.

Before they are examined under the microscope, cells and tissues usually are colored with special dyes called **stains** to aid in viewing. These stains produce the variety of colors you see when looking at pictures of cells and tissues taken under a microscope. The metric unit used for microscopic measurements is the **micrometer** (MI-kro-me-ter). This unit is 1/1000 of a millimeter and is symbolized with the Greek letter for m as μm.

## Structure of the Cell

Just as people may look different but still have certain features in common—two eyes, a nose, and a mouth, for example—all cells share certain characteristics. A "typical" animal cell is shown in Fig. 3-2.

The outer covering of the cell is the **cell membrane,** also called the *plasma membrane* (Fig. 3-3). This membrane is composed of two layers of lipid molecules in which float a variety of different proteins and a small amount of carbohydrates. The proteins may act as receptors, that is, points of attachment for materials coming to the cell in the blood; they may also act as carriers, shuttling materials into or out of the cell. Some form channels through which selected substances can pass. The cell membrane is

very important in regulating what can enter and leave the cell.

The main substance that fills the cell and holds the cell contents is the **cytoplasm** (SI-to-plazm). This is a suspension of nutrients, minerals, enzymes, and other specialized materials in water. Although the composition of the cytoplasm has been analyzed, no one has been able to produce it in a laboratory; there must be something about the organization of these substances that has not yet been discovered.

## The Organelles

Just as the body has different organs to carry out special functions, the cell contains specialized subdivisions that perform different tasks. These structures are called **organelles,** which means "little organs." Each of these will be examined in turn (see Fig. 3-2).

The largest organelle is the **nucleus** (NU-kle-us). This is often called the *control center* of the cell because it contains the genetic material, which governs all the activities of the cell, including cell reproduction. Within the nucleus is a smaller globule called the **nucleolus** (nu-KLE-o-lus), which means "little nucleus." Its functions are not entirely understood, but it is believed to act in the manufacture of proteins within the cell. The actual formation of proteins occurs on small bodies called **ribosomes** (RI-bo-somes). These are attached to a network of membranes throughout the cell called the **endoplasmic reticulum** (en-do-PLAS-mik re-TIK-u-lum). The name literally means "network" (reticulum) "within the cytoplasm" (endoplasmic), but for ease it is almost always called simply the *ER.*

Fairly large and important organelles are the **mitochondria** (mi-to-KON-dre-ah). These are round or bean-shaped structures with folded membranes on the inside. Within the mitochondria, food is converted to energy for the cell in the form of ATP. These are the "power plants" of the cell. Active cells have large numbers of mitochondria. Other organelles in a typical cell include the **centrioles,** rod-shaped bodies near the nucleus that function in cell division; **lysosomes** (LI-so-somes), which contain digestive enzymes; and the **Golgi** (GOL-je) **apparatus,** which formulates special substances, such as mucus, released from cells. All these cell structures are summarized in Table 3-1 for easy study.

*(Text continues on page 27)*

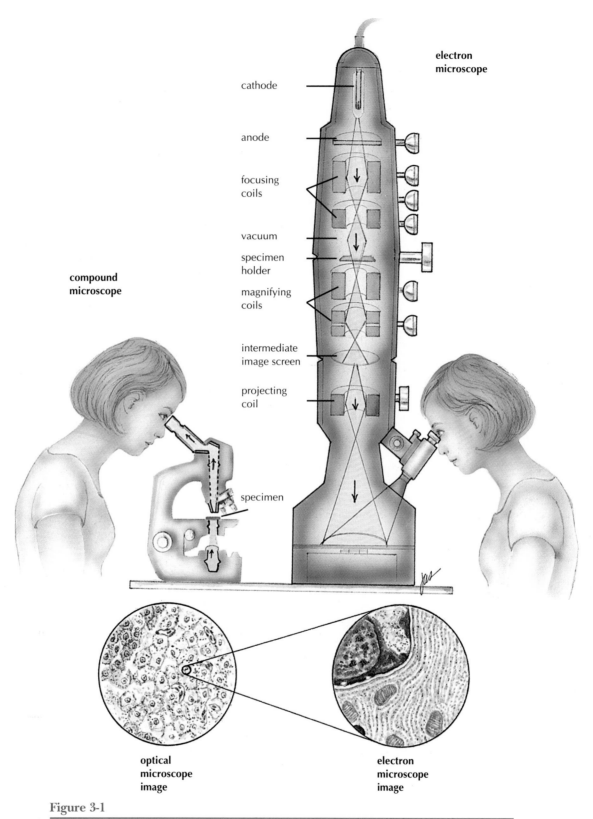

**electron microscope**

cathode

anode

focusing coils

vacuum

specimen holder

magnifying coils

intermediate image screen

projecting coil

**compound microscope**

specimen

optical microscope image

electron microscope image

**Figure 3-1**

A simplified comparison of an optical (compound) microscope and an electron microscope.

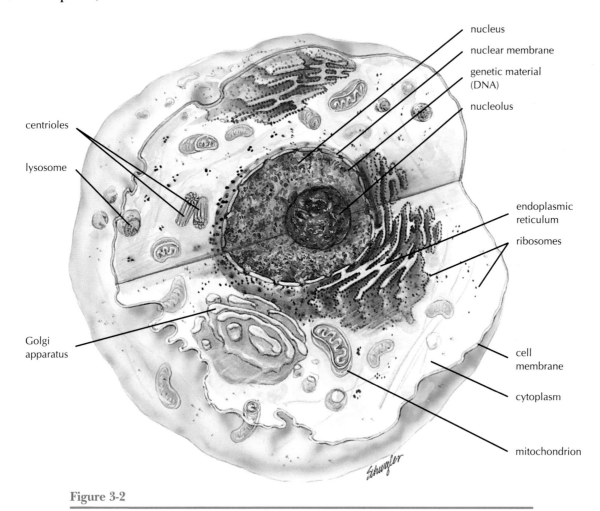

**Figure 3-2**

Diagram of a typical animal cell showing the main organelles.

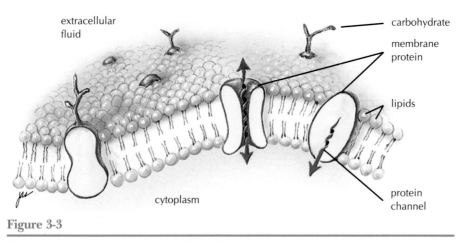

**Figure 3-3**

Current concept of the structure of the cell membrane.

## Cell Junctions

Some cells, such as blood cells, are not in direct contact with other cells. Most cells, however, are held in contact with other cells. What keeps cells together? First, the cell membranes contain complex substances that are sticky and tend to bind cells to each other. Second, there are special intercellular junctions, which may vary according to the functions of the cells.

A *tight junction* is formed when the membranes of adjacent cells actually become fused. This type of junction is found in the lining of the digestive tract and acts to keep digestive juices and other harmful substances from damaging the organs. One substance that can seep through the tight junctions of the stomach and damage the deeper tissues is alcohol.

Connections can form between adjacent cells where tissues are under great stress, for example in the skin. Protein filaments extend from the cells, bridging the space between the membranes and anchoring cells to each other. These junctions are called *desmosomes* (DES-mo-somz).

Finally, *gap junctions* allow materials to pass between cells. Small tubes made of protein connect the cells and act as channels for exchange of chemicals. These are found where cells need to communicate easily, as in the nervous system.

Although the basic structure of all body cells is the same, individual cells may vary widely in size, shape, and composition according to the function of each. In size they may range from the 7 µm of a red blood cell to the 200 µm or more of a muscle cell. A neuron with its long fibers is very different in appearance from a muscle cell or from a transparent cell in the clear lens of the eye. Most human cells have all the organelles described, but these may vary in number. Some cells have small, hairlike projections from the surface called *cilia* (SIL-e-ah), which wave to create movement around the cell. Examples are the cells that line the passageways of the respiratory and reproductive tracts. A long, whiplike extension from the cell is a *flagellum* (flah-JEL-lum). Each human sperm cell has a flagellum that is used for locomotion. Thus, each cell is specialized for its particular function.

## ▶ Cell Functions

### Protein Synthesis

Since protein molecules play an indispensable part in the body's activities, we need to identify the cellular substances that direct the production of protein. In the cytoplasm and in the nucleus are chemicals called *nucleic* (nu-KLE-ik) *acids.* The two nucleic acids important in protein production are *deoxyribonucleic* (de-ok-se-RI-bo-nu-kle-ik) *acid,* or *DNA,* and *ribonucleic* (RI-bo-nu-kle-ik) *acid,* abbreviated *RNA.* Both of these are large, complex molecules composed of subunits called *nucleotides* (NU-kle-o-tides) (Fig. 3-4). There are four nucleotides in DNA and four in RNA, but only three of these are common to both.

### DNA

DNA molecules are found mostly in the nucleus of the cell, where they make up the **chromosomes** (KRO-mo-somes), dark-staining, threadlike bodies. Looking at Figure 3-4, you can see that DNA exists as a double strand. The two strands are matched according to the pattern of the nucleotides. The adenine (A) nucleotide always pairs with the thymine (T) nucleotide; the cytosine (C) nucleotide always pairs with the guanine (G) nucleotide. The two strands of DNA are held together by weak bonds. The doubled strands are then coiled into a spiral, giving DNA the descriptive name of the *double helix.*

Specific regions of the DNA in the chromosomes make up the **genes,** the hereditary factors of each cell. It is the genes that carry the messages for the development of particular inherited characteristics, such as brown eyes or curly hair. Their message is actually contained in the pattern of the four nucleotides in the DNA. These code for the amino acids in the various cellular proteins. DNA is considered the master blueprint of the cell, and, because it contains the DNA, the nucleus is considered the cell's control center.

### THE ROLE OF RNA

A blueprint is only a map. The directions it illustrates must be translated into appropriate actions. RNA is the substance needed for this step. RNA is much like DNA except that it is a single strand and has one nucleotide that is different from those in DNA. By pairing of nucleotides, RNA picks up the message of the DNA, which has by now broken its weak bonds and uncoiled into single strands. This

## TABLE 3-1
## Cell Structures

| Name | Description | Function |
| --- | --- | --- |
| Cell membrane | Outer layer of the cell; composed mainly of lipids and proteins | Limits the cell; regulates what enters and leaves the cell |
| Cytoplasm | Colloidal suspension that fills cell | Holds cell contents |
| Nucleus | Large, dark-staining body near the center of the cell; composed of DNA and proteins | Contains the chromosomes with the genes (the hereditary material that directs all cell activities) |
| Nucleolus | Small body in the nucleus; composed of RNA, DNA, and protein | Needed for protein manufacture |
| Endoplasmic reticulum (ER) | Network of membranes in the cytoplasm | Used for storage and transport; holds ribosomes |
| Ribosomes | Small bodies attached to the ER; composed of RNA and protein | Manufacture proteins |
| Mitochondria | Large organelles with folded membranes inside | Convert energy from nutrients into ATP |
| Golgi apparatus | Layers of membranes | Put together special substances such as mucus |
| Lysosomes | Small sacs of digestive enzymes | Digest substances within the cell |
| Centrioles | Rod-shaped bodies (usually 2) near the nucleus | Help separate the chromosomes in cell division |
| Cilia | Short, hairlike projections from the cell | Create movement around the cell |
| Flagellum | Long, whiplike extension from the cell | Moves the cell |

*messenger RNA* travels into the cytoplasm to the ribosomes attached to the ER. Ribosomes are the organelles responsible for protein manufacture. They are so named because they are composed mainly of RNA. At the ribosomes the genetic message is decoded to produce chains of amino acids, the building blocks of proteins.

### Cell Division

To ensure that every cell in the body has similar genetic information, cell reproduction occurs by a very precise division process (Fig. 3-5). In this process of *mitosis* (mi-TO-sis), each original parent cell divides to form two identical daughter cells. Before mitosis can occur, however, the genetic information in the parent cell must be doubled. This duplication occurs during the period between mitoses that is termed *interphase.* While DNA is uncoiled from its double-stranded form, each strand duplicates itself according to the pattern of the nucleotides.

Mitosis itself is described as occurring in four stages, during which distinct changes can be seen in the dividing cell. In *prophase* (PRO-faze), the doubled strands return to their spiral organization and become visible under the microscope as dark, threadlike chromosomes. Meanwhile, in the cytoplasm, the two centrioles move to opposite ends of the cell, trailing very thin threadlike substances that form a structure resembling a spindle stretched across the cell. In *metaphase* (MET-ah-faze), the chromosomes line up across the threadlike spindle. Then, in *anaphase* (AN-ah-faze), the duplicated chromosomes separate and begin to move toward opposite ends of the cell. As mitosis continues into *telophase* (TEL-o-faze), the nuclear area becomes pinched in the middle until two nuclear regions have formed. A similar change occurs in the cell membrane, making the cell resemble a dumbbell. The midsection between the two halves of the dumbbell becomes progressively smaller until, finally, the cell splits in two. There are now two identical daughter cells, which are themselves identical

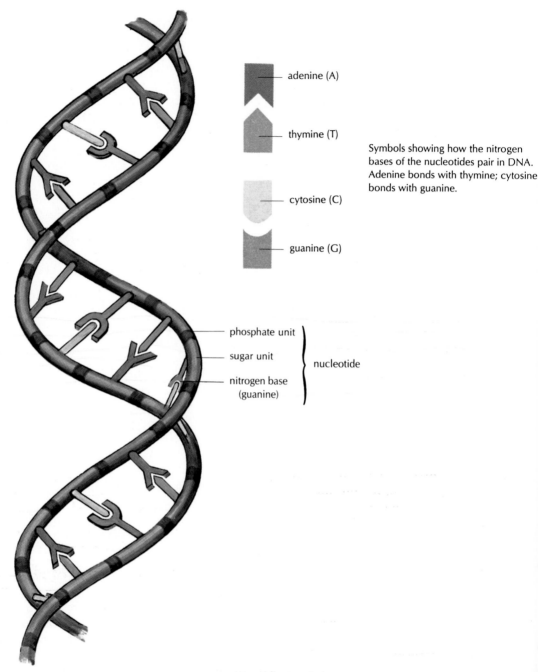

adenine (A)

thymine (T)

Symbols showing how the nitrogen bases of the nucleotides pair in DNA. Adenine bonds with thymine; cytosine bonds with guanine.

cytosine (C)

guanine (G)

phosphate unit

sugar unit

nitrogen base (guanine)

} nucleotide

**Figure 3-4**

Schematic representation of the basic structure of a DNA molecule. Each structural unit, or nucleotide, consists of a phosphorus-containing unit and a sugar unit to which is attached a nitrogen base. There are four different nucleotides. Their arrangement "spells out" the genetic instructions that control all activities of the cell. The symbols in the upper right show how the nucleotides pair in DNA.

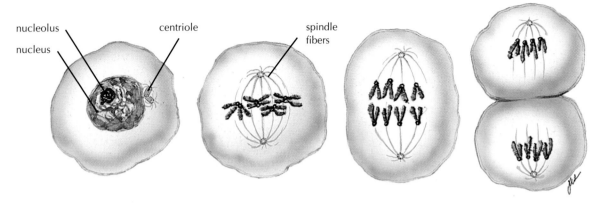

nucleolus          centriole                    spindle
nucleus                                         fibers

**Figure 3-5**

The stages of mitosis. The cell shown is for illustration only. It is not a human cell, which has 46 chromosomes.

to, but smaller than, the parent cell. The two new cells will grow and mature and continue to carry out life functions.

During mitosis, all the organelles, except those needed for the division process, temporarily disappear. After the cell splits, these organelles reappear in each daughter cell. Also at this time, the centrioles usually duplicate in preparation for the next cell division.

Body cells differ in the rate at which they reproduce. Some, such as nerve cells and muscle cells, stop dividing at some point in development and are not replaced if they die. They remain in interphase. Others, such as skin cells, multiply rapidly to replace cells destroyed by injury, disease, or natural wear-and-tear. As a person ages, characteristic changes in the overall activity of his or her body cells take place. One example of these is the slowing down of repair processes. A bone fracture, for example, takes considerably longer to heal in an aged person than in a young one.

### Movement of Substances Across the Cell Membrane

To function, cells require nutrients. How to receive this nourishment would seem to be a problem, since cells are surrounded by a protective membrane. However, if a cell is bathed in a liquid containing dissolved nutrient materials, an interesting thing happens: the liquid with the dissolved nutrient particles passes through the cell membrane. Not only do these molecules pass in, but waste products pass out of the cell in the opposite direction, enabling the cell to perform the function of elimination. The membrane also keeps valuable proteins and other substances from leaving the cell and prevents the admission of undesirable substances. For this reason, the cell membrane is classified as a ***semipermeable*** (sem-e-PER-me-ah-bl) membrane. It is permeable or passable to some molecules but impassable to others.

Water, a tiny molecule, is always able to penetrate the membrane with ease. In contrast, other molecules cannot pass through because they are too large, even though they are readily soluble. It is by the process of digestion that large molecules are split into smaller molecules so that they can travel through the membrane. For example, sucrose (table sugar) is converted to glucose, a smaller molecule that is a major source of cellular energy.

Various physical processes are responsible for exchanges through cell membranes. Some of these processes are:

1. ***Diffusion,*** the constant movement of molecules from a region of relatively higher concentration to one of lower concentration. Molecules, especially those in solution, tend to spread throughout an area until they are equally concentrated in all parts of a container (Figs. 3-6 and 3-7). When substances diffuse through a membrane, such as the intact cell membrane, passage is limited to those particles small enough to pass through spaces in the membrane. A large-scale example is shown in Figure 3-8.

**Figure 3-6**

Diffusion of gaseous molecules throughout a given space. When a solution vaporizes, the molecules tend to spread throughout the area. Common examples are the diffusion of perfumes, oils, and cleaning solutions.

2. ***Osmosis.*** Of all substances, water moves most rapidly through the cell membrane. The term *osmosis* means specifically the diffusion of water through a semipermeable membrane. The water molecules move, as expected, from an area where they are in higher number to an area where they are in lower number. That is, they move from a more dilute solution into a more concentrated solution. The tendency of a solution to pull water into it is called the ***osmotic pressure*** of the solution.

**Figure 3-7**

Diffusion of a solid in a liquid. Here, a solid diffuses into water. The molecules of the solid tend to spread evenly throughout the water.

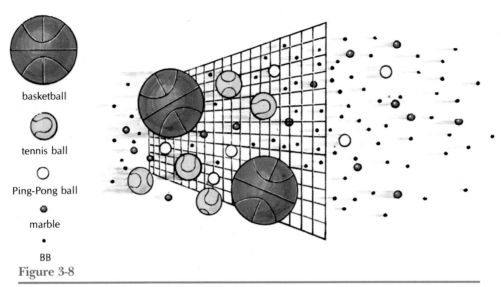

basketball

tennis ball

Ping-Pong ball

marble

BB

**Figure 3-8**

Diffusion through a semipermeable membrane. In this example, large objects (basketballs and tennis balls) cannot pass through the net. In the human body, large particles (proteins and blood cells) do not pass through intact membranes, such as the walls of the blood capillaries.

This force is directly related to concentration: the higher the concentration of a solution, the greater is its tendency to pull water into it.

3. *Filtration,* the passage of water containing dissolved materials through a membrane as a result of a mechanical ("pushing") force on one side (Fig. 3-9). An example of filtration in the human body is the formation of urine in the microscopic functional units of the kidney (see Chap. 19).

The three processes just described do not require cellular energy. They depend on the natural energy of the molecules for movement. The next three processes to be described do require cellular energy.

4. *Active transport.* Often molecules move into or out of a living cell in a direction opposite to the way they would normally flow by diffusion. That is, they move from an area where they are in relatively lower concentration to an area where they are in higher concentration. This movement, because it is against the natural flow, requires energy in the form of ATP. It also requires proteins in the cell membrane that act as *carriers* for the molecules. This process, called *active transport,* is a function of the living cell membrane. It allows the cell to take in what it needs from the surrounding fluids and to release materials from the cell. Because the cell membrane can carry on active trans-

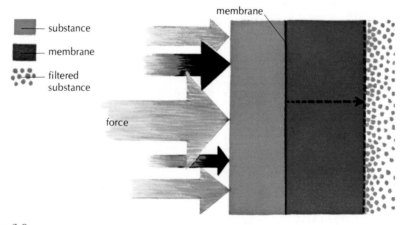

— substance

— membrane

— filtered substance

membrane

force

**Figure 3-9**

Filtration. A mechanical force pushes a substance through a membrane.

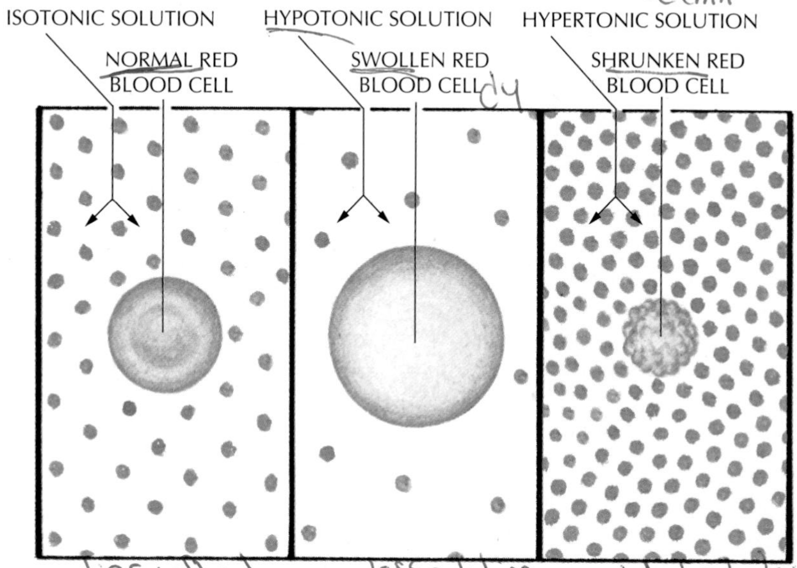

**Figure 3-10**

Osmosis. Water molecules moving through a red blood cell membrane in three different concentrations of fluid. *(Left)* The normal saline solution has a concentration nearly the same as that inside the cell, and water molecules move into and out of the cell at the same rate. *(Center)* The dilute solution causes the cell to swell and eventually hemolyze (burst) because of the large number of water molecules moving into the cell. *(Right)* The concentrated solution causes the water molecules to move out of the cell, leaving it shrunken.

ISOTONIC SOLUTION    HYPOTONIC SOLUTION    HYPERTONIC SOLUTION

NORMAL RED BLOOD CELL    SWOLLEN RED BLOOD CELL    SHRUNKEN RED BLOOD CELL

*[handwritten annotations: edema; tissue fluid blood plasma; less solution cell may burst; lot of solution more concentrated than the cell]*

port, it is most accurately described as **selectively permeable.** It regulates what can enter and leave the cell based on the needs of the cell.

5. **Phagocytosis** (fag-o-si-TO-sis) is the engulfing of relatively large particles by the cell membrane and the movement of these particles into the cell. Certain white blood cells carry out phagocytosis to rid the body of foreign material and dead cells.

6. In **pinocytosis** (pi-no-si-TO-sis), droplets of fluid are engulfed by the cell membrane. This is a way for large protein molecules in suspension to travel into the cell. The word *pinocytosis* means "cell drinking."

### HOW OSMOSIS AFFECTS CELLS

As stated earlier, water moves very easily through the cell membrane. Therefore, for a normal fluid balance to be maintained, all cells must be kept in solutions that have the same concentration of molecules as the fluids within the cell—the intracellular fluids (Fig. 3-10). Such equivalent solutions are described as **isotonic.** Tissue fluids and blood plasma are isotonic for body cells. Manufactured solutions that are isotonic for the cells and can thus be used to replace body fluids include 0.9% salt or **normal saline** and 5% dextrose (glucose).

A solution that is less concentrated than the intracellular fluid is described as **hypotonic.** A cell placed in a hypotonic solution draws water in, swells, and may burst. When this occurs to a red blood cell, the cell is said to **hemolyze** (HE-mo-lize). If a cell is placed in a **hypertonic** solution, which is more concentrated than the cell fluids, it loses water to the surrounding fluids and shrinks. Osmosis affects the total amount and distribution of body fluids, as discussed in Chapter 19.

## SUMMARY

### I. Microscopes
**A.** Types
  1. Compound light microscope
  2. Transmission electron microscope—magnifies up to 1 million times
  3. Scanning electron microscope—gives three-dimensional image

**B.** Stains—dyes used to aid in viewing cells under the microscope

**C.** Micrometer—metric unit commonly used for microscopic measurements

### II. Cell structure
**A.** Cell membrane—regulates what enters and leaves cell

**B.** Cytoplasm—colloidal suspension that holds organelles

**C.** Organelles—subdivisions that carry out special functions

1. Nucleus, nucleolus, ribosomes, centrioles
2. Mitochondria, ER, lysosomes, Golgi apparatus
3. Cilia, flagellum

**III. Cell functions**
   **A.** Protein synthesis—carried out by nucleic acids
      **1.** DNA
         **a.** Double strand of nucleotides
         **b.** Located in the nucleus
         **c.** Carries the genetic message
      **2.** RNA
         **a.** Single strand of nucleotides
         **b.** Located in the cytoplasm
         **c.** Translates DNA message into proteins
         **d.** Ribosomes—site of protein synthesis
   **B.** Cell division
      **1.** Duplication of chromosomes during interphase
      **2.** Mitosis—separation of chromosomes and division into two identical daughter cells
      **3.** Four stages—prophase, metaphase, anaphase, telophase
   **C.** Movement of substances across cell membrane
      **1.** Do not require cell energy
         **a.** Diffusion—molecules move from area of higher concentration to area of lower concentration
         **b.** Osmosis—diffusion of water through semipermeable membrane
         **c.** Filtration—movement of materials through cell membrane under mechanical force
      **2.** Require cell energy
         **a.** Active transport—movement of molecules from area of lower concentration to area of higher concentration
         **b.** Phagocytosis—engulfing of large particles by cell membrane
         **c.** Pinocytosis—intake of droplets of fluid
   **D.** Effect of osmosis on cells
      **1.** Isotonic solution—same concentration as cell fluids; cell remains the same
      **2.** Hypotonic solution—lower concentration than cell fluids; cell swells
      **3.** Hypertonic solution—higher concentration than cell fluids; cell shrinks

---

## QUESTIONS FOR STUDY AND REVIEW

1. Why is the study of cells so important in the study of the body?
2. Define the term *organelle*. List 10 organelles found in cells and give the function of each.
3. Compare DNA and RNA with respect to location in the cell and composition.
4. What are chromosomes? Where are they located in the cell and what is their function?
5. Explain the role of each of the following in protein synthesis: DNA, nucleotide, RNA, ribosomes.
6. Name the stage between one mitosis and the next.
7. Name the process of cell division. Name the stages of cell division and describe what happens during each.
8. Explain why it is necessary to reduce large molecules to smaller molecules by the process of digestion.
9. List and define six methods by which materials cross the cell membrane. Which of these requires cellular energy?
10. What substance moves most rapidly through the cell membrane?
11. Why is the cell membrane described as selectively permeable?
12. What is meant by the term *isotonic*? Name four isotonic solutions.
13. What will happen to a red blood cell placed in a 5.0% salt solution? in distilled water?

# CHAPTER 4

# Tissues, Glands, and Membranes

**Behavioral Objectives**

After careful study of this chapter, you should be able to:

- Name the four main groups of tissues and give the location and general characteristics of each

- Describe the difference between exocrine and endocrine glands and give examples of each

- Give examples of soft, fibrous, hard, and liquid connective tissues

- Name four types of membranes and give the location and functions of each

- Describe two types of epithelial membranes

Memmler, RL, Cohen, BJ, Wood, DL. *STRUCTURE AND FUNCTION OF THE HUMAN BODY, 6/e,*
© 1996 Lippincott-Raven Publishers

Tissues are groups of cells similar in structure, arranged in a characteristic pattern, and specialized for the performance of specific tasks. The study of tissues is known as ***histology*** (his-TOL-o-je). The tissues in our bodies might be compared with the different materials used to construct a building. Think for a moment of the great variety of materials used according to need—wood, stone, steel, plaster, insulation, and so forth. Each of these has different properties, but together they contribute to the building as a whole. The same may be said of tissues in the body.

## ▶ Tissue Classification

The four main groups of tissue are the following:

1. ***Epithelial*** (ep-ih-THE-le-al) ***tissue*** covers surfaces, lines cavities, and forms glands.
2. ***Connective tissue*** supports and forms the framework of all parts of the body.
3. ***Nervous tissue*** conducts nerve impulses.
4. ***Muscle tissue*** contracts and produces movement.

This chapter concentrates mainly on epithelial and connective tissues; the other types of tissues receive more attention in later chapters.

## ▶ Epithelial Tissue

Epithelial tissue, or ***epithelium*** (ep-ih-THE-le-um), forms a protective covering for the body and all the organs. It is the main tissue of the outer layer of the skin. It forms the lining of the intestinal tract, the respiratory and urinary passages, the blood vessels, the uterus, and other body cavities.

Epithelium has many forms and many purposes, and the cells of which it is composed vary accordingly. Epithelial tissue is classified according to the shape and the arrangement of its cells (Fig. 4-1). The cells may be

1. ***Squamous*** (SKWA-mus)—flat and irregular
2. ***Cuboidal***—square
3. ***Columnar***—long and narrow.

They may be arranged in a single layer, described as ***simple,*** or in many layers, termed ***stratified.*** Thus, a single layer of flat, irregular cells would be described as *simple squamous epithelium,* whereas tissue with many layers of these same cells would be described as *stratified squamous epithelium.*

The cells of some kinds of epithelium produce secretions, such as ***mucus*** (MU-kus) (a clear, sticky fluid), digestive juices, sweat, and other substances. The digestive tract is lined with a special kind of epithelium, the cells of which not only produce secretions but also are designed to absorb digested foods. The air that we breathe passes over yet another form of epithelium that lines the respiratory tract. This lining secretes mucus and is provided with tiny hairlike projections called ***cilia.*** Together, the mucus and the cilia help trap bits of dust and other foreign particles that could otherwise reach the lungs and damage them.

Some organs, such as the urinary bladder, must vary a great deal in size during the course of their work; for this purpose there is a special wrinkled, crepe-like tissue, called ***transitional epithelium,*** which is capable of great expansion yet will return to its original form once tension is relaxed—as when, in this case, the bladder is emptied. Certain areas of the epithelium that form the outer layer of the skin are capable of modifying themselves for greater strength whenever they are subjected to unusual wear and tear; the growth of calluses is a good example of this.

Epithelium repairs itself very quickly if it is injured. If, for example, there is a cut, the cells near and around the wound immediately form daughter cells, which grow until the cut is closed. Epithelial tissue reproduces frequently in areas of the body subject to normal wear and tear, such as the skin, the inside of the mouth, and the lining of the intestinal tract.

### Glands

The active cells of many glands are epithelial tissue. A gland is a group of cells specialized to produce a substance that is sent out to other parts of the body. These secretions are manufactured from blood constituents.

Glands are divided into two categories:

1. ***Exocrine*** (EK-so-krin) ***glands*** have ducts or tubes to carry the secretion from the gland to another organ, to a body cavity, or to the body surface. These ***external secretions*** are effective in a limited area near their origins. Examples include the digestive juices, the secretions from the sebaceous (oil) glands of the skin, and tears from the lacrimal glands. These secretions are discussed in the chapters on specific systems.

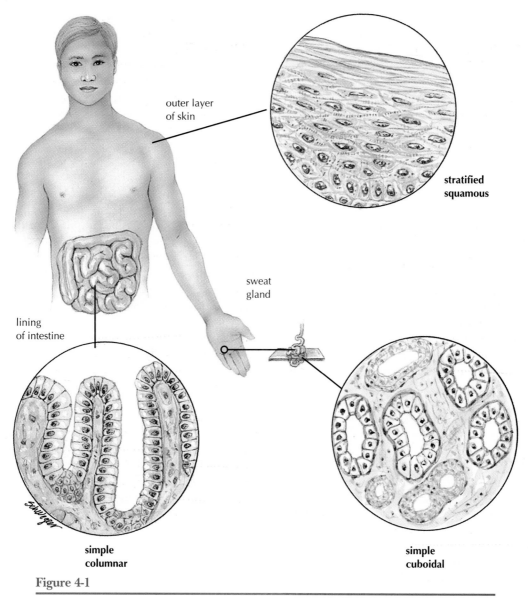

outer layer
of skin

**stratified
squamous**

sweat
gland

lining
of intestine

**simple
columnar**

**simple
cuboidal**

**Figure 4-1**

Three types of epithelium.

2. **Endocrine** (EN-do-krin) **glands** depend on blood flowing through the gland to carry the secretion to another organ. These secretions, called **hormones,** have specific effects on other tissues. Endocrine glands are the so-called "ductless glands."

Exocrine glands vary in design from very simple depressions resembling tiny dimples to more complex structures. Simple tubelike glands are found in the stomach wall and in the intestinal lining. Com-plex glands composed of treelike groups of ducts are found in the liver, the pancreas, and the salivary glands. Most glands are made largely of epithelial tissue with a framework of connective tissue. There may be a tough connective tissue capsule (a fibrous envelope) enclosing the gland, with extensions into the organ that form partitions. Between the partitions are groups of cells that unite to form lobes.

Endocrine glands produce **internal secretions,** which are carried to all parts of the body by the blood. These substances often affect tissues at a considerable distance from the point of origin. Endocrine glands, because they secrete directly into the bloodstream, have an extensive blood vessel network. The organs believed to have the richest blood supply in the body are the tiny adrenal glands located near the upper part of the kidneys. Some glands, such as the pancreas, have both endocrine and exocrine functions and therefore have both ducts and a rich blood supply. Hormones and the glands that produce them are discussed in Chapter 11.

# ▶ Connective Tissue

The supporting fabric of all parts of the body is connective tissue (Fig. 4-2). This is so extensive and widely distributed that if we were able to dissolve all the tissues except connective tissue, we would still be able to recognize the contours of the parts and the organs of the entire body.

Connective tissue has large amounts of nonliving material, called *intercellular* *material*, between the cells. This material contains varying amounts of water, fibers, and hard minerals. Connective tissue may be classified simply according to its degree of hardness:

1. Soft connective tissue—loosely held together with semi-liquid material between the cells; includes adipose (fat) tissue and loose (areolar) connective tissue
2. Fibrous connective tissue—Most connective tissue contains fibers, but this type is densely packed with fibers. Examples are ligaments, tendons, and fascia.
3. Hard connective tissue—has a very firm consistency, as in cartilage, or is hardened, as in bone, by minerals in the intercellular material.
4. Liquid connective tissue—Blood and lymph (the fluid that circulates in the lymphatic system) are examples of liquid connective tissues. Cells are suspended in a fluid environment.

## Soft Connective Tissue

The loose, or *areolar* (ah-RE-o-lar), form of connective tissue is found in membranes around vessels and organs, between muscles, and under the skin. It is the most common type of connective tissue in the body. It contains cells and fibers in a very loose, jellylike background material. *Adipose* (AD-ih-pose) *tissue* contains cells that are able to store large amounts of fat. This tissue is used as a reserve of energy for the body. Fat also serves as a heat insulator and as padding for various structures, such as organs and joints.

## Fibrous Connective Tissue

Fibrous connective tissue is very dense and has large numbers of fibers that give it strength and flexibility. The main type of fiber in this and other connective tissues is *collagen* (KOL-ah-jen), a flexible white protein. Some tissues, such as the tissue that makes up the vocal cords, consist mainly of elastic fibers that allow the tissue to stretch. This type of connective tissue serves as a binding between organs and also as a framework for some organs. Particularly strong forms make up the tough *capsules* around certain organs, such as the kidneys, liver, and glands, the deep fascia around muscles, and the *periosteum* (per-e-OS-te-um) around bones. If the fibers in the connective tissue are all arranged in the same direction, like the strands of a cable, the tissue can pull in one direction. Examples are the cordlike *tendons*, which connect muscles to bones, and the *ligaments*, which connect bones to other bones.

### FASCIA

The word *fascia* (FASH-e-ah) means "band"; hence, fascial membranes are bands or sheets that support the organs and hold them in place. An example of a fascia is the continuous sheet of tissue that underlies the skin. This contains fat (adipose tissue) and is called the *superficial fascia*. *Superficial* refers to a surface; the superficial fascia is closer than any other kind of fascia to the surface of the body.

As we penetrate more deeply into the body, we find examples of the *deep fascia*, which contains no fat and has many different purposes. Deep fascia covers and protects the muscle tissue with coverings known as *muscle sheaths*. The blood vessels and the nerves are sheathed with fascia; the brain and the spinal cord are encased in a multilayered covering called the *meninges* (men-IN-jeze). In addition, fascia serves to anchor muscle tissue to structures such as the bones.

### TISSUE REPAIR

Like epithelial tissue, fibrous connective tissue can repair itself easily. A large, gaping wound requires a correspondingly large growth of this new connective tissue; such new growth is called *scar tissue*. Excess production of collagen in the formation of a scar may result in the development of *keloids*

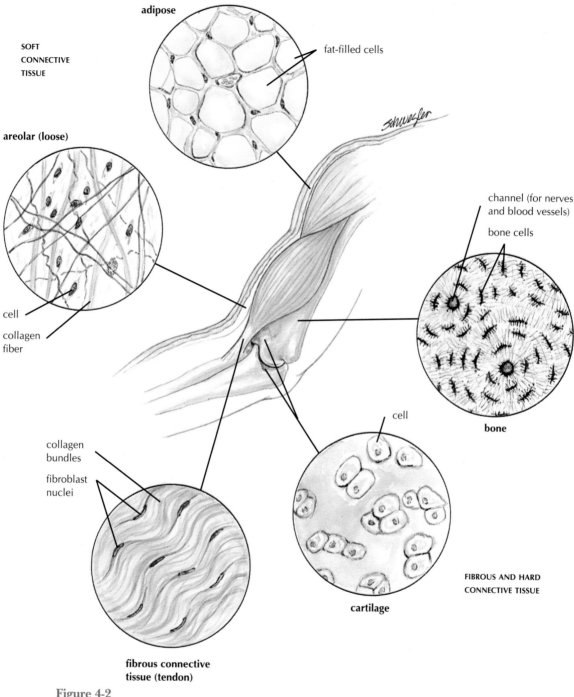

**SOFT CONNECTIVE TISSUE**

adipose

fat-filled cells

areolar (loose)

cell

collagen fiber

channel (for nerves and blood vessels)

bone cells

bone

cell

collagen bundles

fibroblast nuclei

cartilage

**FIBROUS AND HARD CONNECTIVE TISSUE**

fibrous connective tissue (tendon)

**Figure 4-2**

Connective tissue.

(KE-loyds), sharply raised areas on the surface of the skin. These are not dangerous but may be removed for the sake of appearance. The process of repair includes stages in which new blood vessels are formed in the wound, followed by the growth of the scar tissue. Overdevelopment of the blood vessels in the early stages of repair may lead to the formation of excess tissue. Normally, however, the blood vessels are gradually replaced by white fibrous connective tissue, which forms the scar. Suturing (sewing) the

## Collagen: The Body's Scaffolding

The most abundant protein in the body, making up about 25% of total protein, is the tough, flexible, white material known as *collagen*. Its name comes from a Greek word meaning "glue," indicating that it is the main structural substance in the body. Collagen is the major ingredient in all connective tissue.

The arrangement of the fibers in collagen gives this substance its different properties. It may form a fibrous connective tissue, as in the heart valves, a cord of great strength, as in tendons and ligaments, or a tough, transparent tissue, as in the cornea of the eye. It is collagen that gives strength and resilience to the skin.

The varied properties of collagen are evident in the preparation of a gelatin dessert. Gelatin is a collagen extract made by boiling animal bones and other connective tissue. It is a viscous liquid in hot water but forms a semisolid gel on cooling.

edges of a clean wound together, as is done in the case of operative wounds, decreases the amount of scar tissue needed and hence reduces the size of the resulting scar. Such scar tissue may be stronger than the original tissue.

### Hard Connective Tissue

The hard connective tissues, which as the name suggests are more solid than the other groups, include cartilage and bone. A common form of cartilage is the tough, elastic, translucent material popularly called *gristle*. This and other forms of cartilage are found in such places as between the segments of the spine and at the ends of the long bones. In these positions, cartilage acts as a shock absorber and as a bearing surface that reduces friction between moving parts. Cartilage is found in other structures, including the nose, the ear, and parts of the larynx, or "voice box." Except in joint cavities, cartilage is covered by a layer of fibrous connective tissue called *perichondrium* (per-e-KON-dre-um).

The tissue of which bones are made, called *osseous* (OS-e-us) *tissue*, is very much like cartilage in its cellular structure. In fact, the skeleton of the fetus in the early stages of development is made almost entirely of cartilage. This tissue gradually becomes impregnated with the calcium salts that make bones characteristically solid and hard. Within the bones are nerves, blood vessels, bone-forming cells, and a special form of tissue, bone marrow, in which blood cells are manufactured.

The liquid connective tissues, blood and lymph, are discussed in Chapters 12 and 15, respectively.

## ▶ Muscle Tissue

Muscle tissue is designed to produce movement by a forcible contraction. The cells of muscle tissue are threadlike and thus are called *muscle fibers.* If a piece of well-cooked meat is pulled apart, small groups of these muscle fibers may be seen. Muscle tissue is usually classified as follows (Fig. 4-3):

1. *Skeletal muscle,* which combines with connective tissue structures such as tendons (discussed later along with the skeleton) to provide for movement of the body. This type of tissue is also known as *voluntary muscle,* since it can be made to contract by an act of will. In other words, in theory at least, any of your skeletal muscles can be made to contract as you want them to.

The next two groups of muscle tissue are known as *involuntary muscle* because they typically contract independently of the will. In fact, most of the time we do not think of their actions at all. These are:

2. *Cardiac muscle,* which forms the bulk of the heart wall and is known also as *myocardium* (mi-o-KAR-de-um). This is the muscle that produces the regular contractions known as *heartbeats.*
3. *Smooth muscle,* known also as *visceral muscle,* which forms the walls of the *viscera* (VIS-er-ah), or organs of the ventral body cavities (with the exception of the heart). Some examples of visceral muscles are those that move food and waste materials along the digestive tract. Visceral muscles are found in other kinds of structures as well. Many tubular structures contain them, such as the blood vessels and the tubes that carry urine from the kidneys. Even certain structures at the base of body hair have this type of muscle. When these muscles contract, the skin condition we call *gooseflesh* results. Other types of visceral muscles are discussed under the various body systems.

Muscle tissue, like nervous tissue, repairs itself only with difficulty or not at all once an injury has

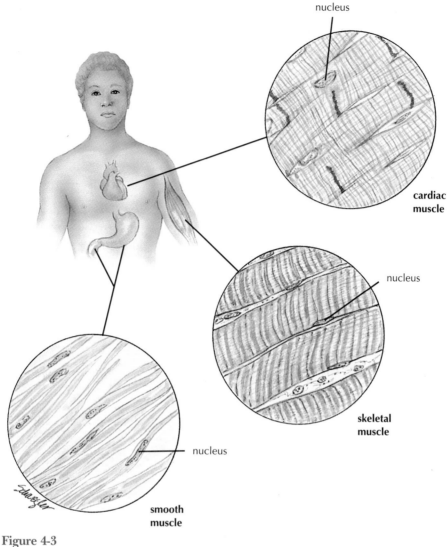

nucleus

cardiac
muscle

nucleus

skeletal
muscle

nucleus

smooth
muscle

**Figure 4-3**

Muscle tissue.

been sustained. When injured, muscle tissue is frequently replaced with connective tissue.

## ▶ Nervous Tissue

The human body is made up of countless structures, both large and small, each of which contributes something to the action of the whole organism. This aggregation of structures might be compared to an army. For all the members of the army to work together, there must be a central coordinating and order-giving agency somewhere; otherwise chaos would ensue. In the body this central agency is the **brain.** Each structure of the body is in direct communication with the brain by means of its own set of telephone wires, called **nerves.** The nerves from even the most remote parts of the body all come together and form a great trunk cable called the **spinal cord,** which in turn leads directly into the central switchboard of the brain. Here, messages come in and orders go out 24 hours a day. This entire communication system, brain and all, is made of nervous tissue.

The basic unit of nervous tissue is the **neuron** (NU-ron), or nerve cell (Fig. 4-4). A neuron consists of a nerve cell body plus small branches, like those of a tree, called *fibers*. One type of fiber, the ***dendrite*** (DEN-drite), carries nerve impulses, or messages, to the nerve cell body. A single fiber, the ***axon*** (AK-son), carries impulses away from the nerve cell body. Neurons may be quite long; their fibers can extend for several feet. A nerve is a bundle of nerve cell fibers held together with connective tissue.

Just as telephone wires are insulated to keep them from being short-circuited, some axons are insulated and protected by a coating of material called ***myelin*** (MI-eh-lin). Groups of myelinated fibers form "white matter," so called because of the color of the myelin, which is very much like fat in appearance and consistency. Not all neurons have myelin, however; some axons are unmyelinated, as are all dendrites and all cell bodies. These areas appear gray in color. Since the outer layer of the brain has large collections of cell bodies and unmyelinated fibers, the brain is popularly termed *gray matter.*

Nervous tissue is supported by ordinary connective tissue everywhere except in the brain and spinal cord. Here, the supporting cells are ***neuroglia*** (nu-ROG-le-ah), which have a protective function as well. Neuroglia, also called *glial cells*, are named from the Greek word *glia* meaning "glue." There is a more detailed discussion of nervous tissue in Chapter 8.

## ▶ Membranes

Membranes are thin sheets of tissue. Their properties vary: some are fragile, others tough; some are transparent, others opaque (*i.e.,* they cannot be seen through). Membranes may cover a surface, may

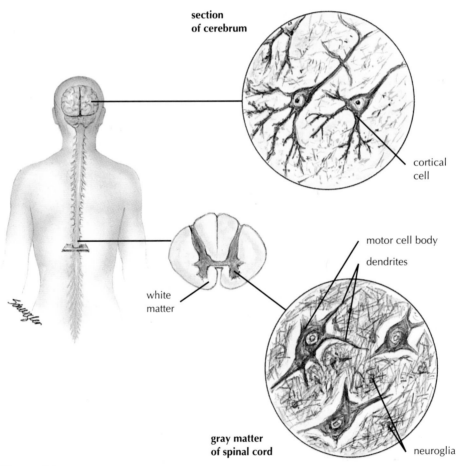

section
of cerebrum

cortical
cell

motor cell body

dendrites

white
matter

gray matter
of spinal cord

neuroglia

**Figure 4-4**

Nervous tissue.

serve as dividing partitions, may line hollow organs and body cavities, or may anchor various organs. They may contain cells that secrete lubricants to ease the movement of organs such as the heart and the movement of joints.

### Epithelial Membranes

*Epithelial membrane* is so named because its outer surface is made of epithelium. Underneath, however, there is a layer of connective tissue that strengthens the membrane, and in some cases there is a thin layer of smooth muscle under that. Epithelial membranes may be divided into two subgroups:

1. *Serous* (SE-rus) *membranes* line the walls of body cavities and are folded back onto the surface of internal organs, forming the outermost layer of many of them.
2. *Mucous* (MU-kus) *membranes* line tubes and other spaces that open to the outside of the body.

Epithelial membranes are made of closely crowded active cells that manufacture lubricants and protect the deeper tissues from invasion by microorganisms. Serous membranes secrete a thin, watery lubricant. Mucous membranes produce a rather thick and sticky substance called *mucus*. (Note that the adjective *mucous* contains an "o," whereas the noun *mucus* does not.)

The special serous membrane lining a closed cavity or covering an organ is called the *serosa* (se-RO-sah). In referring to the mucous membrane of a particular part, the noun *mucosa* (mu-KO-sah) is used.

#### SEROUS MEMBRANES

Serous membranes, unlike mucous membranes, usually do not communicate with the outside of the body. This group lines the closed ventral body cavities. There are three serous membranes:

1. The *pleurae* (PLU-re) or *pleuras* (PLU-rahs) line the thoracic cavity and cover each lung.
2. The *pericardium* (per-ih-KAR-de-um) is a sac that encloses the heart. It fits into a space in the chest between the two lungs.
3. The *peritoneum* (per-ih-to-NE-um) is the largest serous membrane. It lines the walls of the abdominal cavity, covers the organs of the abdomen, and forms supporting and protective structures within the abdomen (see Fig. 17-3).

The epithelium covering serous membranes is of a special kind called *mesothelium* (mes-o-THE-le-um). The mesothelium is smooth and glistening and is lubricated so that movements of the organs can take place with a minimum of friction.

Serous membranes are so arranged that one portion forms the lining of the closed cavity while another part covers the surface of the organs. The serous membrane attached to the wall of a cavity or sac is known as the *parietal* (pah-RI-eh-tal) *layer;* the word *parietal* refers to a wall. Parietal pleura lines the chest wall, and parietal pericardium lines the sac that encloses the heart. Because internal organs are called *viscera*, the membrane attached to the organs is the *visceral layer.* On the surface of the heart is visceral pericardium, and each lung surface is covered by visceral pleura.

Areas lined with serous membrane are examples of *potential spaces* because it is *possible* for a space to exist. Normally the membrane surfaces are in direct contact and a minimal amount of lubricant is present. Only if substances accumulate between the membranes, as when inflammation causes the production of excessive amounts of fluid, is there an actual space.

#### MUCOUS MEMBRANES

Mucous membranes form extensive continuous linings in the digestive, the respiratory, the urinary, and the reproductive systems, all of which are connected with the outside of the body. They vary somewhat in both structure and function. The cells that line the nasal cavities and most parts of the respiratory tract are supplied with tiny, hairlike extensions called *cilia,* which have already been mentioned in describing cells. The microscopic cilia move in a wavelike manner that forces the secretions outward away from the deeper parts of the lungs. In this way, foreign particles such as bacteria and dust become trapped in the sticky mucus and are prevented from causing harm. Ciliated epithelium is also found in certain tubes of both the male and the female reproductive systems.

The mucous membranes that line the digestive tract have special functions. For example, the mucous membrane of the stomach serves to protect the deeper tissues from the action of powerful digestive juices. If for some reason a portion of this membrane is injured, these juices begin to digest a part of the stomach itself—as in the case of peptic ulcers. Mucous membranes located farther along in the digestive system are designed to absorb food materials, which are then transported to all the cells of the body. Other mucous membranes are discussed in more detail later.

### Other Membranes

Another epithelial membrane is the skin, or *cutaneous* (ku-TA-ne-us) *membrane*, described fully in Chapter 5. *Synovial* (sin-O-ve-al) *membranes* are thin connective tissue membranes that line the joint cavities. They secrete a lubricating fluid that reduces friction between the ends of bones, thus permitting free movement of the joints.

With this we conclude our brief introduction to membranes. As we study each system in turn, other membranes will be encountered. They may have unfamiliar names, but they will be either epithelial or connective tissue membranes, and we will become familiar with their general locations; in short, they will be easy to recognize and remember.

## SUMMARY

I. **Tissue**—groups of cells specialized for a specific task
   A. Epithelial tissue—covers surfaces, lines cavities
      1. Squamous, cuboidal, and columnar cells
      2. Simple or stratified cell layers
      3. Glands
         a. Exocrine
            (1) Secrete through ducts
            (2) Produce external secretions
         b. Endocrine
            (1) Secrete into bloodstream
            (2) Produce internal secretions—hormones
   B. Connective tissue—supports, binds, forms framework of body
      1. Soft—jellylike intercellular material
         a. Adipose—stores fat
         b. Areolar (loose)
      2. Fibrous—Dense tissue with collagenous or elastic fibers between cells
         a. Tendons—attach muscle to bone
         b. Ligaments—connect bones
         c. Capsules—around organs
         d. Fascia—bands or sheets that support organs
      3. Hard—firm and solid
         a. Cartilage—found at ends of bones, nose, outer ear, trachea, etc.
         b. Bone—contains mineral salts
      4. Liquid
         a. Blood
         b. Lymph
   C. Muscle tissue—contracts to produce movement

   1. Skeletal muscle—voluntary; moves skeleton
   2. Cardiac muscle—forms the heart
   3. Smooth muscle—involuntary; forms visceral organs
   D. Nervous tissue
      1. Neuron—nerve cell
         a. Cell body—contains nucleus
         b. Dendrite—fiber carrying impulses toward cell body
         c. Axon—fiber carrying impulses away from cell body
      2. Myelin—fatty material that insulates some axons
         a. Myelinated fibers—make up white matter
         b. Unmyelinated cells and fibers—make up gray matter
      3. Neuroglia—support and protect nervous tissue
II. **Membrane**—thin sheet of tissue
   A. Epithelial membrane—outer layer epithelium
      1. Serous membrane—secretes watery fluid
         a. Parietal layer—lines body cavity
         b. Visceral layer—covers internal organs
      2. Mucous membrane
         a. Secretes mucus
         b. Lines tube or space that opens to the outside (*e.g.*, respiratory, digestive, reproductive tracts)
      3. Cutaneous membrane—skin
   B. Synovial membrane—lines a joint

## QUESTIONS FOR STUDY AND REVIEW

1. Define *tissue.* Give a few general characteristics of tissues.
2. Define *epithelium* and give two examples.
3. Describe the difference between endocrine and exocrine glands and give examples of each.
4. Define *connective tissue.* Name the main kinds of connective tissue and give an example of each.
5. What kinds of fibers are found in connective tissue?
6. What is the main purpose of nervous tissue? What is its basic structural unit called?
7. Define *myelin.* Where is it found?
8. Name three kinds of muscle tissue and give an example of each.
9. What is the difference between voluntary and involuntary muscle?
10. Compare how easily the four basic groups of tissues repair themselves.
11. What are some general characteristics of epithelial membranes?
12. Name two types of epithelial membrane. Give their general characteristics and two examples of each.
13. Name and describe the two layers of serous membranes.
14. Give two examples of fascia.

# CHAPTER 5

# The Skin

## Selected Key Terms

The following terms are defined in the Glossary:

**dermis**

**epidermis**

**integument**

**keratin**

**melanin**

**sebaceous**

**sebum**

**subcutaneous**

**sudoriferous**

## Behavioral Objectives

After careful study of this chapter, you should be able to:

- Describe the layers of the integumentary system
- Describe the location and function of the appendages of the skin
- List the main functions of the skin
- Summarize the information to be gained by observation of the skin

Memmler, RL., Cohen, BJ, Wood, DL. *STRUCTURE AND FUNCTION OF THE HUMAN BODY*, 6/e,
© 1996 Lippincott-Raven Publishers

The skin is often considered to be merely a membrane covering the body. However, in both structure and function the skin assumes more complex properties. The skin and its appendages are easily observed, giving us clues to the functioning and health of other body systems. The skin is classified in the following three ways:

1. It may be called an enveloping *membrane* because it is a thin layer of tissue that covers the entire body.
2. It may be referred to as an *organ* (the largest one, in fact) because it contains several kinds of tissue, including epithelial, connective, and nervous tissues.
3. It is most properly known as the *integumentary* (in-teg-u-MEN-tar-e) *system* because it includes glands, vessels, nerves, and a subcutaneous (sub-ku-TA-ne-us) layer that work together as a body system. The name is from the word *integument* (in-TEG-u-ment), which means "covering." The term *cutaneous* (ku-TA-ne-us) also refers to the skin.

## ▶ Structure of the Integumentary System

The integumentary system consists of the skin and the subcutaneous (under the skin) layer, which contains structures extending from the skin. The skin itself consists of two main layers, different from each other in structure and function (Fig. 5-1). These are:

1. The *epidermis* (ep-ih-DER-mis), or outermost layer, which is subdivided into *strata* (STRA-tah), or layers, and is made entirely of epithelial cells with no blood vessels.
2. The *dermis,* or true skin, which has a framework of connective tissue and contains many blood vessels, nerve endings, and glands.

### Epidermis

The epidermis is the surface layer of the skin, the outermost cells of which are constantly lost through wear and tear. Since there are no blood vessels in the epidermis, the only living cells are in its deepest layer, the *stratum germinativum* (jer-min-a-TI-vum), where nourishment is provided by capillaries in the underlying dermis. The cells in this layer are constantly dividing and producing daughter cells, which are pushed upward toward the surface. As the surface cells die from the gradual loss of nourishment,

they undergo changes. Mainly, they develop large amounts of a protein called *keratin* (KER-ah-tin), which serves to thicken and protect the skin.

By the time epidermal cells reach the surface, they have become flat and horny, forming the uppermost layer of the epidermis, the **stratum corneum** (KOR-ne-um). Between the stratum germinativum and the stratum corneum there are additional layers that vary with the thickness of the skin. Cells in the deepest layer of the epidermis produce **melanin** (MEL-ah-nin), a dark pigment that colors the skin; irregular patches of melanin are called *freckles*. Ridges and grooves in the skin of the fingers, palms, toes, and soles form unchanging patterns determined by heredity. These ridges actually serve to prevent slipping, but because they are unique to each individual, fingerprints and footprints are excellent means of identification. The ridges are due to elevations and depressions in the epidermis and the dermis. The deep surface of the epidermis is accurately molded upon the outer part of the dermis, which has raised and depressed areas.

### Dermis

The *dermis,* or *corium* (KO-re-um), the so-called "true skin," has a framework of elastic connective tissue and is well supplied with blood vessels and nerves. The thickness of the dermis and epidermis varies. Some areas, such as the soles of the feet and the palms of the hands, are covered with very thick layers of skin, whereas others, such as the eyelids, are covered with very thin and delicate layers. Most of the appendages of the skin, including the sweat glands, the oil glands, and the hair, are located in the dermis and may extend into the subcutaneous layer. Because of its elasticity, the skin can stretch, even dramatically as in pregnancy, with little damage.

### Subcutaneous Layer

The dermis rests on the subcutaneous layer, sometimes referred to as the *superficial fascia,* which connects the skin to the surface muscles. This layer consists of elastic and fibrous connective tissue and adipose (fat) tissue. The fat serves as insulation and as a reserve store for energy. Continuous bundles of elastic fibers connect the subcutaneous tissue with the dermis, so there is no clear boundary between the two. The major blood vessels that supply the skin run through the subcutaneous layer. The blood vessels that are concerned with temperature regulation

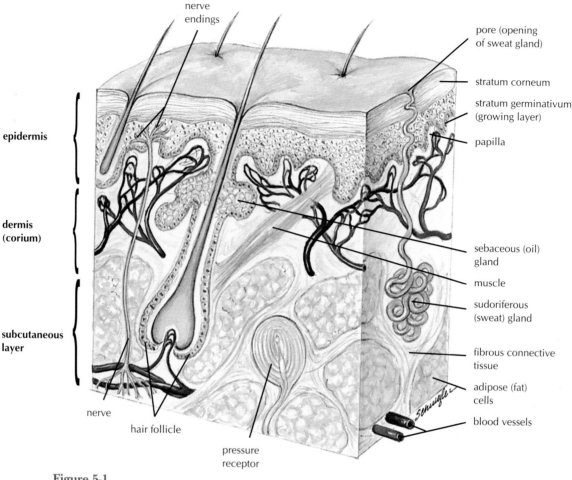

nerve endings

pore (opening of sweat gland)

stratum corneum

stratum germinativum (growing layer)

papilla

epidermis

dermis (corium)

sebaceous (oil) gland

muscle

sudoriferous (sweat) gland

subcutaneous layer

fibrous connective tissue

adipose (fat) cells

blood vessels

nerve

hair follicle

pressure receptor

**Figure 5-1**

Cross section of the skin.

are located here. Some of the appendages of the skin, such as sweat glands and hair roots, extend into the subcutaneous layer. The subcutaneous tissues are rich in nerves and nerve endings, including those that supply the dermis and epidermis. The thickness of the subcutaneous layer varies in different parts of the body, the layer being at its thinnest on the eyelids and at its thickest on the abdomen.

Figure 5-2 is a photograph of the skin and its appendages as seen through a microscope.

## ▶ Appendages of the Skin

### Sweat Glands

The *sudoriferous* (su-do-RIF-er-us) *glands,* or sweat glands, are coiled, tubelike structures located in the dermis and the subcutaneous tissue. Each

gland has an excretory tube that extends to the surface and opens at a pore. The slant at which the excretory tube joins the skin serves as a valve. Sudoriferous glands function to regulate body temperature through the evaporation of sweat from the body surface. Sweat consists of water with small amounts of mineral salts and other substances.

Another type of sweat gland is located mainly in the armpits and the groin area. These glands release their secretions through the hair follicles in response to emotional stress and sexual stimulation. The secretions contain cellular material that is broken down by bacteria, producing body odor.

Several types of glands associated with the skin are modified sweat glands. These are the *ceruminous* (seh-RU-min-us) *glands* in the ear canal that produce ear wax, or *cerumen,* the *ciliary* (SIL-e-er-e) *glands* at the edges of the eyelids, and the *mammary glands.*

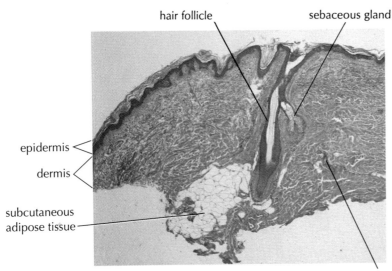

hair follicle          sebaceous gland

epidermis

dermis

subcutaneous
adipose tissue

sweat gland

**Figure 5-2**

Microscopic view of thin skin showing tissue layers and some appendages. (Adapted from Cormack DH: Essential Histology, Plate 12-1B, Philadelphia, JB Lippincott, 1993.)

### Sebaceous Glands

The *sebaceous* (se-BA-shus) *glands* are saclike in structure, and their oily secretion, *sebum* (SE-bum), lubricates the skin and hair and prevents drying. The ducts of the sebaceous glands open into the hair follicles.

Babies are born with a covering produced by these glands that resembles cream cheese; this secretion is called the *vernix caseosa* (VER-niks ka-se-O-sah).

Blackheads consist of a mixture of dried sebum and keratin that may collect at the openings of the sebaceous glands. If these glands become infected, pimples result. If sebaceous glands become blocked by accumulated sebum, a sac of this secretion may form and gradually increase in size. These sacs are referred to as *sebaceous cysts.* Usually it is not difficult to remove such tumorlike cysts by surgery.

### Hair and Nails

Almost all of the body is covered with hair, which in most areas is very soft and fine. Hair is composed mainly of keratin and is not living. Each hair develops within a sheath called a *follicle,* and new hair is formed from cells at the bottom of the follicles (see Fig. 5-1). Attached to most hair follicles is a thin band of involuntary muscle. When this muscle contracts, the hair is raised, forming "goose bumps" on the skin. As it contracts, the muscle presses on the sebaceous gland associated with the hair follicle, causing the release of sebum.

Nails are protective structures made of hard keratin produced by cells that originate in the outer layer of the epidermis (stratum corneum). New cells form continuously at the proximal end of the nail in an area called the *nail root.* Nails of both the toes and the fingers are affected by general health.

## ▶ Functions of the Skin

Although the skin has several functions, the three that are by far the most important are the following:

1. Protection of deeper tissues against drying and against invasion by pathogenic organisms or their toxins
2. Regulation of body temperature by dissipation of heat to the surrounding air
3. Receipt of information about the environment by means of the many nerve endings distributed throughout the skin. Much of this sensory information has a protective function. For example, it prompts one to withdraw from harmful stimuli, such as a hot stove.

The outermost layer of the skin, the stratum corneum, protects the body against invasions by pathogens and against dehydration. These dry, dead cells, composed of keratin, are found in a tight, interlocking pattern that is impervious to penetration by organisms and by water. The outermost cells are constantly being shed, causing the mechanical removal of pathogens. The function of epidermis as a

## Vitamin D

One function of the skin is the synthesis of vitamin D. This "vitamin" is actually a group of related fat-soluble compounds that aids in the absorption of calcium from foods. It is manufactured from a substance in the skin that is activated by the ultraviolet (UV) radiation in sunlight. Enzymes in the liver and kidneys then modify it into its most active form. Since it is manufactured in the body and carried in the blood to the tissues on which it acts, vitamin D is now considered to be a hormone.

A lack of vitamin D results in bone malformation. There is no dietary requirement for the vitamin because even moderate exposure to sunlight is enough to satisfy our needs. Exposure of the arms and face for only 1 hour per week is enough to avoid a deficiency. As a safeguard, all milk sold in the United States is fortified with vitamin D.

water barrier is vital to provide the wet environment required by cells.

The regulation of body temperature, both the loss of excess heat and the protection from cold, is a very important function of the skin. Indeed, most of the blood that flows through the skin is concerned with temperature regulation. The skin forms a large surface for radiating body heat to the air. When the blood vessels dilate (enlarge), more blood is brought to the surface so that heat can be dissipated. The activation of sweat glands, and the evaporation of sweat from the surface of the body also helps cool the body. In cold conditions, the flow of small amounts of blood in veins deep in the subcutaneous tissue serves to heat the skin and protect deeper tissues from excess heat loss. Special vessels that directly connect arteries and veins in the skin of the ears, nose, and other exposed locations provide the volume of blood flow needed to prevent freezing. As is the case with so many body functions, temperature regulation is complex and involves several parts of the body, including certain centers in the brain.

Another important function of the skin is obtaining information from the environment. Because of the many nerve endings and other special receptors for pain, touch, pressure, and temperature, which are located mostly in the dermis, the skin may

be regarded as one of the chief sensory organs of the body. Many of the reflexes that make it possible for humans to adjust themselves to the environment begin as sensory impulses from the skin. Here, too, the skin works with the brain and the spinal cord to make these important functions possible.

The functions of absorption and excretion are minimal in the skin. Most medicated ointments used on the skin are for the treatment of local conditions only, and the injection of medication into the subcutaneous tissues is limited by the slow absorption that occurs here. The skin excretes a mixture of water and mineral salts in perspiration. Some nitrogen-containing wastes are also eliminated through the skin, but even in disease the amount of waste products excreted by the skin is small.

The human skin does not "breathe." The pores of the epidermis serve only as outlets for perspiration and oil from the sweat glands and sebaceous glands.

The skin has the ability to regenerate after injury. Skin that is injured can recover, even to the extent of growing a new epidermis from the cells of the hair follicle. During healing, a *scab,* or crust, made of dried pus and blood may form over the wound.

## ▶ Observation of the Skin

What can the skin tell you? What do its color, texture, and other attributes indicate? Is there any damage? Much can be learned by an astute observer. The skin reflects the general state of one's health.

The color of the skin depends on a number of factors, including the following:

1. The amount of pigment in the epidermis
2. The quantity of blood circulating in the surface blood vessels
3. The composition of the circulating blood
   a. Presence or absence of oxygen
   b. Concentration of hemoglobin
   c. Presence of bile, silver compounds, or other chemicals

### Pigment

The main pigment of the skin, as we have noted, is called *melanin.* This pigment is also found in the hair, the middle coat of the eyeball, the iris of the eye, and certain tumors. Melanin is common to all races, but darker people have a much larger quantity in their tissues. The melanin in the skin helps to protect against damaging ultraviolet radiation from the sun. Thus, skin that is exposed to the sun shows a normal increase in this pigment, a response we call *tanning*.

### Effects of Aging on Skin, Hair, and Nails

As people age, wrinkles, or crow's feet, develop around the eyes and mouth owing to the loss of fat and collagen in the underlying tissues. The dermis becomes thinner, and the skin may become quite transparent and lose its elasticity, the effect of which is so-called "parchment skin." The hair does not replace itself as rapidly as before and thus becomes thinner. The formation of pigment also decreases with age, causing hair to become gray or white. However, there may be localized areas of extra pigmentation in the skin with the formation of brown spots, especially on areas exposed to the sun (*e.g.*, the hands). The sweat glands decrease in number, so there is less output of perspiration. The fingernails may flake, become brittle, or develop ridges. Toenails may become discolored or abnormally thickened.

### Hair Testing for Genetic Identification

The living cells of the hair root can be used as sources of genetic material for the creation of a genetic fingerprint. Each person's genetic information (DNA) is easily separated into segments of varying lengths. The pattern of these segments is unique to each individual. Sorted, and then put in order by length, the resultant pattern looks much like a supermarket bar code. Comparisons of the codes of related individuals will show many matching bars. Pictures of the pattern can be stored and used later for identification (*e.g.*, of criminals or missing children).

## ▶ Care of the Skin

The most important factors in caring for the skin are those that ensure good general health. Proper nutrition and adequate circulation are vital to the maintenance of the skin. Regular cleansing removes dirt and dead skin debris and sustains the slightly acid environment that inhibits bacterial growth on the skin. Careful handwashing with soap and water, with attention to the undernail areas, is a simple measure that reduces the spread of disease.

The skin needs protection from continued exposure to sunlight to prevent premature aging and cancerous changes. Appropriate applications of sunscreens before and during time spent in the sun can prevent skin damage.

---

## SUMMARY

I. **Classification of the skin**
   A. Membrane (cutaneous)
   B. Organ
   C. System (integumentary)
II. **Structure of the integumentary system**
   A. Epidermis—surface layer of the skin
   B. Dermis—deeper layer of the skin
   C. Subcutaneous layer—under the skin
III. **Appendages of the skin**
   A. Sweat (sudoriferous) glands
   B. Sebaceous (oil) glands
   C. Hair
   D. Nails
IV. **Functions of the skin**
   A. Protection
   B. Regulation of body temperature
   C. Sensory perception
V. **Observation of the skin**
   A. Pigment
   B. Effects of aging
   C. Hair testing for genetic identification
VI. **Care of the skin**

## QUESTIONS FOR STUDY AND REVIEW

1. What characteristics of the skin classify it as a membrane? as an organ? as a system?
2. Of what type of cells is the epidermis composed?
3. Explain how the outermost cells of the epidermis are replaced.
4. Describe the structure of the dermis.
5. Describe the contents and functions of the subcutaneous layer.
6. Describe the location and function of the skin glands.
7. Explain the three most important functions of the skin.
8. What are the most important contributors to the color of the skin, normally?
9. Define melanin and explain its function.
10. What changes may occur in the skin with age?
11. List some factors involved in proper care of the skin.

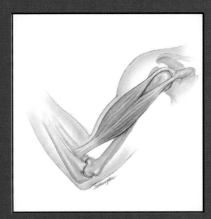

# UNIT II

# Movement and Support

This unit deals with the skeletal and muscular systems. It covers the functions of the skeletal system, going beyond support purposes and including those related to blood formation and to the storage and metabolism of certain mineral salts. The important muscles and their functions in various movements, as well as their ability to produce heat and to aid in the circulation of body fluids, are noted in Chapter 7, The Muscular System.

# CHAPTER 6

# The Skeleton: Bones and Joints

## Selected Key Terms

The following terms are defined in the Glossary:

**amphiarthrosis**

**bursa**

**circumduction**

**diaphysis**

**diarthrosis**

**endosteum**

**epiphysis**

**fontanelle**

**joint**

**osteoblast**

**osteoclast**

**osteocyte**

**periosteum**

**resorption**

**synarthrosis**

**synovial**

## Behavioral Objectives

After careful study of this chapter, you should be able to:

- Name the three different types of bone cells and describe the functions of each
- Differentiate between compact bone and spongy bone with respect to structure and location
- Describe the structure of a long bone
- Explain how a long bone grows
- Differentiate between red and yellow marrow with respect to function and location
- Name and describe various markings found on bones
- List the bones in the axial skeleton
- List the bones in the appendicular skeleton
- Describe how the skeleton changes with age
- Describe the three types of joints
- Describe the structure of a synovial joint and give six examples of synovial joints
- Define six types of movement that occur at synovial joints

Memmler, RL, Cohen, BJ, Wood, DL. *STRUCTURE AND FUNCTION OF THE HUMAN BODY*, 6/e, © 1996 Lippincott-Raven Publishers

The bones are the framework of the body. They are a combination of several kinds of tissue and contain blood vessels and nerves. Bones are attached to each other at joints. The combination of bones and joints together with related connective tissues forms the skeletal system.

## ▶ The Bones

### Bone Structure

The bones are composed chiefly of bone tissue, called *osseous* (OS-e-us) *tissue.* Bones are not lifeless. Even though the spaces between the cells of bone tissue are permeated with stony deposits of calcium salts, these cells themselves are very much alive. Bones are organs, with their own system of blood vessels, lymphatic vessels, and nerves.

In the embryo (the early developmental stage of a baby) most of the bones-to-be are composed of cartilage. (Portions of the skull develop from fibrous connective tissue.) Bone formation begins during the second and third months of embryonic life. At this time, bone-building cells, called *osteoblasts* (OS-te-o-blasts), become very active. First, they manufacture a substance, the *intercellular* (in-ter-SEL-u-lar) *material,* which is located between the cells and contains large quantities of a protein called *collagen.* Then, with the help of enzymes, calcium compounds are deposited within the intercellular material. Once the intercellular material has hardened around these cells, they are called *osteocytes.* They are still living and continue to maintain the bone, but they do not produce new bone tissue. Other cells, called *osteoclasts* (OS-te-o-klasts), are responsible for the process of *resorption,* or the breakdown of bone. Enzymes also implement this process.

The complete bony framework of the body, known as the *skeleton* (Fig. 6-1), consists of bones of many different shapes. They may be flat (rib, skull), cube shaped (wrist, ankle), or irregular (vertebrae). However, typical bones, which make up most of the arms and legs, are described as *long bones.* This type of bone has a long shaft, called the *diaphysis* (di-AF-ih-sis), which has a central marrow cavity and two irregular ends, each called an *epiphysis* (eh-PIF-ih-sis) (Fig. 6-2).

There are two types of bone tissue. One type is *compact bone,* which is hard and dense. This makes up the main shaft of a long bone and the outer layer of other bones. The osteocytes in this type of bone are located in rings of bone tissue around a central *Haversian* (ha-VER-shan) *canal* containing nerves and blood vessels. The second type, called *spongy bone,* has more spaces than compact bone. It is made of a meshwork of small, bony plates filled with red marrow and is found at the ends of the long bones and at the center of other bones.

Bones contain two kinds of marrow: *red marrow,* found at the end of the long bones and at the center of other bones, which manufactures blood cells; and *yellow marrow* of the "soup bone" type, found chiefly in the central cavities of the long bones. Yellow marrow is largely fat.

Bones are covered on the outside (except at the joint region) by a membrane called the *periosteum* (per-e-OS-te-um). The inner layer of this membrane contains osteoblasts, which are essential in bone formation, not only during growth but also in the repair of fractures. Blood vessels and lymphatic vessels in the periosteum play an important role in the nourishment of bone tissue. Nerve fibers in the periosteum make their presence known when one suffers a fracture, or when one receives a blow, such as on the shinbone. A thinner membrane, the *endosteum* (en-DOS-te-um), lines the marrow cavity of a bone; it too contains cells that aid in the growth and repair of bone tissue.

### Bone Growth and Repair

In a long bone, the transformation of cartilage into bone begins at the center of the shaft. Later, secondary bone-forming centers develop across the ends of the bones. The long bones continue to grow in length at these centers through childhood and into the late teens. Finally, by the late teens or early 20s, the bones stop growing in length. Each bone-forming region hardens and can be seen in x-ray films as a thin line across the end of the bone. Physicians can judge the future growth of a bone by the appearance of these lines on x-ray films.

As a bone grows in length, the shaft is remodeled so that it grows wider as the central marrow cavity increases in size. Thus, alterations in the shape of the bone are a result of the addition of bone tissue to some surfaces and its resorption from others.

The processes of bone formation and bone resorption continue throughout life, more rapidly at some times than at others. The bones of small

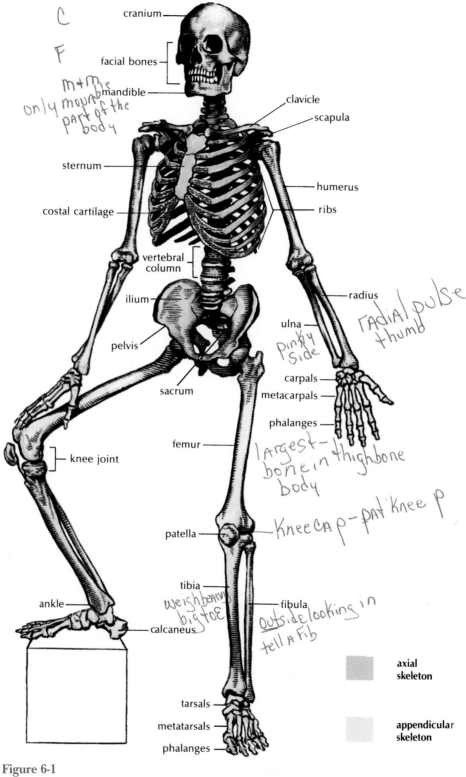

C
F
M+Me
only mouab part of the
body

radial pulse
thumb

pinky side

largest-bone in body    thighbone

kneecap - pat knee p

weight bearing
big toe

outside looking in
tell a Fib

**Figure 6-1**

The skeleton.

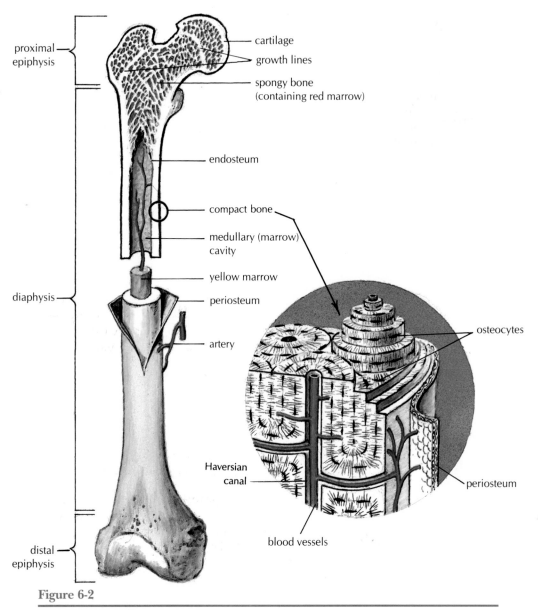

**Figure 6-2**

The structure of a long bone; the composition of compact bone.

children are relatively pliable because they contain a larger proportion of cartilage and are undergoing active bone formation. In the elderly, there is a slowing of the processes that continually renew bone tissue. As a result, the bones are weaker and more fragile. The elderly also have a decreased ability to form the protein framework on which calcium salts are deposited. Fractures in old people heal more slowly due to these decreases in bone metabolism.

## Main Functions of Bones

Bones have a number of functions, many of which are not at all obvious. Some of these are listed below:

1. To serve as a firm framework for the entire body
2. To protect such delicate structures as the brain and the spinal cord
3. To serve as levers, working with attached muscles to produce movement

## The Timetable of Bone Formation

 The conversion of cartilage to bone, called *ossification,* begins in the fetus by the 3rd month. At birth, evidence of ossification should be noted in most of the bones. The long bones begin to ossify in the center of the shaft (diaphysis). At birth, secondary ossification centers are apparent at each end (epiphyses). The short bones usually have only one center of ossification.

During the first 12 years of life, bone formation continues throughout the entire skeleton. By 17 to 20 years of age the bones of the upper extremities and the scapulae, but not the clavicles, are ossified. By age 18 to 23, the lower limbs and the pelvis are converted to bone tissue. By age 23 to 25 the clavicles, the sternum, and the vertebrae are totally ossified, so that by 25 years of age bone formation is complete.

4. To serve as a storehouse for calcium salts, which may be resorbed into the blood if there is not enough calcium in the diet
5. To produce blood cells (in the red marrow).

### Bone Markings

In addition to their general shape, bones have other distinguishing features called **bone markings.** These markings include raised areas and depressions that help to form joints or serve as points for muscle attachments, and various holes that allow the passage of nerves and blood vessels. Some of these identifying features are:

1. *Projections*
   a. *Head,* a rounded knob-like end separated from the rest of the bone by a slender region, the neck
   b. *Process,* a large projection of a bone, such as the upper part of the ulna in the forearm that creates the elbow
   c. *Crest,* a distinct border or ridge, often rough, such as over the top of the hip bone
   d. *Spine,* a sharp projection from the surface of a bone, such as the spine of the scapula (shoulder blade)
2. *Depressions* or *holes*
   a. *Foramen* (fo-RA-men), a hole that allows a vessel or a nerve to pass through or between

bones. The plural is *foramina* (fo-RAM-ih-nah).
   b. *Sinus* (SI-nus), an air space found in some skull bones
   c. *Fossa* (FOS-sah), a depression on a bone surface. The plural is fossae (FOS-se).
   d. *Meatus* (me-A-tus), a short channel or passageway, such as the channel in the temporal bone of the skull that leads to the inner ear.

Examples of these and other markings will be seen on the bones illustrated in this chapter.

### Divisions of the Skeleton

The skeleton may be divided into two main groups of bones (see Fig. 6-1):

1. The **axial** (AK-se-al) **skeleton,** which includes the bony framework of the head and the trunk
2. The **appendicular** (ap-en-DIK-u-lar) **skeleton,** which forms the framework for the **extremities** (limbs) and for the shoulders and hips.

#### FRAMEWORK OF THE HEAD

The bony framework of the head, called the **skull,** is subdivided into two parts: the cranium and the facial portion. Refer to Figures 6-3 through 6-6, which show different views of the skull, as you study the following descriptions. The numbers in brackets refer to the numbers in Figure 6-3.

A. The **cranium** is a rounded box that encloses the brain; it is composed of eight distinct cranial bones.
   1. The **frontal bone** [1] forms the forehead, the front of the skull's roof, and part of the roof over the eyes and the nasal cavities. The **frontal sinuses** (air spaces) communicate with the nasal cavities. These sinuses and others near the nose are described as **paranasal sinuses.**
   2. The two **parietal** (pah-RI-eh-tal) bones [2] form most of the top and the side walls of the cranium.
   3. The two **temporal bones** [4] form part of the sides and some of the base of the skull. Each one contains **mastoid sinuses** as well as the ear canal, the eardrum, and the entire middle and internal ears. The **mastoid process** of the temporal bone projects downward immediately behind the external part of the ear. It contains the mastoid air cells and serves as a place for muscle attachment.

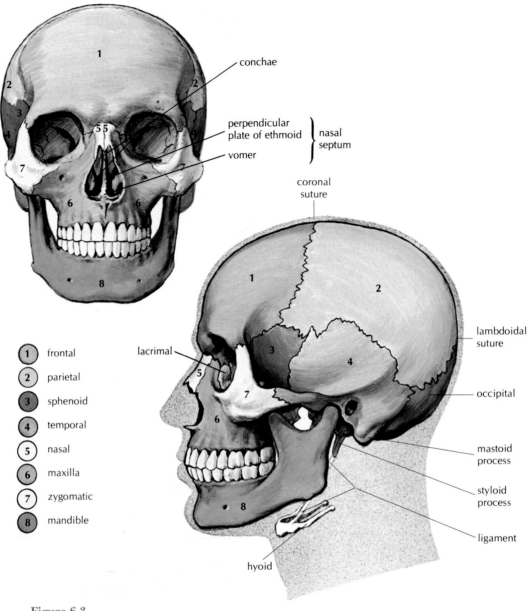

**Figure 6-3**

The skull, from the front and from the left.

4. The ***ethmoid*** (ETH-moyd) ***bone*** is a very light, fragile bone located between the eyes. It forms a part of the medial wall of the eye sockets, a small portion of the cranial floor, and most of the nasal cavity roof. It contains several air spaces—comprising some of the paranasal sinuses. A thin, platelike, downward extension of this bone (the perpendicular plate) forms much of the nasal septum, a midline partition in the nose.

5. The ***sphenoid*** (SFE-noyd) ***bone*** [3], when seen from above, resembles a bat with its wings extended. It lies at the base of the skull in front of the temporal bones.

6. The ***occipital*** (ok-SIP-ih-tal) ***bone*** forms the back and a part of the base of the skull. The ***foramen magnum***, located at the base of the occipital bone, is a large opening through which the spinal cord communicates with the brain (Figs. 6-4 and 6-5).

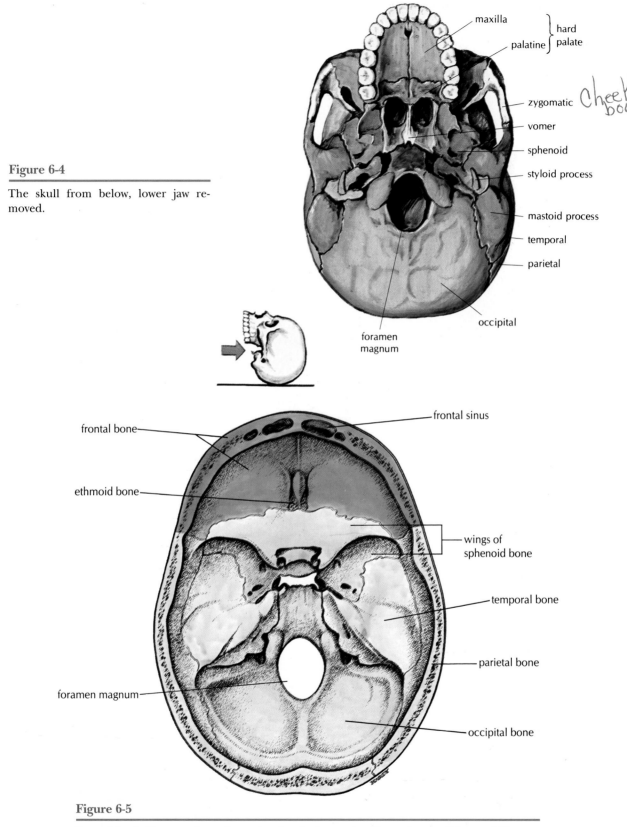

maxilla

hard palate

palatine

zygomatic — Cheek bone

vomer

sphenoid

styloid process

mastoid process

temporal

parietal

occipital

foramen magnum

**Figure 6-4**

The skull from below, lower jaw removed.

frontal sinus

frontal bone

ethmoid bone

wings of sphenoid bone

temporal bone

parietal bone

foramen magnum

occipital bone

**Figure 6-5**

Base of the skull as seen from above, showing the internal surfaces of some of the cranial bones. (Chaffee EE, Lytle IM: Basic Physiology and Anatomy, 4th ed, p. 95. Philadelphia, JB Lippincott, 1980)

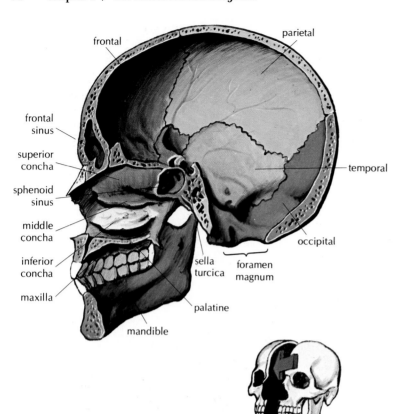

frontal

parietal

frontal
sinus

superior
concha

sphenoid
sinus

middle
concha

inferior
concha

maxilla

temporal

occipital

sella
turcica

foramen
magnum

palatine

mandible

**Figure 6-6**

The skull, internal view.

B. The *facial portion* of the skull is composed of 14 bones.
1. The *mandible* (MAN-dih-bl) [8], or lower jaw bone, is the only movable bone of the skull.
2. The two *maxillae* (mak-SIL-e) [6] fuse in the midline to form the upper jaw bone, including the front part of the hard palate (roof of the mouth). Each maxilla contains a large air space, called the *maxillary sinus,* that communicates with the nasal cavity.
3. The two *zygomatic* (zi-go-MAT-ik) *bones* [7], one on each side, form the prominences of the cheeks.
4. Two slender *nasal bones* [5] lie side by side, forming the bridge of the nose.
5. The two *lacrimal* (LAK-rih-mal) *bones,* each about the size of a fingernail, lie near the inside corner of the eye in the front part of the medial wall of the orbital cavity.
6. The *vomer* (VO-mer), shaped like the blade of a plow, forms the lower part of the nasal septum.

7. The paired *palatine bones* form the back part of the hard palate.
8. The two *inferior nasal conchae* (KON-ke) extend horizontally along the lateral wall (sides) of the nasal cavities. The paired superior and middle conchae are part of the ethmoid bone.

In addition to the bones of the cranium and the facial bones, there are three tiny bones, or *ossicles* (OS-sik-ls), in each middle ear (see Chap. 10) and a single horseshoe, or U-shaped, bone just below the skull proper, called the *hyoid* (HI-oyd) *bone,* to which the tongue and other muscles are attached.

Openings in the base of the skull provide spaces for the entrance and exit of many blood vessels, nerves, and other structures. Projections and slightly elevated portions of the bones provide for the attachment of muscles. Some portions contain delicate structures, such as the part of the temporal bone that encloses the middle and internal sections of the ear. The air sinuses provide lightness and serve as resonating chambers for the voice.

## THE INFANT SKULL

The skull of the infant has areas in which the bone formation is incomplete, leaving so-called "soft spots." Although there are a number of these, the largest and best known is near the front at the junction of the two parietal bones with the frontal bone. It is called the ***anterior fontanelle*** (fon-tah-NEL), and it usually does not close until the child is about 18 months old (Fig. 6-7).

## FRAMEWORK OF THE TRUNK

The bones of the trunk include the ***vertebral*** (VER-teh-bral) ***column*** and the bones of the chest, or ***thorax*** (THO-raks). The vertebral column is made of a series of irregularly shaped bones. These number 33 or 34 in the child, but because of fusions that occur later in the lower part of the spine, there usually are just 26 separate bones in the adult spinal column (Figs. 6-8 and 6-9). Each of these ***vertebrae*** (VER-teh-bre), except the first two cervical vertebrae, has a drum-shaped body (centrum) located toward the front (anteriorly) that serves as the weight-bearing part; disks of cartilage between the vertebral bodies act as shock absorbers and provide flexibility.

In the center of each vertebra is a large hole, or foramen. When all the vertebrae are linked in series by strong connective tissue bands (ligaments), these spaces form the spinal canal, a bony cylinder that protects the spinal cord. Projecting backward from the bony arch that encircles the spinal cord is the ***spinous process***, which usually can be felt just under the skin of the back. When viewed from the side the vertebral column can be seen to have a series of ***intervertebral foramina*** through which spinal nerves emerge as they leave the spinal cord.

The bones of the vertebral column are named and numbered from above downward, on the basis of location:

1. The ***cervical*** (SER-vih-kal) ***vertebrae,*** seven in number, are located in the neck. The first vertebra, called the ***atlas,*** supports the head; when one nods the head, the skull rocks on the atlas at the occipital bone. The second cervical vertebra, called the ***axis,*** serves as a pivot when the head is turned from side to side.
2. The ***thoracic vertebrae,*** 12 in number, are located in the thorax. The posterior ends of the 12 pairs of ribs are attached to these vertebrae.
3. The ***lumbar vertebrae,*** five in number, are located in the small of the back. They are larger and heav-

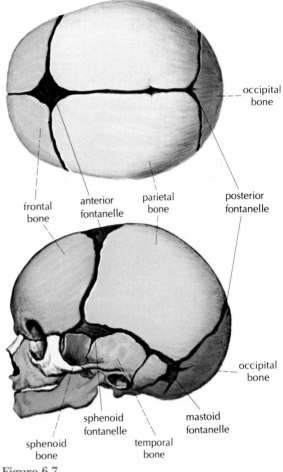

**Figure 6-7**

Infant skull showing fontanelles.

ier than the other vertebrae to support more weight.
4. The ***sacral*** (SA-kral) ***vertebrae*** are five separate bones in the child. However, they eventually fuse to form a single bone, called the ***sacrum*** (SA-krum), in the adult. Wedged between the two hip bones, the sacrum completes the posterior part of the bony pelvis.
5. The ***coccyx*** (KOK-siks), or tailbone, consists of four or five tiny bones in the child. These fuse to form a single bone in the adult.

When viewed from the side, the vertebral column can be seen to have four curves, corresponding to the four groups of vertebrae (see Fig. 6-8). In the newborn infant, the entire column is concave forward (curves away from a viewer facing the infant), as seen in Figure 6-10. This is the primary

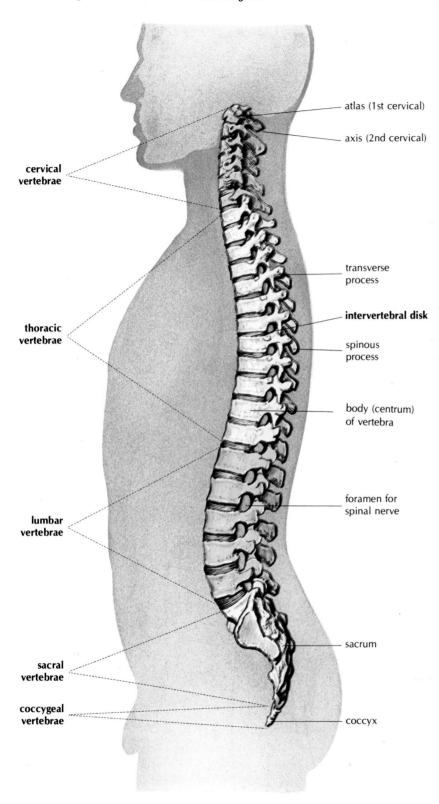

atlas (1st cervical)

axis (2nd cervical)

**cervical vertebrae**

transverse process

**intervertebral disk**

spinous process

**thoracic vertebrae**

**Figure 6-8**

Vertebral column from the side.

body (centrum) of vertebra

foramen for spinal nerve

**lumbar vertebrae**

sacrum

**sacral vertebrae**

**coccygeal vertebrae**

coccyx

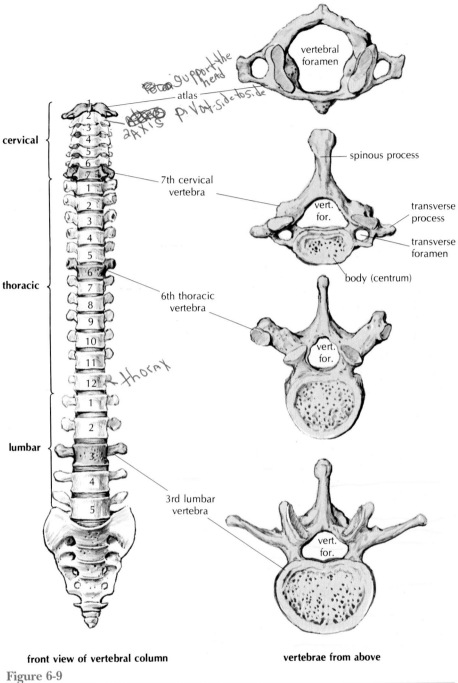

**front view of vertebral column**          **vertebrae from above**

**Figure 6-9**

Front view of the vertebral column; vertebrae from above.

curve. When the infant begins to assume an erect posture, secondary curves, which are convex (curve toward the viewer) may be noted. For example, the cervical curve appears when the infant begins to hold up the head at about 3 months of age; the lumbar curve appears when he or she begins to walk.

The thoracic and sacral curves remain the two primary curves. These curves of the vertebral column provide some of the resilience and spring so essential in balance and movement.

The bones of the ***thorax*** (Fig. 6-11) form a cone-shaped cage. Twelve pairs of ***ribs*** form the bars of this

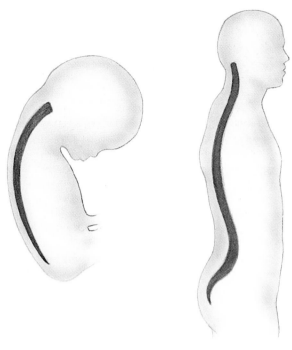

**Figure 6-10**

Curves of the infant spine and adult spine.

cage, completed by the **sternum** (STER-num), or breastbone, anteriorly.

The top portion of the sternum is the broadly T-shaped **manubrium** (mah-NU-bre-um) that joins at the top on each side with the clavicle (collarbone) and joins laterally with the first pair of ribs. The **body** of the sternum is long and bladelike. It joins along each side with ribs two through seven. Where the manubrium joins the body of the sternum, there is a slight elevation, the **sternal angle,** which easily can be felt as a surface landmark.

The lower end of the sternum consists of a small tip that is made of cartilage in youth but becomes bone in the adult. This is the **xiphoid** (ZIF-oyd) **process.** It is used as a landmark for CPR (cardiopulmonary resuscitation) to locate the region for chest compression. The thorax protects the heart, lungs, and other organs.

All 24 of the ribs are attached to the vertebral column posteriorly. However, variations in the *anterior* attachment of these slender, curved bones have led to the following classification:

1. **True ribs,** the first seven pairs, are those that attach directly to the sternum by means of individual extensions called **costal** (KOS-tal) **cartilages.**

2. **False ribs** are the remaining five pairs. Of these, the 8th, 9th, and 10th pairs attach to the cartilage of the rib above. The last two pairs have no anterior attachment at all and are known as **floating ribs.**

The spaces between the ribs, called **intercostal spaces,** contain muscles, blood vessels, and nerves.

Some changes occur in the vertebral column with age that lead to a loss in height. Approximately 1.2 cm (about 0.5 inches) are lost each 20 years beginning at age 40, owing primarily to a thinning of the intervertebral disks (between the bodies of the vertebrae). Even the vertebral bodies themselves may lose height in later years. The costal (rib) cartilages become calcified and less flexible, and the chest may decrease in diameter by 2 to 3 cm (about 1 inch), mostly in the lower part.

## BONES OF THE APPENDICULAR SKELETON

The appendicular skeleton may be considered in two divisions: upper and lower. The upper division on each side includes the shoulder, the arm (between the shoulder and the elbow), the forearm (between the elbow and the wrist), the wrist, the hand, and the fingers. The lower division includes the hip (part of the pelvic girdle), the thigh (between the hip and the knee), the leg (between the knee and the ankle), the ankle, the foot, and the toes.

The bones of the upper division may be divided into two groups for ease of study:

A. The **shoulder girdle** (Fig. 6-12) consists of two bones:
   1. The **clavicle** (KLAV-ih-kl), or collarbone, is a slender bone with two shallow curves. It joins the sternum anteriorly and the scapula laterally and helps to support the shoulder. Because it often receives the full force of falls on outstretched arms or of blows to the shoulder, it is the most frequently broken bone.
   2. The **scapula** (SKAP-u-lah), or shoulder blade, is shown from a posterior view in Figure 6-13. The **spine** of the scapula is the raised ridge that can be felt behind the shoulder in the upper portion of the back. Muscles that move the arm attach to the fossae (depressions) above and below the scapular spine. The **acromion** (ah-KRO-me-on) is the process that joins the clavicle. This can be felt as the highest point of the shoulder. Below the acromion there is a shallow socket, the **glenoid cavity,** that forms

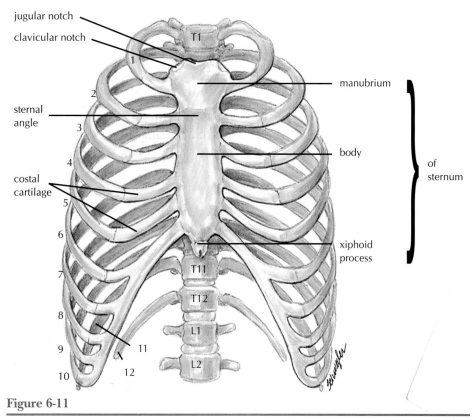

**Figure 6-11**

Bones of the thorax (anterior view). The first seven pairs of ribs are the true ribs; pairs 8 through 12 are the false ribs, of which the last two pairs are floating ribs.

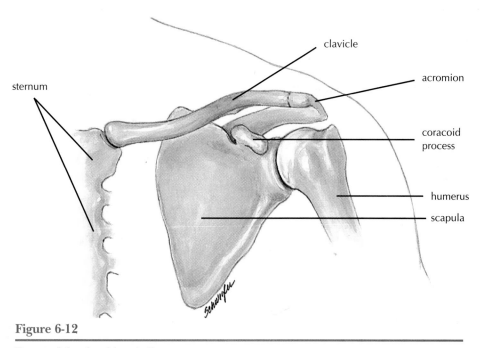

**Figure 6-12**

Bones of the shoulder girdle.

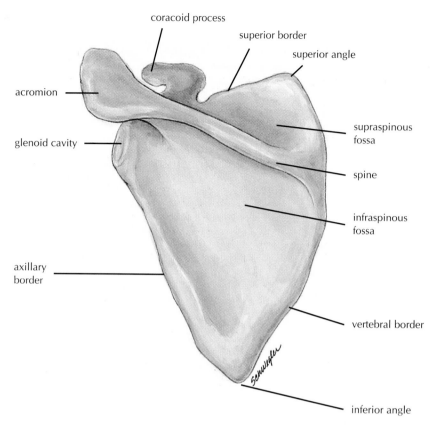

coracoid process

superior border

superior angle

acromion

glenoid cavity

supraspinous
fossa

spine

infraspinous
fossa

axillary
border

vertebral border

inferior angle

**Figure 6-13**

Left scapula (posterior view).

a ball-and-socket joint with the arm bone (humerus). Medial to the glenoid cavity is the ***coracoid process*** to which muscles attach.

B. Each ***upper extremity***, or upper limb, consists of the following bones:

1. The proximal bone, the ***humerus*** (HU-mer-us), or arm bone, forms a joint with the scapula above and with the two forearm bones at the elbow (Fig. 6-14).

2. The forearm bones are the ***ulna*** (UL-nah), which lies on the medial, or little finger, side, and the ***radius*** (RA-de-us), on the lateral, or thumb, side. When the palm is up, or forward, the two bones are parallel; when the palm is turned down, or back, the lower end of the radius moves around the ulna so that the shafts of the two bones are crossed. In this position, a distal projection of the ulna (the styloid process) pops up at the outside of the wrist. The top of the ulna has the large ***olecranon*** (o-LEK-rah-non) that forms the point of the elbow. The pulley-like ***trochlea*** (TROK-le-ah) of the distal humerus fits into

the deep ***trochlear notch*** of the ulna, allowing a hinge action at the elbow joint.

3. The wrist contains eight small ***carpal*** (KAR-pal) ***bones*** arranged in two rows of four each. The names of these eight different bones are given in Figure 6-15.

4. Five ***metacarpal bones*** are the framework for the palm of each hand. Their rounded distal ends form the knuckles.

5. There are 14 ***phalanges*** (fah-LAN-jeze), or finger bones, in each hand, two for the thumb and three for each finger. Each of these bones is called a ***phalanx*** (FA-lanx). They are identified as the first, or proximal, which is attached to a metacarpal; the second, or middle; and the third, or distal. Note that the thumb has only two phalanges, a proximal and a distal.

The bones of the lower division are grouped together in a similar fashion:

A. The bony pelvis supports the trunk and the organs in the lower abdomen, or pelvic cavity, including the urinary bladder, the internal repro-

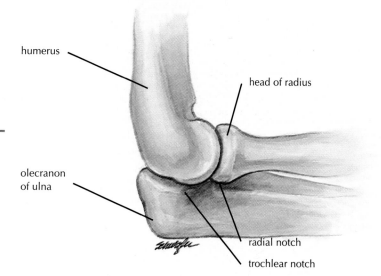

**Figure 6-14**

Lateral view of the right elbow.

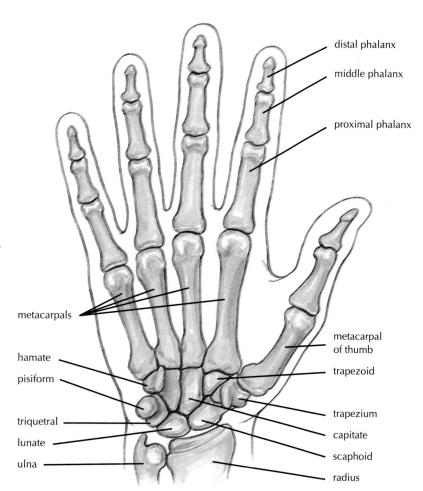

**Figure 6-15**

Bones of the right hand, anterior view.

ductive organs, and parts of the intestine. The female pelvis is adapted for pregnancy and childbirth; it is broader and lighter than the male pelvis (Fig. 6-16).

The *pelvic girdle* is a strong bony ring that forms the walls of the pelvis. It is composed of two hipbones, which form the front and the sides of the ring, and the sacrum, which joins with the hipbones to complete the ring at the back (posteriorly). Each hipbone, or *os coxae,* begins its development as three separate bones which later fuse:

1. The *ilium* (IL-e-um), which forms the upper, flared portion. The *iliac* (IL-e-ak) *crest* is the curved rim along the upper border of the ilium. It can be felt just below the waist. At either end of the crest are two bony projections. The most prominent of these is the *anterior superior iliac spine*, which is often used as a surface landmark in diagnosis and treatment.

2. The *ischium* (IS-ke-um), which is the lowest and strongest part. The *ischial* (IS-ke-al) *spine* at the back of the pelvic outlet is used as a point of reference during childbirth to indicate the progress of the presenting part (usually the baby's head) down the birth canal. Just below this spine is the large *ischial tuberosity*, which helps support the weight of the trunk when one sits down.

3. The *pubis* (PU-bis), which forms the anterior part. The joint formed by the union of the two hip bones anteriorly is called the *symphysis* (SIM-fih-sis) *pubis.*

   Portions of all three bones contribute to the formation of the *acetabulum* (as-eh-TAB-u-lum), the deep socket that holds the head of the femur (thigh bone) to form the hip joint. The largest foramina in the entire body are found in the pelvic girdle, near the front of each hipbone, one on each side of the symphysis pubis. Called the *obturator* (OB-tu-ra-tor) *foramina*, these are partially covered by a membrane.

B. Each *lower extremity*, or lower limb, consists of the following bones:
1. The thigh bone, called the *femur* (FE-mer), is the longest and strongest bone in the body. Proximally it has a large ball-shaped head that joins the os coxae (Fig. 6-17). The large lateral projection at the top of the femur is the *greater trochanter* (tro-KAN-ter), used as a surface landmark. The *lesser trochanter*, a smaller elevation, is located on the medial side.

2. The *patella* (pah-TEL-lah), or kneecap, is embedded in the tendon of the large anterior thigh muscle, the quadriceps femoris, where it crosses the knee joint. It is an example of a *sesamoid* (SES-ah-moyd) *bone*, a type of bone that develops within a tendon or a joint capsule.

3. There are two bones in the leg. Medially (on the great toe side), the *tibia,* or shin bone, is the longer, weight-bearing bone. It has a sharp anterior ridge that can be felt at the surface. Laterally, the slender *fibula* (FIB-u-lah) does not reach the knee joint; thus, it is not a weight-bearing bone. The *medial malleolus* (mal-LE-o-lus) is a downward projection at the lower end of the tibia; it forms the prominence on the inner aspect of the ankle (Fig. 6-18). The *lateral malleolus,* at the lower end of the fibula, forms the prominence on the outer aspect of the ankle.

4. The structure of the foot is similar to that of the hand. However, the foot supports the weight of the body, so it is stronger and less mobile than the hand. There are seven *tarsal bones* associated with the ankle and foot. These are named and illustrated in Figure 6-18. The largest of these is the *calcaneus* (kal-KA-ne-us), or heel bone. Five *metatarsal bones* form the framework of the instep, and the heads of these bones form the ball of the foot.

5. The *phalanges* of the toes are counterparts of those in the fingers. There are three of these in each toe except for the great toe, which has only two.

## ▶ The Joints

An *articulation,* or *joint,* is an area of junction or union between two or more bones.

### Kinds of Joints

Joints are classified into three main types on the basis of the material between the bones (Fig. 6-19).

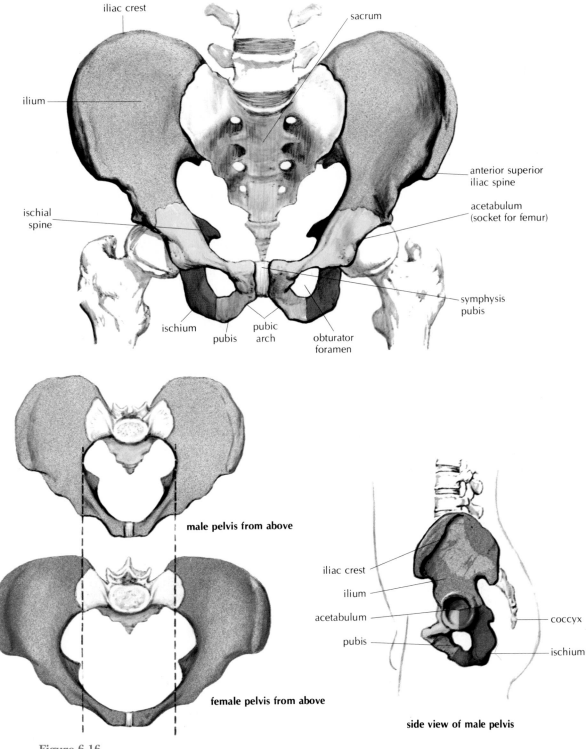

iliac crest

sacrum

ilium

anterior superior
iliac spine

ischial
spine

acetabulum
(socket for femur)

symphysis
pubis

ischium

pubis

pubic
arch

obturator
foramen

**male pelvis from above**

**female pelvis from above**

iliac crest

ilium

acetabulum

pubis

coccyx

ischium

**side view of male pelvis**

**Figure 6-16**

Pelvic girdle showing male pelvis and female pelvis.

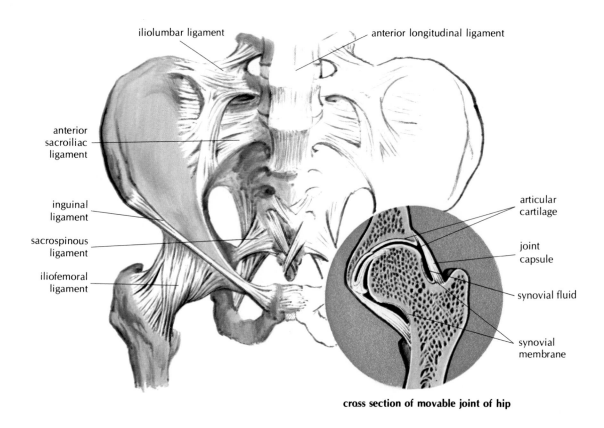

iliolumbar ligament

anterior longitudinal ligament

anterior
sacroiliac
ligament

inguinal
ligament

sacrospinous
ligament

iliofemoral
ligament

articular
cartilage

joint
capsule

synovial fluid

synovial
membrane

**cross section of movable joint of hip**

Hip joint and pelvis, showing ligaments.

They may also be classified according to the degree of movement permitted:

1. ***Fibrous joint.*** The bones in this type of joint are held together by fibrous connective tissue. An example is a ***suture*** (SU-chur) between bones of the skull. This type of joint is immovable and is termed a ***synarthrosis*** (sin-ar-THRO-sis).

2. ***Cartilaginous joint.*** The bones in this type of joint are connected by cartilage. Examples are the joint between the pubic bones of the pelvis and the joints between the bodies of the vertebrae. This type of joint is slightly movable and is termed an ***amphiarthrosis*** (am-fe-ar-THRO-sis).

3. ***Synovial*** (sin-O-ve-al) ***joint.*** The bones in this type of joint have a potential space between them. It is called the ***joint cavity,*** and it contains a small amount of thick colorless fluid that resembles uncooked egg white. This lubricant, ***synovial fluid,*** is secreted by the membrane that lines the joint cavity. This type of joint is freely movable and is

termed a ***diarthrosis*** (di-ar-THRO-sis). Most joints are synovial joints; they are described in more detail below.

Table 6-1 gives a summary of joint types.

## Structure of Synovial Joints

The bones in freely movable joints are held together by ***ligaments,*** bonds of fibrous connective tissue. Additional ligaments reinforce and help stabilize the joints at various points (see Fig. 6-17). Also, for strength and protection there is a ***joint capsule*** of connective tissue that encloses each joint and is continuous with the periosteum of the bones. The bone surfaces in freely movable joints are protected by a smooth layer of gristle called the ***articular*** (ar-TIK-u-lar) ***cartilage.***

Near some joints are small sacs called ***bursae*** (BER-se), which are filled with synovial fluid. These lie in areas subject to stress and help ease movement over and around the joints.

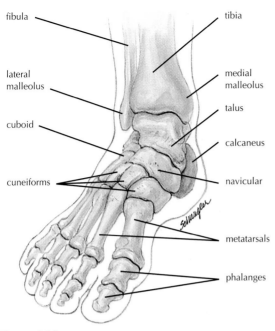

fibula

tibia

lateral malleolus

medial malleolus

talus

cuboid

calcaneus

cuneiforms

navicular

metatarsals

phalanges

**Figure 6-18**

Bones of the right foot.

### Types of Synovial Joints

Synovial joints may be classified according to the types of movement they allow as follows:

1. *Gliding* joint, in which the bone surfaces slide over one another. Examples are joints in the wrist and ankle.
2. *Hinge* joint, which allows movement in one direction, changing the angle of the bones at the joint. Examples are the elbow joint and the joints between the phalanges of the fingers and toes.
3. *Pivot* joint, which allows rotation around the length of the bone. Examples are the joint between the first and second cervical vertebrae and the joint at the proximal ends of the radius and ulna.
4. *Condyloid* joint, which allows movement in two directions. Examples are the joint between the wrist and the bones of the forearm; the joint between the occipital bone of the skull and the first cervical vertebra (the atlas).
5. *Saddle* joint, which is like a condyloid joint, but with deeper articulating surfaces. An example is the joint between the wrist and the metacarpal bone of the thumb.
6. *Ball-and-socket* joint, which allows movement in many directions around a central point. This type of joint gives the greatest freedom of movement. Examples are the shoulder and hip joints (see Fig. 6-17).

The knee is illustrated as an example of a synovial joint in Figure 6-20. Although basically a hinge joint, the knee is very complex and has some of the properties of a gliding joint as well. An x-ray picture of the knee is shown in Figure 6-21.

### Function of Synovial Joints

The chief function of the freely movable joints is to allow for changes of position and so provide

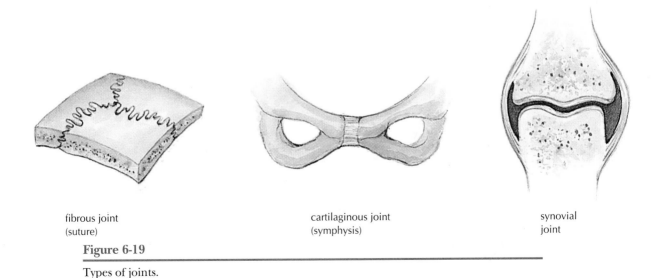

fibrous joint
(suture)

cartilaginous joint
(symphysis)

synovial joint

**Figure 6-19**

Types of joints.

## TABLE 6-1
## Types of Joints

| Name | Movement | Material Between the Bones | Examples |
|---|---|---|---|
| Fibrous | Immovable (Synarthrosis) | No joint cavity; fibrous connective tissue between bones | Sutures between bones of skull |
| Cartilaginous | Slightly movable (Amphiarthrosis) | No joint cavity; cartilage between bones | Pubic symphysis; joints between bodies of vertebrae |
| Synovial | Freely movable (Diarthrosis) | Joint cavity containing synovial fluid | Gliding, hinge, pivot, condyloid, saddle, ball-and-socket joints |

for motion. These movements are given names to describe the nature of the changes in the positions of the body parts (Fig. 6-22). For example, there are four kinds of angular movement, or movement that changes the angle between bones, as listed below:

1. *Flexion* (FLEK-shun) is a bending motion that decreases the angle between bones, as in bending the fingers to close the hand.
2. *Extension* is a straightening motion that increases the angle between bones, as in straightening the fingers to open the hand.

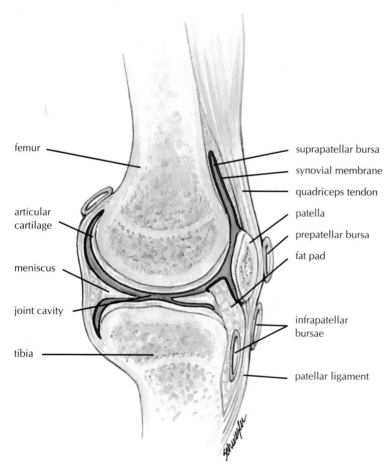

femur

articular cartilage

meniscus

joint cavity

tibia

suprapatellar bursa

synovial membrane

quadriceps tendon

patella

prepatellar bursa

fat pad

infrapatellar bursae

patellar ligament

**Figure 6-20**

The knee joint (sagittal section).

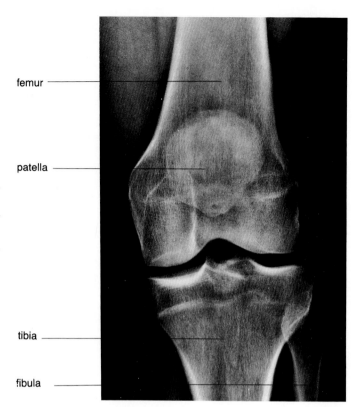

femur

patella

tibia

fibula

**Figure 6-21**

X-ray (radiograph) of the knee joint showing the articulating bones. (Greenspan A: Orthopedic Radiology, p. 6.2. New York: Gower 1988)

3. *Abduction* (ab-DUK-shun) is movement away from the midline of the body, as in moving the arms straight out to the sides.
4. *Adduction* is movement toward the midline of the body, as in bringing the arms back to their original position beside the body.

A combination of these angular movements enables one to execute a movement referred to as *circumduction* (ser-kum-DUK-shun). To perform this movement, stand with your arm outstretched and draw a large imaginary circle in the air. Note the smooth combination of flexion, abduction, extension, and adduction that makes circumduction possible. *Rotation* refers to a twisting or turning of a bone on its own axis, as in turning the head from side to side to say "no."

There are special movements that are characteristic of the forearm and the ankle:

1. *Supination* (su-pin-A-shun) is the act of turning the palm up or forward; *pronation* (pro-NA-shun) turns the palm down or backward.
2. *Inversion* (in-VER-zhun) is the act of turning the sole inward, so that it faces the opposite foot; *eversion* (e-VER-zhun) turns the sole outward, away from the body.
3. In *dorsiflexion* (dor-sih-FLEK-shun) the foot is bent upward at the ankle, narrowing the angle between the leg and the top of the foot; in *plantar flexion* the toes point downward, as in toe dancing, flexing the arch of the foot.

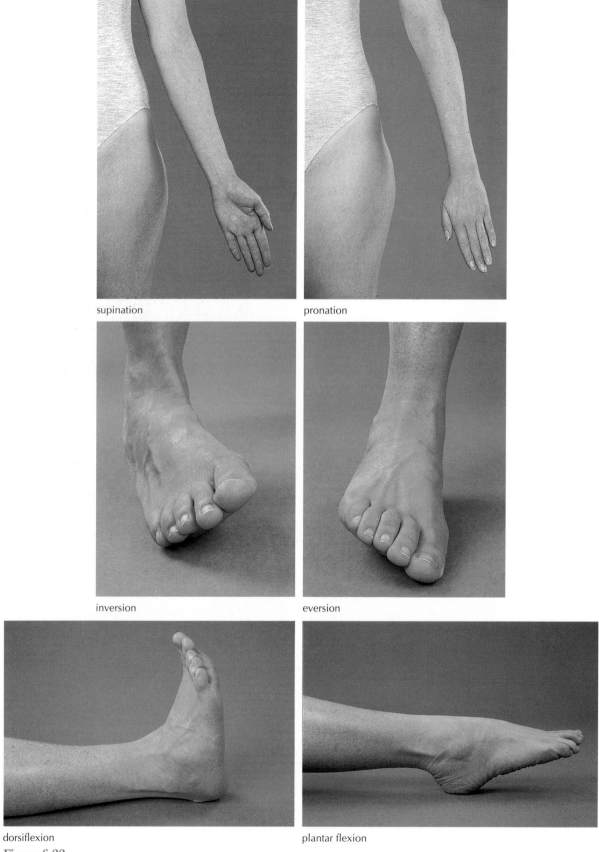

supination

pronation

inversion

eversion

dorsiflexion

plantar flexion

**Figure 6-22**

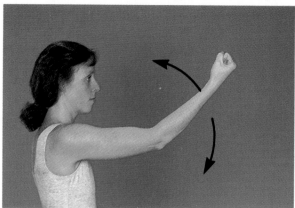

flexion/extension

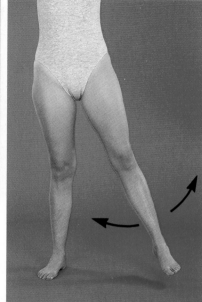

abduction/adduction

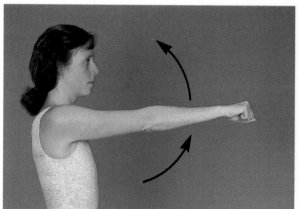

circumduction

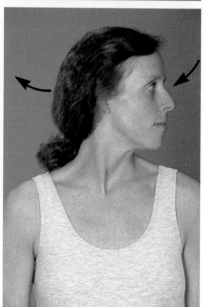

rotation

**Figure 6-22**

Movements at synovial joints.

# SUMMARY

**I. Bones**
  **A.** Structure
    **1.** Cells
      **a.** Osteoblasts—bone-forming cells
      **b.** Osteocytes—mature bone cells that maintain bone
      **c.** Osteoclasts—bone cells that break down (resorb) bone
    **2.** Tissue
      **a.** Compact—shaft (diaphysis) of long bones; outside of other bones
      **b.** Spongy—end (epiphysis) of long bones; center of other bones
    **3.** Marrow
      **a.** Red—ends of long bones, center of other bones
      **b.** Yellow—center of long bones
    **4.** Membranes—contain bone-forming cells
      **a.** Periosteum—covers bone
      **b.** Endosteum—lines marrow cavity
  **B.** Long bone growth—begins in center of shaft and continues at ends of bone; growing area forms line across epiphysis
  **C.** Bone functions—serve as body framework; protect organs; serve as levers; store calcium salts; form blood cells
  **D.** Bone markings
    **1.** Projections—head, process, crest, spine
    **2.** Depressions and holes—foramen, sinus, fossa, meatus

**II. Divisions of the skeleton**
  **A.** Axial
    **1.** Head
      **a.** Cranial—frontal, parietal, temporal, ethmoid, sphenoid, occipital
      **b.** Facial—mandible, maxillae, zygomatic, nasal, lacrimal, vomer, palatine, inferior nasal conchae
      **c.** Other—ossicles (of ear), hyoid
      **d.** Infant skull—fontanelles (soft spots)
    **2.** Trunk
      **a.** Vertebral column—cervical, thoracic, lumbar, sacral, coccygeal
      **b.** Thorax
        **(1)** Sternum—manubrium, body, xiphoid process
        **(2)** Ribs
          **(a)** True—first seven pairs
          **(b)** False—remaining five pairs, including two floating ribs
  **B.** Appendicular
    **1.** Shoulder girdle—clavicle, scapula
    **2.** Upper extremity—humerus, ulna, radius, carpals, metacarpals, phalanges
    **3.** Pelvic girdle—os coxae: ilium, ischium, pubis
    **4.** Lower extremity—femur, patella, tibia, fibula, tarsals, metatarsals, phalanges

**III. Joints**
  **A.** Types
    **1.** Fibrous—immovable (synarthrosis)
    **2.** Cartilaginous—slightly movable (amphiarthrosis)
    **3.** Synovial—freely movable (diarthrosis)
  **B.** Structure of synovial joints
    **1.** Joint cavity—contains synovial fluid
    **2.** Ligaments—hold joint together
    **3.** Joint capsule—strengthens and protects joint
    **4.** Articular cartilage—covers ends of bones
    **5.** Bursae—fluid-filled sacs near joints; cushion and protect joints and surrounding tissue
  **C.** Types of synovial joints—gliding, hinge, pivot, condyloid, saddle, ball-and-socket
  **D.** Movement at synovial joints—flexion, extension, abduction, adduction, circumduction, rotation, supination, pronation, inversion, eversion, dorsiflexion, plantar flexion

## QUESTIONS FOR STUDY AND REVIEW

1. Explain the difference between the terms in each of the following pairs:
   a. *osteoblast* and *osteoclast*
   b. *red marrow* and *yellow marrow*
   c. *periosteum* and *endosteum*
   d. *compact bone* and *spongy bone*
   e. *epiphysis* and *diaphysis*
2. Define *resorption* as it applies to bone.
3. Name five general functions of bones.
4. Name the main cranial and facial bones.
5. What is a sinus? Name several bones in which there are sinuses.
6. Define *fontanelle* and name the largest fontanelle.
7. What are the main divisions of the vertebral column? the ribs?
8. Name the bones of the upper and lower appendicular skeleton.
9. Name the bones on which you would find the following bone markings: mastoid process, acromion, olecranon, iliac crest, trochanter, medial malleolus, lateral malleolus, trochlea, ischial tuberosity, acetabulum, glenoid cavity, xiphoid process.

10. What is a foramen? Give at least two examples of foramina.
11. Describe the effects of aging on the skeletal system.
12. List three types of joints. Describe the amount of movement that occurs at each type. Find an example of each type in the chapter illustrations.
13. Describe where each of the following is found in a synovial joint: joint capsule, articular cartilage, and synovial fluid.
14. Name six types of synovial joints and give one example of the location of each. Find one example of each in the chapter illustrations.
15. Differentiate between the terms in each of the following pairs:
    a. *flexion* and *extension*
    b. *abduction* and *adduction*
    c. *supination* and *pronation*
    d. *inversion* and *eversion*
    e. *circumduction* and *rotation*
    f. *dorsiflexion* and *plantar flexion*

# CHAPTER 7

# The Muscular System

## Behavioral Objectives

After careful study of this chapter, you should be able to:

- List the characteristics of skeletal muscle
- Briefly describe how muscles contract
- List the substances needed in muscle contraction and describe the function of each
- Define the term *oxygen debt*
- List several compounds that delay the onset of oxygen debt
- Cite the effects of exercise on muscles
- Compare isotonic and isometric contractions
- Explain how muscles work in pairs to produce movement
- Compare the workings of muscles and bones to lever systems
- Explain how muscles are named
- Name some of the major muscles in each muscle group and describe the main function of each
- Describe how muscles change with age

Memmler, RL, Cohen, BJ, Wood, DL. *STRUCTURE AND FUNCTION OF THE HUMAN BODY*, 6/e,
© 1996 Lippincott-Raven Publishers

There are three kinds of muscle tissue: skeletal, smooth, and cardiac muscle. Smooth muscle makes up the walls of the hollow body organs as well as those of the blood vessels and respiratory passageways. It moves involuntarily, producing the wavelike motions of peristalsis that move substances through a system. Cardiac muscle, also involuntary, makes up the wall of the heart and creates the pulsing action of that organ. Skeletal muscle is attached to the bones and produces movement of the skeleton. This chapter is concerned with the skeletal muscles, which comprise the largest amount of muscle tissue of the body. There are more than 650 individual muscles in this muscular system; together they make up about 40% of the total body weight. Although each muscle is a distinct structure, muscles usually act in groups to execute body movements. Skeletal muscle is also known as *voluntary muscle*, because it is usually under conscious control.

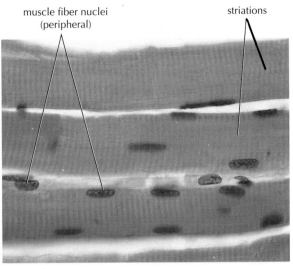

**Figure 7-1**

Microscopic view of skeletal muscle fibers (cells) showing multiple nuclei and striations. (Cormack DH: Essential Histology. Philadelphia, JB Lippincott, 1993)

## ▶ Characteristics of Skeletal Muscle

### Structure

When seen through a microscope (Fig. 7-1), individual skeletal muscle cells are long and threadlike; therefore they are often called *muscle fibers*. An extraordinary feature of these cells is that each one has many nuclei and so is said to be *multinucleated*. Because of the striped or banded appearance of the cells, skeletal muscle is also described as *striated muscle*.

In forming whole muscles, these fibers are arranged in bundles, or *fascicles* (FAS-ih-kls), held together by connective tissue (Fig. 7-2). Deeper layers of connective tissue surround each individual fiber in the bundle. The entire muscle is then encased in a tough connective tissue sheath that is part of the *deep fascia*. All these supporting tissues merge to form the tendon that attaches the muscle to a bone. Each muscle also has a rich supply of blood vessels and nerves.

### Muscles in Action

Like all cells, muscle fibers show *irritability*, which is the capacity to respond to a stimulus. Skeletal muscles usually are stimulated by the electrical energy of nerve impulses coming from the brain and the spinal cord (see Chap. 8). Nerve fibers carry impulses to the muscles, each fiber branching to supply from a few to more than 100 individual muscle cells. The point at which a nerve fiber contacts a muscle cell is called the *neuromuscular junction* (Fig. 7-3). It is here that chemicals called *neurotransmitters* are released from the neuron to stimulate the muscle fiber. The stimulus received by way of the nerve ending results in an electrical current, called an *action potential*, that is transmitted along the cell membrane.

*Contractility* is another important property of muscle tissue. This is the capacity of a muscle fiber to undergo shortening and to change its shape, becoming thicker. Studies of muscle chemistry and observation of cells under the powerful electron microscope have given a concept of how muscle cells work. These studies reveal that each skeletal muscle fiber contains many threads, or filaments, made of two kinds of proteins called *actin* (AK-tin) and *myosin* (MI-o-sin). Filaments made of actin are thin and light; those made of myosin are heavy and dark. These filaments are present in alternating bundles within the muscle cell (Fig. 7-4). It is the alternating bands of light actin and heavy myosin filaments that give skeletal muscle its striated appearance. They also give a view of what occurs when muscles contract.

In movement, the myosin filaments "latch on" to the actin filaments by means of many paddle-like

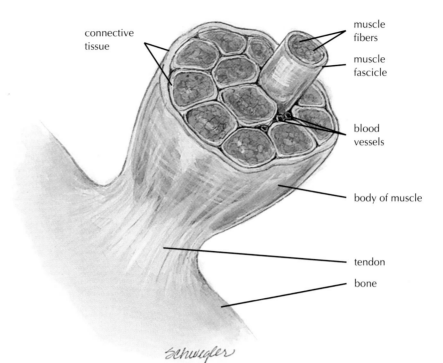

connective tissue

muscle fibers

muscle fascicle

blood vessels

body of muscle

tendon

bone

Schweigler

**Figure 7-2**

Structure of a skeletal muscle showing connective tissue coverings.

extensions, These attachments are described as *cross-bridges* between the two types of filaments. Using the energy of ATP for repeated movements, these cross-bridges, like the oars of a boat, push all the actin molecules closer together. As the overlapping filaments slide together, the muscle fiber contracts, becoming shorter and thicker (see Fig. 7-4).

It is well known that the mineral calcium is needed for muscle contraction. What calcium does is to uncover points on the actin where cross-bridges can form between the filaments so that the sliding action can begin. The calcium needed for contraction is stored within the endoplasmic reticulum (ER) of the muscle cell and is released into the cytoplasm when the cell is stimulated by a nerve fiber. Muscles relax when nerve stimulation stops and the calcium is pumped back into the ER.

### Muscles and Energy

As noted earlier, all muscle contraction requires energy in the form of ATP. The source of this energy is the oxidation, or "burning," of nutrients within the cells.

To produce ATP, muscle cells must have an adequate supply of oxygen and glucose or other nutri-

ents. These substances are constantly brought to the cells by the circulation, but muscle cells also store a small reserve supply of each. Additional oxygen is stored in the form of a compound similar to the blood's hemoglobin but located specifically in muscle cells, as indicated by the prefix *myo-* in its name, **myoglobin** (mi-o-GLO-bin). Additional glucose is stored as a compound that is built from glucose molecules and is called **glycogen** (GLI-ko-jen). Glycogen can be broken down into glucose when needed by the muscle cells, for example during vigorous exercise.

Oxygen plays an important role in preventing the accumulation of **lactic acid,** a waste product of metabolism that causes muscle fatigue. During strenuous activity, a person may not be able to breathe in oxygen rapidly enough to meet the needs of the hard-working muscles. The increased demands are met at first by the energy-rich compounds that are stored in the tissues. However, continual exercise depletes these stores. For a short time glucose may be used without the benefit of oxygen. In this case lactic acid is formed in the cells by an alternate pathway of metabolism. This anaerobic process (occurring without oxygen) permits greater magnitude of

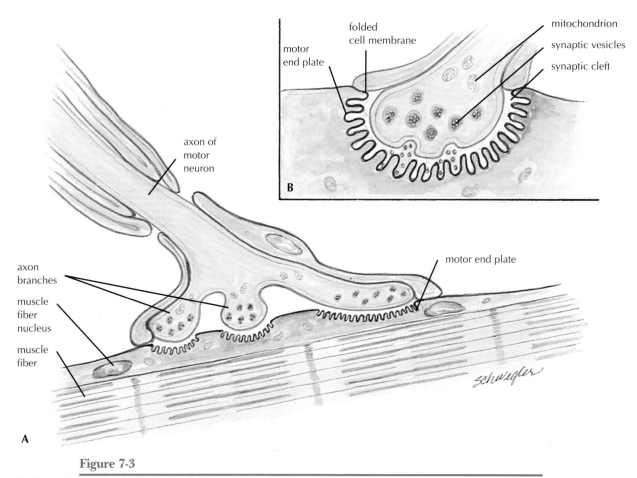

**Figure 7-3**

Neuromuscular junction. (**A**) The branched end of a motor neuron makes contact with the membrane of a muscle fiber (cell). (**B**) Enlarged view.

activity than would otherwise be possible, as, for example, allowing sprinting instead of jogging.

Anaerobic metabolism can continue until build-up of lactic acid causes the muscle to fatigue. As lactic acid accumulates, the person is said to develop an ***oxygen debt***. After stopping exercise, he or she must continue to take in more oxygen by continued rapid breathing (panting) until the debt is paid in full. That is, enough oxygen must be taken in to convert the lactic acid to other substances that can be used in metabolism. In addition, the energy-rich compounds that are stored in the cells must be replenished.

### Effects of Exercise

Changes that occur as muscle metabolism increases during exercise cause ***vasodilation*** (vas-o-di-LA-shun)—an increase in the diameter of blood vessels—thereby allowing blood to flow more easily to these tissues. As muscles work, more blood is pumped back to the heart. The temporarily increased load on the heart acts to strengthen the heart muscle and to improve its circulation. The chambers of the heart gradually enlarge to accommodate more blood. Regular exercise also improves respiratory efficiency. Circulation in the capillaries surrounding the alveoli, or air sacs, is increased, and this brings about enhanced gas exchange and deeper breathing. Athletic training leads to more efficient distribution and use of oxygen so that the onset of oxygen debt is delayed. Even moderate regular exercise has the additional benefits of weight control, strengthening of the bones, decreased blood pressure, and decreased risk of heart attacks. The effects of exercise on the body are studied in the fields of sports medicine and exercise physiology.

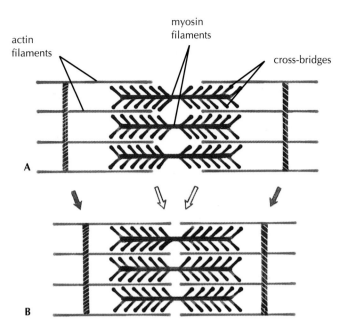

## Figure 7-4

The sliding filament mechanism of skeletal muscle contraction (**A**) Muscle is in a relaxed state. (**B**) Muscle has contracted and the cross-bridges have moved to a new site to prepare for another "push."

### Types of Muscle Contractions

Muscle *tone* refers to a partially contracted state of the muscles that is normal even when the muscles are not in use. The maintenance of this tone, or **tonus** (TO-nus), is due to the action of the nervous system in keeping the muscles in a constant state of readiness for action. Muscles that are little used soon become flabby, weak, and lacking in tone.

In addition to the partial contractions that are responsible for muscle tone, there are two other types of contractions on which the body depends:

### Rigor Mortis

After death, muscles enter a stage of rigidity known as *rigor mortis*. The calcium that triggers contraction escapes from its storage areas and causes the muscle filaments to slide together. Because all metabolism has ceased, however, there is no ATP to power further movement of the cross-bridges, and the filaments remain in a contracted position. This stiffness lasts for about 24 hours, then gradually fades as body tissues begin to decay. The rate at which these changes occur varies with the individual and the environment.

1. *Isotonic* (i-so-TON-ik) *contractions* are those in which the tone or tension within the muscle remains the same but the muscle as a whole shortens, producing movement. Lifting weights, walking, running, or any other activity in which the muscles become shorter and thicker (forming bulges) are isotonic contractions.
2. *Isometric* (i-so-MET-rik) *contractions* are those in which there is no change in muscle length but there is a great increase in muscle tension. For example, if you push against a brick wall, there is no movement, but you can feel the increased tension in your arm muscles.

Most muscles contract either isotonically or isometrically, but most movements of the body involve a combination of the two types of contraction.

## ▶ The Mechanics of Muscle Movement

### Attachments of Skeletal Muscles

Most muscles have two or more attachments to the skeleton. The method of attachment varies. In some instances the connective tissue of the muscle ties directly to the periosteum of the bone. In other cases the connective tissue sheath and partitions within the muscle extend to form specialized structures that aid in attaching the muscle to the bone.

Such an extension may take the form of a cord, called a *tendon* (Fig. 7-5); alternatively, a broad sheet called an *aponeurosis* (ap-o-nu-RO-sis) may attach muscles to bones or to other muscles, as in the abdomen (Fig. 7-6) or across the top of the skull (Fig. 7-7).

Whatever the nature of the muscle attachment, the principle remains the same: to furnish a means of harnessing the power of the muscle contractions. A muscle has two (or more) attachments, one of which is connected to a more freely movable part than the other. The less movable (more fixed) attachment is called the *origin;* the attachment to the part of the body that the muscle puts into action is called the *insertion.* When a muscle contracts, it pulls on both points of attachment, bringing the more movable insertion closer to the origin and thereby causing movement of the body part (see Fig. 7-5).

### Muscles Work Together

Many of the skeletal muscles function in pairs. A movement is performed by muscle called the ***prime mover;*** the muscle that produces an opposite movement is known as the ***antagonist.*** Clearly, for any given movement, the antagonist must relax when the prime mover contracts. For example, when the biceps brachii at the front of the arm contracts to flex the arm, the triceps brachii at the back must relax; when the triceps brachii contracts to extend the arm, the biceps brachii must relax. In addition to prime movers and antagonists, there are also muscles that serve to steady body parts or to assist prime movers.

As the muscles work together, body movements are coordinated, and a large number of complicated movements can be carried out. At first, however, the nervous system must learn to coordinate any new, complicated movement. Think of a child learning to walk or to write, and consider the number of muscles she uses unnecessarily or forgets to use when the situation calls for them.

### Levers and Body Mechanics

Proper body mechanics help conserve energy and ensure freedom from strain and fatigue; con-

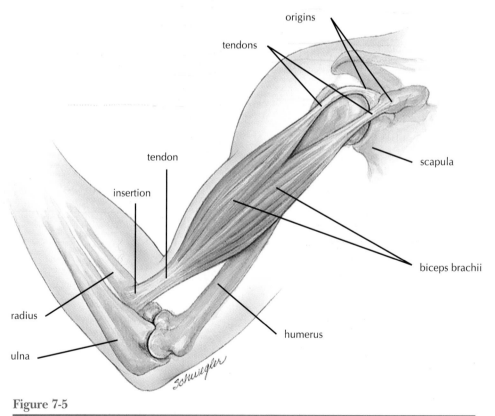

**Figure 7-5**

Diagram of a muscle showing three attachments to bones—two origins and one insertion.

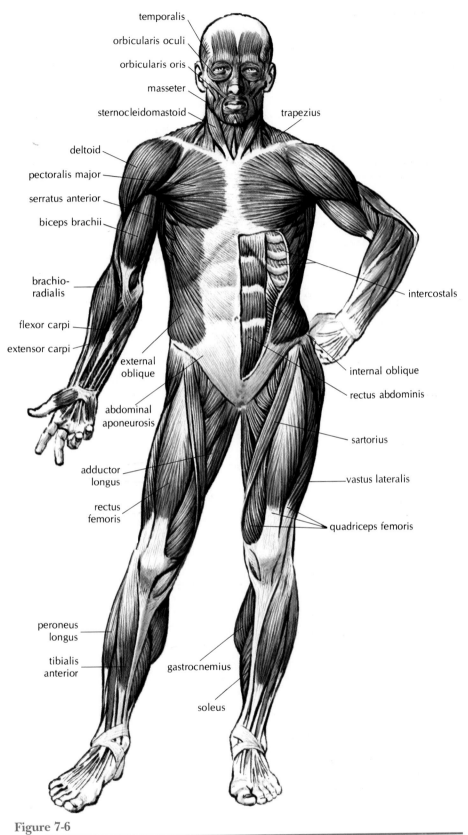

temporalis
orbicularis oculi
orbicularis oris
masseter
sternocleidomastoid
trapezius
deltoid
pectoralis major
serratus anterior
biceps brachii
brachio-
radialis
intercostals
flexor carpi
extensor carpi
external
oblique
internal oblique
rectus abdominis
abdominal
aponeurosis
sartorius
adductor
longus
vastus lateralis
rectus
femoris
quadriceps femoris
peroneus
longus
tibialis
anterior
gastrocnemius
soleus

**Figure 7-6**

Superficial muscles, anterior (front) view.

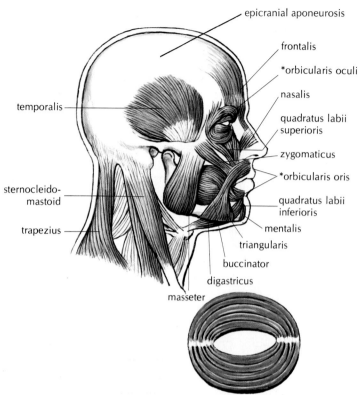

**Figure 7-7**

Muscles of the head.

*orbicularis or ring-shaped muscle

versely, such ailments as lower back pain—a common complaint—can be traced to poor body mechanics. Maintaining the body segments in correct relation to one another has a direct effect on the working capacity of the vital organs that are supported by the skeleton.

If you have had a course in physics, recall your study of levers. A lever is simply a rigid bar that moves about a fixed point, the fulcrum. There are three classes of levers, which differ only in the location of the fulcrum, the effort (force), and the resistance (weight). In a first-class lever, the fulcrum is located between the resistance and the effort; scissors, which you probably use every day, are an example of this class. The second-class lever has the resistance located between the fulcrum and the effort; a mattress lifted at one end is an illustration of this class. In the third-class lever, the effort is between the resistance and the fulcrum. A forceps or a tweezers is an example of this type of lever. The effort is applied in the center of the tool, between the fulcrum where the pieces join and the resistance at the tip.

The musculoskeletal system can be considered a system of levers, in which the bone is the lever, the joint is the fulcrum, and the force is applied by a muscle. Most lever systems in the body are of the third-class type. A muscle usually inserts over a joint and exerts force between the fulcrum and the resistance (see Fig. 7-5), in which the resistance is in the forearm and hand. By understanding and applying knowledge of levers to body mechanics, the health worker can improve his or her skill in carrying out numerous clinical maneuvers and procedures.

## ▶ Skeletal Muscle Groups

The study of muscles is made simpler by grouping them according to body regions. Knowing how muscles are named can also help in remembering them.

### Naming of Muscles

A number of different characteristics are used in naming muscles. These include:

1. Location, near a bone, for example, lateral or medial, internal or external

2. Size, using terms such as maximus, major, minor, longus, brevis
3. Shape, such as circular (orbicularis), triangular (deltoid), trapezoid
4. Direction of fibers, including straight (rectus), angled (oblique)
5. Number of heads (attachment points) as indicated by the suffix -*ceps*, as in biceps, triceps, quadriceps
6. Action, as in flexor, extensor, adductor, abductor, levator.

Often, more than one feature is used in naming. Refer to Figures 7-6 and 7-8 as you study the locations and functions of some of the skeletal muscles and try to figure out why each has the name that it does. Although they are described in the singular, most of the muscles are present on both sides of the body. The information is summarized in Table 7-1.

## Muscles of the Head

The principal muscles of the head are those of facial expression and of mastication (chewing) (see Fig. 7-7).

The muscles of facial expression include ring-shaped ones around the eyes and the lips, called the *orbicularis* (or-bik-u-LAH-ris) *muscles* because of their shape (think of "orbit"). The muscle surrounding each eye is called the *orbicularis oculi* (OK-u-li), while the muscle of the lips is the *orbicularis oris.* These muscles, of course, all have antagonists. For example, the *levator palpebrae* (PAL-pe-bre) *superioris,* or lifter of the upper eyelid, is the antagonist for the orbicularis oculi.

One of the largest muscles of expression forms the fleshy part of the cheek and is called the *buccinator* (BUK-se-na-tor). Used in whistling or blowing, it is sometimes referred to as the *trumpeter's muscle.* You can readily think of other muscles of facial expression: for instance, the antagonists of the orbicularis oris can produce a smile, a sneer, or a grimace. There are a number of scalp muscles by means of which the eyebrows are lifted or drawn together into a frown.

There are four pairs of muscles of mastication, all of which insert on the mandible and move it. The largest are the *temporal* (TEM-po-ral), located above and near the ear, and the *masseter* (mas-SE-ter) at the angle of the jaw.

The tongue has two groups of muscles. The first group, called the *intrinsic muscles,* are located entirely within the tongue. The second group, the *extrinsic muscles,* originate outside the tongue. It is because of these many muscles that the tongue has such remarkable flexibility and can perform so many different functions. Consider the intricate tongue motions involved in speaking, chewing, and swallowing.

## Muscles of the Neck

The neck muscles tend to be ribbon-like and extend up and down or obliquely in several layers and in a complex manner. The one you will hear of most frequently is the *sternocleidomastoid* (ster-no-kli-do-MAS-toyd), sometimes referred to simply as the *sternomastoid.* There are two of these strong muscles, which extend from the sternum upward, across either side of the neck, to the mastoid process. Working together, they bring the head forward on the chest (flexion). Working alone, each muscle tilts and rotates the head so as to orient the face toward the side opposite that muscle. A portion of the trapezius muscle (described later) is located in the back of the neck where it helps hold the head up (extension). Other larger deep muscles are the chief extensors of the head and neck.

## Muscles of the Upper Extremities
### MOVEMENT OF THE SHOULDER AND ARM

The position of the shoulder depends to a large extent on the degree of contraction of the *trapezius* (trah-PE-ze-us), a triangular muscle that covers the back of the neck and extends across the back of the shoulder to insert on the clavicle and scapula. The trapezius muscles enable one to raise the shoulders and pull them back. The upper portion of each trapezius can also extend the head and turn it from side to side.

The *latissimus* (lah-TIS-ih-mus) *dorsi* originates from the vertebral spine in the middle and lower back and covers the lower half of the thoracic region. The fibers of each muscle converge to a tendon that inserts on the humerus. The latissimus dorsi powerfully extends the arm, bringing it down forcibly as, for example, in swimming.

A large *pectoralis* (pek-to-RAL-is) *major* is located on either side of the upper part of the chest at the front of the body. This muscle arises from the sternum, the upper ribs, and the clavicle and forms the anterior "wall" of the arm pit or axilla; it inserts on the upper part of the humerus. The pectoralis major

*(Text continues on page 94)*

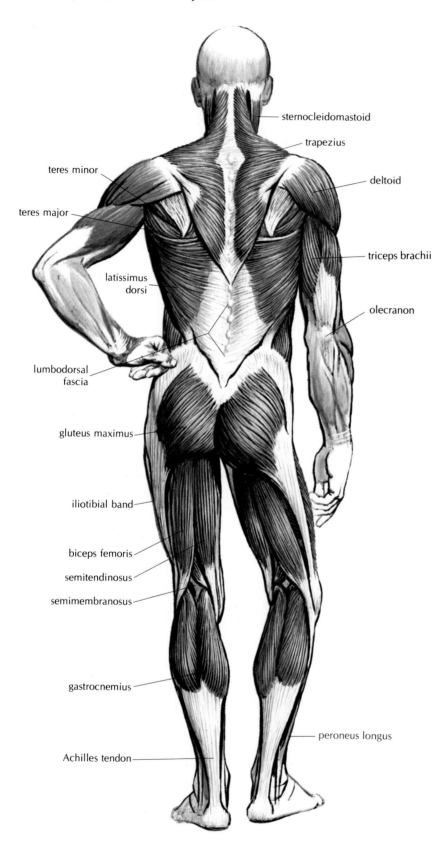

sternocleidomastoid

trapezius

teres minor

deltoid

teres major

triceps brachii

latissimus
dorsi

olecranon

lumbodorsal
fascia

gluteus maximus

iliotibial band

biceps femoris

semitendinosus

semimembranosus

gastrocnemius

peroneus longus

Achilles tendon

**Figure 7-8**

Superficial muscles, posterior
(back) view.

# TABLE 7-1
## Review of Muscles

| Name | Location | Function |
| --- | --- | --- |
| **Muscles of the Head and Neck** | | |
| Orbicularis oculi | Encircles eyelid | Closes eye |
| Levator palpebrae superioris | Back of orbit to upper eyelid | Opens eye |
| Orbicularis oris | Encircles mouth | Closes lips |
| Buccinator | Fleshy part of cheek | Flattens cheek; helps in eating, whistling, and blowing wind instruments |
| Temporal | Above and near ear | Closes jaw |
| Masseter | At angle of jaw | Closes jaw |
| Sternocleidomastoid | Along side of neck, to mastoid process | Flexes head; rotates head toward opposite side from muscle |
| **Muscles of the Upper Extremities** | | |
| Trapezius | Back of neck and upper back, to clavicle scapula | Raises shoulder and pulls it back; extends head |
| Latissimus dorsi | Middle and lower back, to humerus | Extends and adducts arm behind back |
| Pectoralis major | Upper, anterior chest, to humerus | Flexes and adducts arm across chest; pulls shoulder forward and downward |
| Serratus anterior | Below axilla on side of chest to scapula | Moves scapula forward; aids in raising arm |
| Deltoid | Covers shoulder joint, to lateral humerus | Abducts arm |
| Biceps brachii | Anterior arm, to radius | Flexes forearm and supinates hand |
| Triceps brachii | Posterior arm, to ulna | Extends forearm |
| Flexor and extensor carpi groups | Anterior and posterior forearm, to hand | Flex and extend hand |
| Flexor and extensor digitorum groups | Anterior and posterior forearm, to fingers | Flex and extend fingers |
| **Muscles of the Trunk** | | |
| Diaphragm | Dome-shaped partition between thoracic and abdominal cavities | Dome descends to enlarge thoracic cavity from top to bottom |
| Intercostals | Between ribs | Elevate ribs and enlarge thoracic cavity |
| External and internal oblique; transversus and rectus abdominis | Anterolateral abdominal wall | Compress abdominal cavity and expel substances from body; flex spinal column |
| Levator ani | Pelvic floor | Aids defecation |
| Sacrospinalis | Deep in back, vertical mass | Extends vertebral column to produce erect posture |
| **Muscles of the Lower Extremities** | | |
| Gluteus maximus | Superficial buttock, to femur | Extends thigh |
| Gluteus medius | Deep buttock, to femur | Abducts thigh |
| Iliopsoas | Crosses front of hip joint, to femur | Flexes thigh |
| Adductor group | Medial thigh, to femur | Adduct thigh |
| Sartorius | Winds down thigh, ilium to tibia | Flexes thigh and leg ( to sit cross-legged) |
| Quadriceps femoris | Anterior thigh, to tibia | Extends leg |
| Hamstring group | Posterior thigh, to tibia and fibula | Flexes leg |
| Gastrocnemius | Calf of leg, to calcaneus | Extends foot (as in tiptoeing) |
| Tibialis anterior | Anterior and lateral shin, to foot | Dorsiflexes foot (as in walking on heels); inverts foot (sole inward) |
| Peroneus longus | Lateral leg, to foot | Everts foot (sole outward) |
| Flexor and extensor digitorum groups | Posterior and anterior leg, to toes | Flex and extend toes |

flexes and adducts the arm, pulling it across the chest.

Below the axilla, on the side of the chest, is the *serratus* (ser-RA-tus) *anterior.* It originates on the upper eight or nine ribs on the side and the front of the thorax and inserts in the scapula on the side toward the vertebrae. The serratus anterior moves the scapula forward when, for example, one is pushing something. It also aids in raising the arm above the horizontal level.

The *deltoid* covers the shoulder joint and is responsible for the roundness of the upper part of the arm just below the shoulder. Arising from the shoulder girdle (clavicle and scapula), the deltoid fibers converge to insert on the lateral side of the humerus. Contraction of this muscle abducts the arm, raising it laterally to the horizontal position.

The shoulder joint allows for a very wide range of movement. This freedom of movement is possible because of the shallow socket (glenoid cavity of the scapula), which requires the support of four deep muscles and their tendons. In certain activities, such as swinging a golf club, playing tennis, or pitching a baseball, these four muscles, called the *rotator cuff,* may be injured, even torn. Surgery is often required for repair of the rotator cuff.

### MOVEMENT OF THE FOREARM AND HAND

The *biceps brachii* (BRA-ke-i), located on the front of the arm, is the muscle most often displayed when you "flex your muscles" to show your strength. It inserts on the radius and serves to flex the forearm. It is a supinator of the hand (see Fig. 7-5).

The *triceps brachii,* located on the back of the arm, inserts on the olecranon of the ulna. The triceps has been called the *boxer's muscle,* since it straightens the elbow when a blow is delivered. It is also important in pushing because it converts the arm and forearm into a sturdy rod.

Most of the muscles that move the hand and fingers originate from the radius and the ulna; some of them insert on the carpal bones of the wrist, whereas others have long tendons that cross the wrist and insert on bones of the hand and the fingers. The *flexor carpi* and the *extensor carpi muscles* are responsible for many movements of the hand. Muscles that produce finger movements are the several *flexor digitorum* (dij-e-TO-rum) and the *extensor digitorum muscles.* Special groups of muscles in the fleshy parts of the hand are responsible for the intricate movements that can be performed with the

thumb and the fingers. The freedom of movement of the thumb has been one of the most useful endowments of humans.

### Muscles of the Trunk
### MUSCLES OF RESPIRATION

The most important muscle involved in the act of breathing is the *diaphragm.* This dome-shaped muscle forms the partition between the thoracic cavity above and the abdominal cavity below (Fig. 7-9). When the diaphragm contracts, the central dome-shaped portion is pulled downward, thus enlarging the thoracic cavity from top to bottom. The *intercostal muscles* are attached to and fill the spaces between the ribs. The external and internal intercostals run at angles in opposite directions. Contraction of the intercostal muscles serves to elevate the ribs, thus enlarging the thoracic cavity from side to side and from front to back. The mechanics of breathing are described in Chapter 16.

### MUSCLES OF THE ABDOMEN AND PELVIS

The wall of the abdomen has three layers of muscle that extend from the back (dorsally) and around the sides (laterally) to the front (ventrally). They are the *external abdominal oblique* on the outside, the *internal abdominal oblique* in the middle, and the *transversus abdominis,* the innermost. The connective tissue from these muscles extends forward and encloses the vertical *rectus abdominis* of the anterior abdominal wall. The fibers of these muscles, as well as their connective tissue extensions (aponeuroses), run in different directions, resembling the layers in plywood and resulting in a very strong abdominal wall. The midline meeting of the aponeuroses forms a whitish area called the *linea alba* (LIN-e-ah Al-ba) which is an important landmark on the abdomen. It extends from the tip of the sternum to the pubic joint.

These four pairs of abdominal muscles act together to protect the internal organs and compress the abdominal cavity, as in coughing, emptying the bladder (urination) and bowel (defecation), sneezing, vomiting, and childbirth (labor). The two oblique muscles and the rectus abdominis help bend the trunk forward and sideways.

The pelvic floor, or *perineum* (per-ih-NE-um), has its own form of diaphragm, shaped somewhat like a shallow dish. One of the principal muscles of this pelvic diaphragm is the *levator ani* (le-VA-tor A-ni), which acts on the rectum and thus aids in defeca-

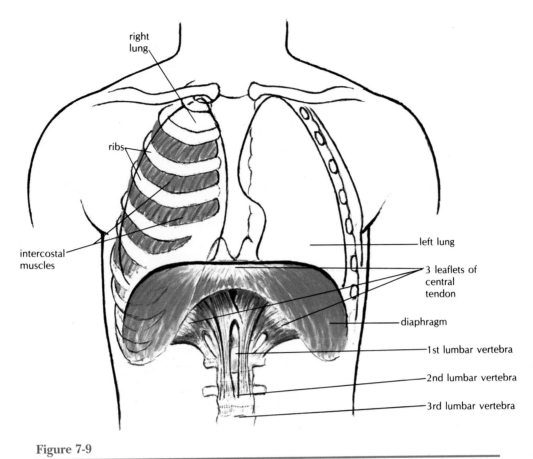

right lung

ribs

intercostal muscles

left lung

3 leaflets of central tendon

diaphragm

1st lumbar vertebra

2nd lumbar vertebra

3rd lumbar vertebra

**Figure 7-9**

The diaphragm forms the partition between the thoracic cavity and the abdominal cavity.

tion. The superficial and deep muscles of the female perineum are shown in Figure 7-10.

### DEEP MUSCLES OF THE BACK

The deep muscles of the back, which act on the vertebral column itself, are thick vertical masses that lie under the trapezius and latissimus dorsi. The longest muscle is the ***sacrospinalis*** (sa-kro-spin-A-lis), which helps maintain the vertebral column in an erect posture.

## Muscles of the Lower Extremities

The muscles in the lower extremities, among the longest and strongest muscles in the body, are specialized for locomotion and balance.

### MOVEMENT OF THE THIGH AND LEG

The ***gluteus maximus*** (GLU-te-us MAK-sim-us), which forms much of the fleshy part of the buttock, is relatively large in humans because of its support function when a person is standing in the erect position. This muscle extends the thigh and is very important in walking and running. The ***gluteus medius,*** which is partially covered by the gluteus maximus, serves to abduct the thigh, and is one of the sites used for intramuscular injections.

The ***iliopsoas*** (il-e-o-SO-as) arises from the ilium and the bodies of the lumbar vertebrae; it crosses the front of the hip joint to insert on the femur. It is a powerful flexor of the thigh and helps keep the trunk from falling backward when one is standing erect.

The ***adductor muscles*** are located on the medial part of the thigh. They arise from the pubis and ischium and insert on the femur. These strong muscles press the thighs together, as in grasping a saddle between the knees when riding a horse.

The ***sartorius*** (sar-TO-re-us) is a long, narrow muscle that begins at the iliac spine, winds downward and inward across the entire thigh, and ends

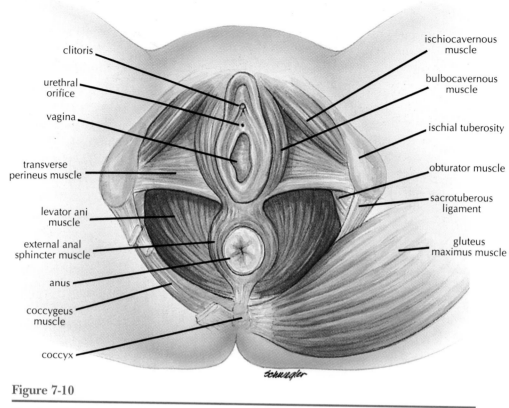

clitoris

urethral orifice

vagina

transverse perineus muscle

levator ani muscle

external anal sphincter muscle

anus

coccygeus muscle

coccyx

ischiocavernous muscle

bulbocavernous muscle

ischial tuberosity

obturator muscle

sacrotuberous ligament

gluteus maximus muscle

**Figure 7-10**

Muscles of the female perineum (pelvic floor).

on the upper medial surface of the tibia. It is called the *tailor's muscle* because it is used in crossing the legs in the manner of tailors, who in days gone by sat cross-legged on the floor.

The front and sides of the femur are covered by the **quadriceps femoris** (KWOD-re-seps FEM-or-is), a large muscle that has four heads of origin. The individual parts are: in the center, covering the anterior thigh, the **rectus femoris**; on either side the **vastus medialis** and **vastus lateralis**; deeper in the center, the **vastus intermedius.** One of these muscles (rectus femoris) originates from the ilium, and the other three are from the femur, but all four have a common tendon of insertion on the tibia. You may remember that this is the tendon that encloses the knee cap, or patella. This muscle extends the leg, as in kicking a ball.

The **hamstring muscles** are located in the posterior part of the thigh. Their tendons can be felt behind the knee as they descend to insert on the tibia and fibula. The hamstrings flex the leg on the thigh, as in kneeling. Individually, moving from lateral to me-

dial position, they are the **biceps femoris,** the **semimembranosus,** and the **semitendinosus.**

### MOVEMENT OF THE FOOT

The **gastrocnemius** (gas-trok-NE-me-us) is the chief muscle of the calf of the leg. It has been called the *toe dancer's muscle* because it is used in standing on tiptoe. It ends near the heel in a prominent cord called the **Achilles tendon,** which attaches to the calcaneus (heel bone). The Achilles tendon is the largest tendon in the body.

Another leg muscle that acts on the foot is the **tibialis** (tib-e-A-lis) **anterior,** located on the front of the leg. This muscle performs the opposite function of the gastrocnemius. Walking on the heels will use the tibialis anterior to raise the rest of the foot off the ground (dorsiflexion). This muscle is also responsible for inversion of the foot. The muscle for eversion of the foot is the **peroneus** (per-o-NE-us) **longus,** located on the lateral side of the leg. The long tendon of this muscle crosses under the foot, forming a sling that supports the transverse (metatarsal) arch.

The toes, like the fingers, are provided with flexor and extensor muscles. The tendons of the extensor muscles are located in the top of the foot and insert on the superior surface of the toe bones (phalanges). The flexor digitorum tendons cross the sole of the foot and insert on the undersurface of the toe bones.

## ▶ Effects of Aging on Muscles

Beginning at about 40 years of age, there is a gradual loss of muscle cells with a resulting decrease in the size of each individual muscle. There is also a loss of power, notably in the extensor muscles, such as the large sacrospinalis near the vertebral column. This causes the "bent over" appearance of a hunchback (kyphosis), which in women is often referred to as the *dowager's hump*. Sometimes there is a tendency to bend (flex) the hips and knees. In addition to the previously noted changes in the vertebral column (see Chap. 6), these effects on the extensor muscles result in a further decrease in the elderly person's height. Activity and exercise throughout life delay and decrease these undesirable effects of aging.

## SUMMARY

I. **Structure of skeletal muscle tissue**
   A. Muscle fibers (cells): large, multinucleated, striated (banded)
   B. Connective tissue
      1. Around individual fibers
      2. Around fascicles (bundles)
      3. Around whole muscle

II. **Characteristics of skeletal muscles**
   A. Irritability—capacity to respond to nerve impulses transmitted at neuromuscular junction
   B. Contractility—ability to shorten
      1. Contracting filaments—slide together to shorten muscle
         a. Actin—thin and light
         b. Myosin—heavy and dark
      2. ATP—supplies energy
      3. Myoglobin—stores energy
      4. Glycogen—stores glucose
      5. Oxygen debt—develops during strenuous exercise
         a. Anaerobic metabolism
         b. Yields lactic acid
   C. Muscle contractions
      1. Tonus—partially contracted state
      2. Isotonic contractions—muscle shortens to produce movement
      3. Isometric contractions—tension increases but muscle does not shorten

III. **Mechanics of movement**
   A. Attachments of skeletal muscles
      1. Tendon—cord of connective tissue that attaches muscle to bone
         a. Origin—attached to more fixed part
         b. Insertion—attached to moving part
      2. Aponeurosis—broad band of connective tissue that attaches muscle to bone or other muscle
   B. Cooperation of muscles
      1. Prime mover—initiates movement
      2. Antagonist—produces opposite movement
      3. Others—steady body parts and assist prime mover
   C. Body mechanics—muscles function with skeleton as lever systems
      1. Lever—bone
      2. Fulcrum—joint
      3. Force—muscle contraction

IV. **Skeletal muscle groups**
   A. Naming of muscles—location, size, shape, direction of fibers, number of heads, action
   B. Muscle groups
      1. Muscles of the head and neck
      2. Muscles of the upper extremities
      3. Muscles of the trunk
      4. Muscles of the lower extremities

V. **Effects of aging on muscles**
   A. Decrease in size of muscles
   B. Weakening of muscles, especially extensors

## QUESTIONS FOR STUDY AND REVIEW

1. Describe the structure of muscles from fiber to deep fascia.
2. List the main characteristics of skeletal muscle cells.
3. Differentiate between the terms in each of the following pairs:
   a. *tendon* and *aponeurosis*
   b. *muscle origin* and *muscle insertion*
   c. *prime mover* and *antagonist*
   d. *isometric contraction* and *isotonic contraction*
4. Explain the role of each of the following in muscle contraction: actin and myosin, calcium, ATP, myoglobin, glycogen.
5. When does oxygen debt occur? What is the role of lactic acid in oxygen debt? How is oxygen debt eliminated?
6. What are some valuable effects of exercise?
7. What are levers and how do they work? The forceps is an example of which class of lever? When muscles and bones act as lever systems, what part of the body is the fulcrum?
8. List six characteristics according to which muscles are named.
9. Name and describe the functions of the principal muscles of the head and neck, upper extremities, trunk, and lower extremities.
10. What effect does aging have on muscles?

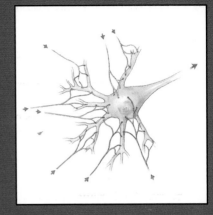

# Coordination and Control

**O**f the four chapters in this unit, two describe the nervous system and some of its many parts and complex functions. The organs of special sense (eye, ear, taste buds, and so forth) are allotted an additional separate chapter. The fourth chapter in this unit discusses hormones and the organs that produce them. Working with the nervous system, these hormones play an important role in coordination and control.

# CHAPTER 8

# The Nervous System: The Spinal Cord and Spinal Nerves

## Behavioral Objectives

After careful study of this chapter, you should be able to:

- Describe the organization of the nervous system according to structure and function
- Describe the structure of a neuron
- Briefly describe the transmission of a nerve impulse
- Explain the role of myelin in nerve conduction
- Briefly describe transmission at a synapse
- Define *neurotransmitter* and give several examples of neurotransmitters
- Name three types of nerves and explain how they differ from each other
- List the components of a reflex arc
- Describe the spinal cord and name several of its functions
- Define a reflex and give several examples
- Describe and name the spinal nerves and three of their main plexuses
- Compare the location and functions of the sympathetic and parasympathetic nervous systems

Memmler, RL, Cohen, BJ, Wood, DL. *STRUCTURE AND FUNCTION OF THE HUMAN BODY, 6/e,*
© 1996 Lippincott-Raven Publishers

None of the body systems is capable of functioning alone. All are interdependent and work together as one unit so that normal conditions (homeostasis) within the body may prevail. The nervous system serves as the chief coordinating agency. Conditions both within and outside the body are constantly changing; one purpose of the nervous system is to respond to these internal and external changes (known as *stimuli*) and so cause the body to adapt itself to new conditions. It is through the instructions and directions sent to the various organs by the nervous system that a person's internal harmony and the balance between that person and the environment are maintained. The nervous system has been compared to a telephone exchange, in that the brain and the spinal cord act as switching centers and the nerve trunks act as cables for carrying messages to and from these centers.

## The Nervous System as a Whole

The parts of the nervous system may be grouped according to structure or function. The anatomic, or structural, divisions of the nervous system are as follows (Fig. 8-1):

1. The *central nervous system* (*CNS*) includes the brain and spinal cord.
2. The *peripheral* (per-IF-er-al) *nervous system* (*PNS*) is made up of all the nerves outside the CNS. It includes all the *cranial* and *spinal nerves*. Cranial nerves are those that carry impulses to and from the brain. Spinal nerves are those that carry messages to and from the spinal cord.

From the standpoint of structure, the CNS and PNS together include most of the nerve tissue in the body. However, certain peripheral nerves have a special function, and for this reason they are grouped together under the designation *autonomic* (aw-to-NOM-ik) *nervous system*. The reason for this separate classification is that the autonomic nervous system governs activities that go on automatically (involuntarily). This system carries impulses from the CNS to the glands, to the smooth muscles found in the walls of tubes and hollow organs, and to the cardiac muscle of the heart. Both cranial and spinal nerves carry autonomic nervous system impulses. The system is subdivided into the *sympathetic* and *parasympathetic nervous systems*, both of which are explained later in this chapter.

The autonomic nervous system makes up a part of the *involuntary*, or *visceral*, *nervous system*, which controls smooth muscle, cardiac muscle, and glands. The *voluntary*, or *somatic* (so-MAH-tik), *nervous system* consists of all of the nerves that control the action of the skeletal muscles (described in Chap. 7), which are under conscious control.

## Neurons and Their Functions

The functional cells of the nervous system are highly specialized cells called *neurons* (Fig. 8-2). The cell body of each neuron contains the nucleus and other organelles. Extending from the cytoplasm of the cell body are threadlike fibers. Nerve cell fibers are of two kinds:

1. *Dendrites* (DEN-drites), which conduct impulses *to* the cell body
2. *Axons* (AK-sons), which conduct impulses *away from* the cell body.

Many dendrites have a highly branched, treelike appearance. In fact the name comes from a Greek word meaning "tree." The dendrites of sensory neurons, however, those which carry impulses toward the CNS, are different from those of other neurons: they may be long (as long as 1 meter) or short, but they are usually single instead of branched. Each of these dendrites functions as a *receptor*, where a stimulus is received and the sensory impulse begins. In the case of the special senses, such as hearing and sight (see Chap. 10), the dendrite is in contact with a modified cell that acts as the receptor.

Some axons are covered with a fatty material called *myelin* that insulates and protects the fiber (Fig. 8-3). In the peripheral nervous system this covering is produced by special myelin-containing cells, called *Schwann* (shwan) *cells,* that wrap around the axon like a jelly roll. When the sheath is complete, small spaces still remain between the cells. These tiny gaps, called *nodes,* are important in speeding the conduction of nerve impulses.

The outermost membranes of the Schwann cells form a thin coating known as the *neurilemma* (nu-rih-LEM-mah). This covering is a part of the mechanism by which some peripheral nerves repair themselves when injured. Under some circumstances, damaged nerve cell fibers may regenerate by growing into the sleeve formed by the neurilemma. Cells of the brain and the spinal cord are myelinated,

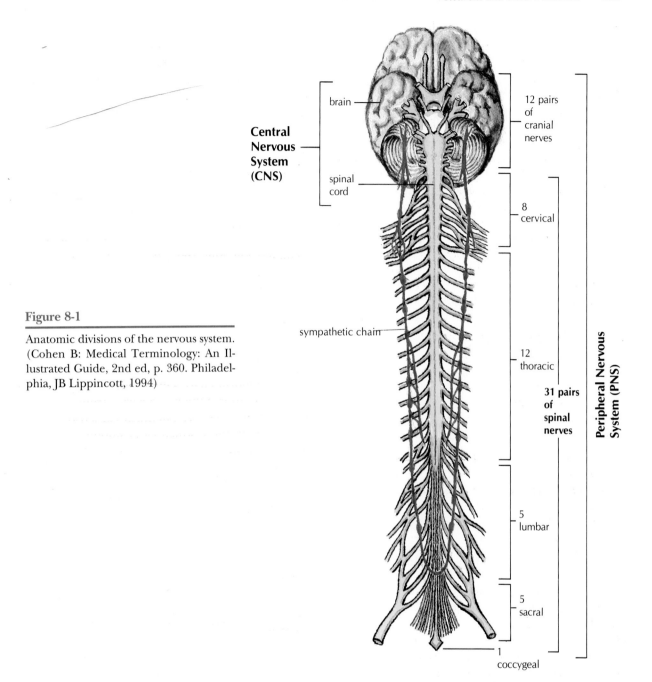

**Figure 8-1**

Anatomic divisions of the nervous system. (Cohen B: Medical Terminology: An Illustrated Guide, 2nd ed, p. 360. Philadelphia, JB Lippincott, 1994)

not by Schwann cells, but by neuroglia, the connective tissue cells of the CNS. As a result, they have no neurilemma. If they are injured, the injury is permanent. Even in the peripheral nerves, however, repair is a slow and uncertain process.

Axons covered with myelin are called *white fibers* and are found in the *white matter* of the brain and spinal cord as well as in the nerve trunks in all parts of the body. The fibers and cell bodies of the *gray matter* are not covered with myelin.

The job of neurons in the peripheral nervous system is to constantly relay information either to or from the central nervous system (CNS). Neurons that conduct impulses *to* the spinal cord and brain are described as *sensory,* or *afferent, neurons.* Those cells that carry impulses *from* the CNS out to muscles

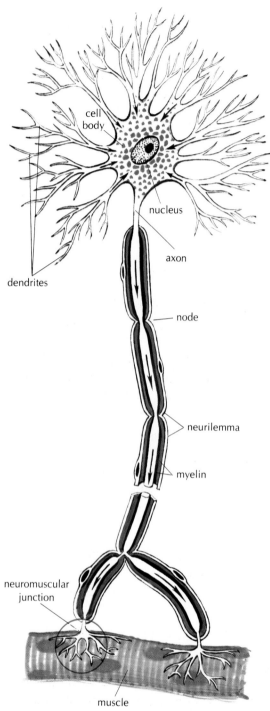

cell body

nucleus

axon

dendrites

node

neurilemma

myelin

neuromuscular junction

muscle

**Figure 8-2**

Diagram of a motor neuron. The break in the axon denotes length. The arrows show the direction of the nerve impulse.

and glands are *motor*, or *efferent, neurons*. Any organ, such as a muscle or gland, activated by a motor neuron is the *effector*.

### The Nerve Impulse

The cell membrane of an unstimulated (resting) neuron carries an electric charge. Because of positive and negative ions concentrated on either side of the membrane, the inside of the membrane at rest is negative as compared with the outside. A *nerve impulse* is a local reversal in the charge on the nerve cell membrane that then spreads along the membrane like an electric current. This sudden electrical change in the membrane is called an *action potential*. A stimulus, then, is any force that can start an action potential. This electric change results from rapid shifts in sodium and potassium ions across the cell membrane. The reversal occurs very rapidly (in less than one thousandth of a second) and is followed by a rapid return of the membrane to its original state so that it can be stimulated again.

A myelinated nerve fiber conducts impulses more rapidly than an unmyelinated fiber of the same size because the electrical impulse "jumps" from node (space) to node in the myelin sheath instead of traveling continuously along the fiber.

### The Synapse

Each neuron is a separate unit, and there is no anatomic connection between neurons. How then is it possible for neurons to communicate? In other words, how does the axon of one neuron make functional contact with the membrane of another neuron? This is accomplished by the *synapse* (SIN-aps), from a Greek word meaning "to clasp." Synapses are points of junction for the transmission of nerve impulses (Fig. 8-4).

Within the branched endings of the axon are small bubbles (vesicles) containing a type of chemical known as a *neurotransmitter*. When stimulated, the axon releases its neurotramsmitter into the narrow gap, the *synaptic cleft*, between the cells. The neurotransmitter then acts as a chemical signal to stimulate the next cell, described as the *postsynaptic cell*. On the receiving membrane, usually that of a dendrite, but sometimes another part of the cell, there are special sites, or *receptors*, ready to pick up and respond to specific neurotransmitters. Receptors in the cell membrane influence how or if that cell will respond to a given neurotransmitter.

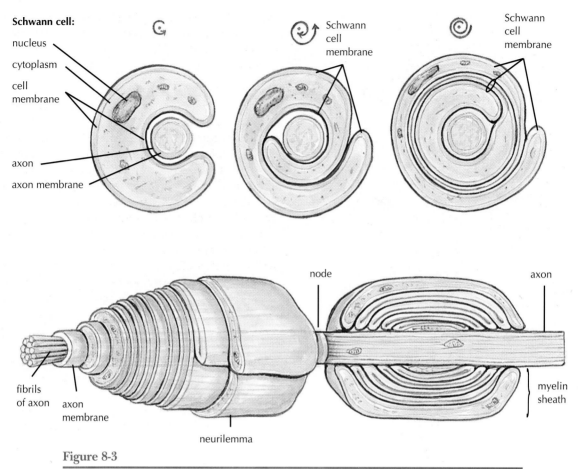

**Figure 8-3**

Formation of the myelin sheath. (**A**) Schwann cells wrap around the axon creating a myelin coating. (**B**) Outermost layer of the Schwann cell forms the neurilemma. Spaces between the cells are the nodes.

Although there are many known neurotransmitters, the main ones are *epinephprine* (ep-ih-NEF-rin), also called *adrenaline;* a related compound, *norepinephrine* (nor-ep-ih-NEF-rin), or *noradrenaline;* and *acetylcholine* (as-e-til-KO-lene). Acetylcholine (ACh) is the neurotransmitter released at the neuromuscular junction, the synapse between a neuron and a muscle cell. All three of the above neurotransmitters function in the autonomic nervous system. It is common to think of neurotransmitters as stimulating the cells they reach; in fact, they have been described as such in this discussion. Note, however, that some of these chemicals act to inhibit the postsynaptic cell and keep it from reacting.

The connections between neurons can be quite complex. One cell can branch to stimulate many receiving cells, or a single cell may be stimulated by a number of different axons (Fig. 8-5). The cell's response is based on the total effects of all the neurotransmitters it receives at any one time.

### Nerves

A *nerve* is a bundle of nerve cell fibers located *outside* the CNS. Bundles of nerve cell fibers *within* the CNS are *tracts.* Tracts are located within the brain and also within the spinal cord to conduct impulses to and from the brain. A nerve or tract can be compared to an electric cable made up of many wires. As with muscles, the "wires," or nerve cell fibers in a nerve, are bound together with connective tissue.

A few of the cranial nerves have only sensory fibers for conducting impulses toward the brain.

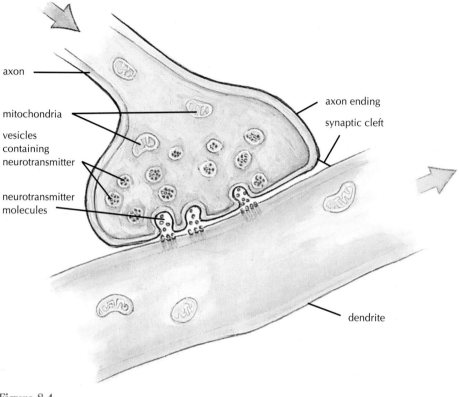

axon

mitochondria

vesicles
containing
neurotransmitter

neurotransmitter
molecules

axon ending

synaptic cleft

dendrite

**Figure 8-4**

Close-up view of a synapse. The axon ending has vesicles containing neurotransmitter, which is released across the synaptic cleft to the membrane of the next cell.

These are described as *sensory,* or *afferent, nerves.* A few of the cranial nerves contain only motor fibers for conducting impulses away from the brain and are classified as *motor,* or *efferent, nerves.* However, the remainder of the cranial nerves and *all* of the spinal nerves contain both sensory *and* motor fibers and are referred to as *mixed nerves.*

### The Reflex Arc

As the nervous system functions, both external and internal stimuli are received, interpreted, and acted on. A complete pathway through the nervous system from stimulus to response is termed a *reflex arc* (Fig. 8-6). This is the basic functional pathway of the nervous system. The parts of a typical reflex arc are:

1. *Receptor*—the end of a dendrite or some specialized receptor cell, as in a special sense organ, that detects a stimulus

2. *Sensory neuron,* or afferent neuron—a cell that transmits impulses *toward* the CNS
3. *Central neuron*—a cell or cells within the CNS. These neurons may carry impulses to and from the brain, may function within the brain, or may distribute impulses to different regions of the spinal cord.
4. *Motor neuron*—or efferent neuron—a cell that carries impulses *away from* the CNS
5. *Effector*—a muscle or a gland outside the CNS that carries out a response.

At its simplest, a reflex arc can involve just two neurons, one sensory and one motor, with a synapse in the CNS. There are very few reflex arcs that require only this minimal number of neurons. The knee jerk reflex described below is one of the few examples in humans. Most reflex arcs involve many more, even hundreds, of connecting neurons within the central nervous system. The many intricate pat-

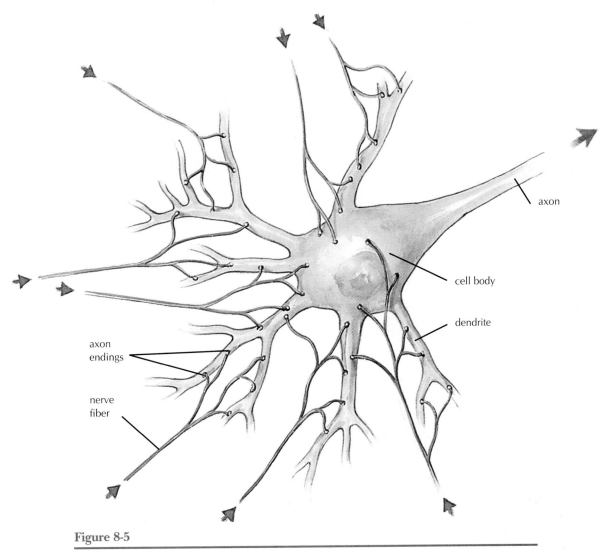

axon

cell body

dendrite

axon
endings

nerve
fiber

**Figure 8-5**

A single neuron is stimulated by axons of many other neurons.

terns that make the nervous system so responsive and adaptable also make it difficult to study, and investigation of the nervous system is one of the most active areas of research today.

## ▶ The Spinal Cord

### Location of the Spinal Cord

In the embryo, the spinal cord occupies the entire spinal canal and so extends down into the tail portion of the vertebral column. However, the column of bone grows much more rapidly than the nerve tissue of the cord, so that eventually the end of the cord no longer reaches the lower part of the spinal canal. This disparity in growth continues to increase; in the adult the cord ends in the region just below the area to which the last rib attaches (between the first and second lumbar vertebrae).

### Structure of the Spinal Cord

The spinal cord (see Fig. 8-6) has a small, irregularly shaped internal section that consists of gray matter (nerve cell bodies) and a larger area surrounding this gray part that consists of white matter

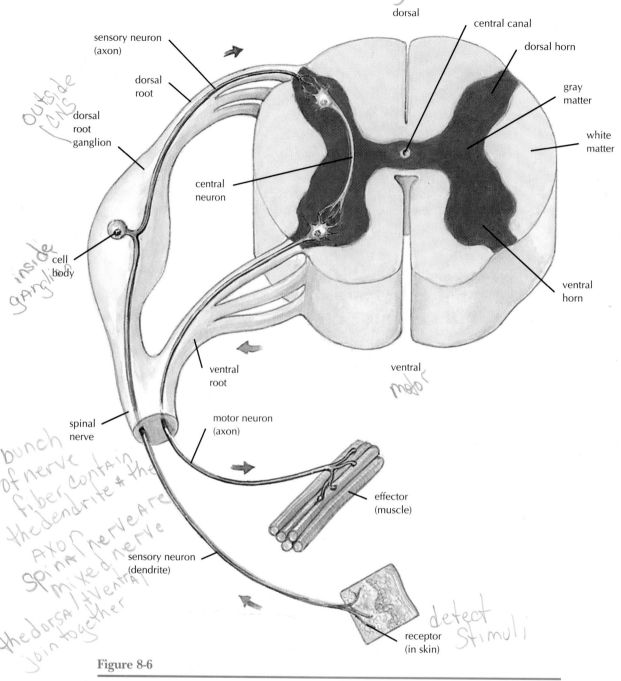

**Figure 8-6**

Reflex arc showing the pathway of impulses and a cross section of the spinal cord.

(nerve cell fibers). The gray matter is so arranged that a column of cells extends up and down dorsally, one on each side; another column is found in the ventral region on each side. These two pairs of columns, called the *dorsal* and *ventral horns,* give the gray matter an H-shaped appearance in cross sec-

tion. In the center of the gray matter is a small channel, the *central canal*, that contains cerebrospinal fluid, the liquid that circulates around the brain and spinal cord. The white matter consists of thousands of nerve cell fibers arranged in three areas external to the gray matter on each side.

## Functions of the Spinal Cord

The spinal cord is a key link between the spinal nerves and the brain. It is also a place where simple responses, known as *reflexes* can be coordinated even without involving the brain. The functions of the spinal cord may be divided into three categories:

1. *Conduction of sensory impulses* upward through ascending tracts to the brain
2. *Conduction of motor impulses* from the brain down through descending tracts to the efferent neurons that supply muscles or glands
3. *Reflex activities.* A reflex is a simple, rapid, and automatic response involving very few neurons.

When you fling out an arm or leg to catch your balance, withdraw from a painful stimulus, or blink to avoid an object approaching your eyes, you are experiencing reflex behavior. A reflex pathway that passes through the spinal cord alone and does not involve the brain is termed a *spinal reflex.* The *stretch reflex,* in which a muscle is stretched and responds by contracting, is one example. If you tap the tendon below the kneecap (the patellar tendon), the muscle of the anterior thigh (quadriceps femoris) contracts, eliciting the knee jerk. Such stretch reflexes may be evoked by appropriate tapping of most large muscles (such as the triceps brachii in the arm and the gastrocnemius in the calf of the leg). Because reflexes occur automatically, they are used in physical examinations to test the condition of the nervous system.

## ▶ Spinal Nerves

### Location and Structure of the Spinal Nerves

There are 31 pairs of spinal nerves, each pair numbered according to the level of the spinal cord from which it arises. Each nerve is attached to the spinal cord by two roots, the *dorsal root* and the *ventral root* (see Fig. 8-6). On each dorsal root is a marked swelling of gray matter called the *dorsal root ganglion,* which contains the cell bodies of the sensory neurons. A *ganglion* (GANG-le-on) is any collection of nerve cell bodies located outside the central nervous system. Nerve cell fibers from sensory receptors throughout the body lead to these dorsal root ganglia.

Whereas sensory fibers form the dorsal roots, the ventral roots of the spinal nerves are a combination of motor (efferent) nerve fibers supplying voluntary muscles, involuntary muscles, and glands. The cell bodies of these neurons are located in the ventral gray matter (ventral horns) of the cord. Because the dorsal (sensory) and ventral (motor) roots are combined to form the spinal nerve, all spinal nerves are mixed nerves.

### Branches of the Spinal Nerves

Each spinal nerve continues only a very short distance away from the spinal cord and then branches into small posterior divisions and rather large anterior divisions. The larger anterior branches interlace to form networks called *plexuses* (PLEK-sus-eze), which then distribute branches to the body parts. The three main plexuses are described as follows:

1. The *cervical plexus* supplies motor impulses to the muscles of the neck and receives sensory impulses from the neck and the back of the head. The phrenic nerve, which activates the diaphragm, arises from this plexus.
2. The *brachial* (BRA-ke-al) *plexus* sends numerous branches to the shoulder, arm, forearm, wrist, and hand. The radial nerve emerges from the brachial plexus.
3. The *lumbosacral* (lum-bo-SA-kral) *plexus* supplies nerves to the lower extremities. The largest of these branches is the *sciatic* (si-AT-ik) *nerve,* which leaves the dorsal part of the pelvis, passes beneath the gluteus maximus muscle, and extends down the back of the thigh. At its beginning it is nearly 1 inch thick, but it soon branches to the thigh muscles; near the knee it forms two subdivisions that supply the leg and the foot.

## ▶ The Autonomic Nervous System

### Parts of the Autonomic Nervous System

Although the internal organs such as the heart, lungs, and stomach contain nerve endings and nerve fibers for conducting sensory messages to the brain and cord, most of these impulses do not reach consciousness. These *afferent* impulses from the viscera are translated into reflex responses without reaching the higher centers of the brain; the sensory neurons from the organs are grouped with those that come from the skin and voluntary muscles.

## Dermatomes

Sensory neurons from all over the skin, except for the skin of the face and scalp, feed information into the spinal cord through the spinal nerves. The skin surface can be mapped into distinct regions that are supplied by a single spinal nerve. Each of these regions is called a ***dermatome*** (DER-mah-tome) (see illustration).

Stimulation of the skin within a given dermatome will be carried over the corresponding spinal nerve.

This information can be used in diagnosis to identify the spinal nerve or spinal segment that is involved in an injury. In some areas, the dermatomes are not absolutely distinct. Some dermatomes may share a nerve supply with neighboring regions. For this reason, it is necessary to numb several adjacent dermatomes to achieve successful anesthesia.

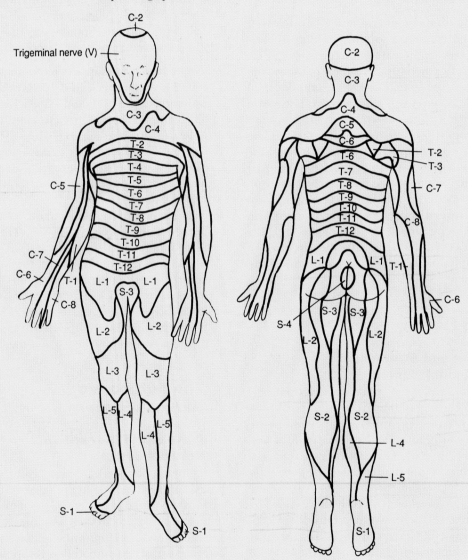

(From Jones SA, Weigel A, White RD, McSwain Jr. NE, Breiler M. Advanced Emergency Care for Paramedic Practice, Fig. 22-8, p. 545. Philadelphia, JB Lippincott, 1992)

In contrast, the *efferent* neurons, which supply the glands and the involuntary muscles, are arranged very differently from those that supply the voluntary muscles. This variation in the location and arrangement of the *visceral efferent* neurons has led to their classification as part of a separate division called the **autonomic nervous system** (**ANS**) (Fig. 8-7). The ANS itself is comprised of **sympathetic** and **parasympathetic** divisions.

The autonomic nervous system has many ganglia that serve as relay stations. In these ganglia each message is transferred at a synapse from the first neuron to a second one and from there to the muscle or gland cell. This differs from the voluntary

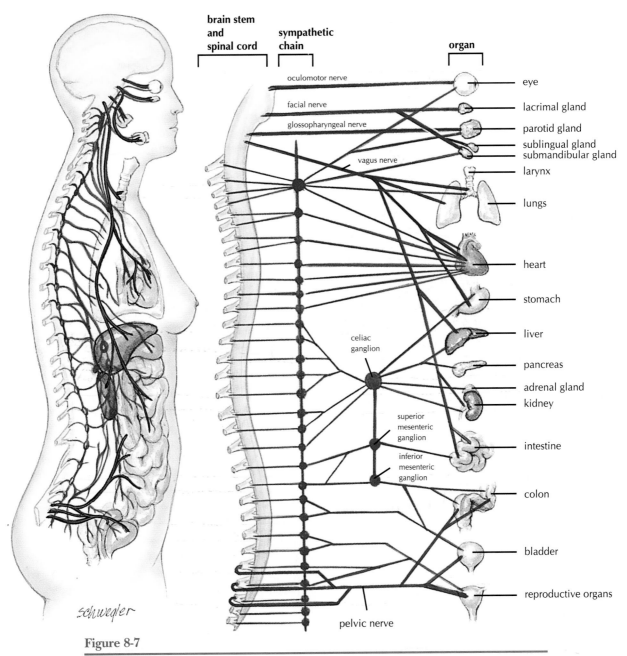

**Figure 8-7**

Autonomic nervous system (only one side is shown). The sympathetic system is shown in green; the parasympathetic system is shown in blue.

(somatic) nervous system, in which each motor nerve fiber extends all the way from the spinal cord to the skeletal muscle with no intervening synapse. Some of the autonomic fibers are within the spinal nerves; some are within the cranial nerves (see Chap. 9). The distribution of the two divisions of the ANS is as follows:

1. The sympathetic pathways begin in the spinal cord with cell bodies in the thoracic and lumbar regions, the **thoracolumbar** (tho-rah-ko-LUM-bar) area. The sympathetic fibers arise from the spinal cord at the level of the first thoracic nerve down to the level of the second lumbar spinal nerve. From this part of the cord, nerve fibers extend to ganglia where they synapse with a second set of neurons, the fibers of which extend to the glands and involuntary muscle tissues. Many of the sympathetic ganglia form the **sympathetic chains,** two cordlike strands of ganglia that extend along either side of the spinal column from the lower neck to the upper abdominal region. The nerves that supply the organs of the abdominal and pelvic cavities synapse in three single ganglia farther from the spinal cord. The second neurons of the sympathetic nervous system act on the effectors by releasing the neurotransmitter epinephrine (adrenaline). This system is therefore described as **adrenergic,** which means "activated by adrenaline."

2. The parasympathetic pathways begin in the **craniosacral** (kra-ne-o-SAK-ral) areas, with fibers arising from cell bodies of the midbrain, medulla, and lower (sacral) part of the spinal cord. From these centers the first set of fibers extends to autonomic ganglia that are usually located near or within the walls of the effector organs. The pathways then continue along a second set of neurons that stimulate the involuntary tissues. These neurons release the neurotransmitter acetylcholine, leading to the description of this system as **cholinergic** (activated by acetylcholine).

## Functions of the Autonomic Nervous System

The autonomic nervous system regulates the action of the glands, the smooth muscles of hollow organs and vessels, and the heart muscle. These actions are all carried on automatically; whenever any changes occur that call for a regulatory adjustment,

the adjustment is made without conscious awareness. The sympathetic part of the autonomic nervous system tends to act as an accelerator for those organs needed to meet a stressful situation. It promotes what is called the **fight-or-flight response.** If you think of what happens to a person who is frightened or angry, you can easily remember the effects of impulses from the sympathetic nervous system:

1. Stimulation of the central portion of the adrenal gland. This produces hormones, including epinephrine, that prepare the body to meet emergency situations in many ways (see Chap. 11). The sympathetic nerves and hormones from the adrenal gland reinforce each other.
2. Dilation of the pupil and decrease in focusing ability (for near objects)
3. Increase in the rate and force of heart contractions
4. Increase in blood pressure due partly to the more effective heartbeat and partly to constriction of small arteries in the skin and the internal organs
5. Dilation of blood vessels to skeletal muscles, bringing more blood to these tissues
6. Dilation of the bronchial tubes to allow more oxygen to enter
7. Increase in metabolism.

The sympathetic system also acts as a brake on those systems not directly involved in the response to stress, such as the urinary and digestive systems. If you try to eat while you are angry, you may note that your saliva is thick and so small in amount that you can swallow only with difficulty. Under these circumstances, when food does reach the stomach, it seems to stay there longer than usual.

The parasympathetic part of the autonomic nervous system normally acts as a balance for the sympathetic system once a crisis has passed. The parasympathetic system brings about constriction of the pupils, slowing of the heart rate, and constriction of the bronchial tubes. It also stimulates the formation and release of urine and activity of the digestive tract. Saliva, for example, flows more easily and profusely and its quantity and fluidity increase.

Most organs of the body receive both sympathetic and parasympathetic stimulation, the effects of the two systems on a given organ generally being opposite. Table 8-1 shows some of the actions of these two systems.

## TABLE 8-1
### Effects of the Sympathetic and Parasympathetic Systems on Selected Organs

*Fight or Flight response* (handwritten)

| Effector | Sympathetic System | Parasympathetic System |
|---|---|---|
| Pupils of eye | Dilation — ↑ *large / more* (handwritten) | Constriction ↓ |
| Sweat glands | Stimulation | None ↓ |
| Digestive glands | Inhibition *Decrease* (handwritten) | Stimulation *Increase* (handwritten) |
| Heart | Increased rate and strength of beat | Decreased rate and strength of beat |
| Bronchi of lungs | Dilation *increase* ↑ (handwritten) | Constriction *small* (handwritten) |
| Muscles of digestive system | Decreased contraction (peristalsis) ↓ | Increased contraction |
| Kidneys | Decreased activity ↓ | None |
| Urinary bladder | Relaxation | Contraction and emptying |
| Liver | Increased release of glucose | None |
| Penis | Ejaculation | Erection |
| Adrenal medulla *Stimulation* (handwritten) | Stimulation | None |
| Blood vessels to | | |
|   Skeletal muscles | Dilation ↑ | Constriction ↓ |
|   Skin | Constriction ↓ | None |
|   Respiratory system | Dilation ↑ | Constriction |
|   Digestive organs | Constriction ↓ | Dilation |

## SUMMARY

I. **Organization of the nervous system**
  A. Structural (anatomic) divisions
    1. Central nervous system (CNS)—brain and spinal cord
    2. Peripheral nervous system (PNS)—spinal and cranial nerves
  B. Functional (physiologic) divisions
    1. Somatic (voluntary) nervous system—supplies skeletal muscles
    2. Visceral (involuntary) nervous system—supplies smooth muscle, cardiac muscle, glands
      a. Autonomic nervous system—involuntary motor nerves

II. **Function of the nervous system**
  A. Neuron—nerve cell
    *Nerve fiber* (handwritten)
    1. Cell body
    2. Cell fibers
      a. Dendrite—carries impulses to cell body
      b. Axon—carries impulses away from cell body
    3. Myelin sheath
      a. Covers some axons
      b. Protects axon and allows more rapid conduction
      c. Made by Schwann cells in PNS; neuroglia in CNS
      d. Neurilemma—outermost layer of Schwann cell; aids neuron repair
      e. White matter—myelinated tissue; gray matter—unmyelinated tissue
    4. Types of neurons
      a. Sensory (afferent)—carry impulses toward CNS
      b. Motor (efferent)—carry impulses away from CNS
  B. Nerve impulse (action potential)—electric current that spreads along nerve fiber
  C. Synapse—junction between neurons where a nerve impulse is transmitted from one neuron to the next
    1. Neurotransmitter—carries impulse across synapse
    2. Postsynaptic cell—picks up neurotransmitter at receptors in membrane

    **D.** Nerve—bundle of nerve cell fibers outside the CNS
       **1.** Sensory (afferent) nerve—contains only fibers that carry impulses toward the CNS (from the receptor)
       **2.** Motor (efferent) nerve—contains only fibers that carry impulses away from the CNS (to the effector)
       **3.** Mixed nerve—contains both sensory and motor fibers
    **E.** Reflex arc—pathway through the nervous system
       **1.** Receptor—detects stimulus
       **2.** Sensory neuron—receptor to CNS
       **3.** Central neuron(s)—in CNS
       **4.** Motor neuron—CNS to effector
       **5.** Effector—muscle or gland that responds

**III. Spinal cord**
    **A.** Structure—H-shaped area of gray matter surrounded by white matter
    **B.** Function
       **1.** Conduction of sensory impulses to brain—ascending tracts
       **2.** Conduction of motor impulses from brain—descending tracts
       **3.** Reflex activities
          **a.** Reflex—simple, rapid, automatic response using few neurons

          **b.** Examples—stretch reflex, eye blink, withdrawal reflex
          **c.** Spinal reflex—coordinated in spinal cord

**IV. Spinal nerves**—31 pairs
    **A.** Roots
       **1.** Sensory (dorsal)
       **2.** Motor (ventral)
    **B.** Plexuses—networks formed by anterior branches
       **1.** Cervical plexus
       **2.** Brachial plexus
       **3.** Lumbosacral plexus

**V. Autonomic nervous system**—motor (efferent) division of the visceral (involuntary) nervous system
    **A.** Divisions
       **1.** Sympathetic system
          **a.** Thoracolumbar
          **b.** Adrenergic—uses adrenaline
          **c.** Stimulates response to stress
       **2.** Parasympathetic system
          **a.** Craniosacral
          **b.** Cholinergic—uses acetylcholine
          **c.** Reverses stress response
    **B.** Effectors—Involuntary muscles (smooth muscle and cardiac muscle) and glands

## QUESTIONS FOR STUDY AND REVIEW

1. What is the main function of the nervous system?
2. Name the two structural divisions of the nervous system.
3. Name the organs or structures controlled by the somatic nervous system; by the visceral nervous system.
4. Define the following terms: *neuron, myelin, synapse, ganglion,* and *Schwann cell.*
5. Describe a nerve impulse. How does conduction along a myelinated fiber differ from conduction along an unmyelinated fiber?
6. What are neurotransmitters? Give several examples of neurotransmitters.
7. Differentiate between the terms in each of the following pairs:
   **a.** *axon* and *dendrite*
   **b.** *gray matter* and *white matter*
   **c.** *receptor* and *effector*
   **d.** *afferent* and *efferent*

   **e.** *sensory* and *motor*
   **f.** *nerve* and *tract*
8. Without the label, how would you know that the neuron in Figure 8-2 is a motor neuron? Is this neuron a part of the somatic or visceral nervous system?
9. What is a mixed nerve? Give several examples.
10. Name the components of a reflex arc.
11. Locate and describe the spinal cord. Name three of its functions.
12. Describe a reflex. Give several examples of reflexes.
13. Differentiate between the dorsal and ventral roots of a spinal nerve.
14. Define a *plexus.* Name the three main plexuses of the spinal nerves.
15. What are the functions of the sympathetic part of the autonomic nervous system, and how do these compare with those of the parasympathetic nervous system?

# CHAPTER 9

# The Nervous System: The Brain and Cranial Nerves

## Selected Key Terms

The following terms are defined in the Glossary:

**brain stem**

**cerebellum**

**cerebral cortex**

**cerebrospinal fluid**

**cerebrum**

**diencephalon**

**hypothalamus**

**medulla oblongata**

**meninges**

**midbrain**

**pons**

**thalamus**

**ventricle**

## Behavioral Objectives

After careful study of this chapter, you should be able to:

- Give the location and functions of the four main divisions of the brain

- Name and locate the three subdivisions of the brain stem

- Name and describe the three meninges

- Cite the function of cerebrospinal fluid and describe where and how this fluid is formed

- Cite one function of the cerebral cortex in each lobe of the cerebrum

- Cite the names and functions of the 12 cranial nerves

Memmler, RL, Cohen, BJ, Wood, DL. *STRUCTURE AND FUNCTION OF THE HUMAN BODY*, 6/e, © 1996 Lippincott-Raven Publishers

# The Brain and its Protective Structures

## Main Parts of the Brain

The brain occupies the cranial cavity and is covered by membranes, fluid, and the bones of the skull. Although the various regions of the brain are in communication and may function together, the brain may be divided into distinct areas for ease of study (Fig. 9-1):

1. The *cerebrum* (SER-e-brum) is the largest part of the brain. It is divided into right and left *cerebral* (SER-e-bral) *hemispheres* by a deep groove called the *longitudinal fissure*.
2. The *diencephalon* (di-en-SEF-ah-lon) is the area between the cerebral hemispheres and the brain stem. It includes the thalamus and the hypothalamus.
3. The *brain stem* connects the cerebrum and diencephalon with the spinal cord. The upper portion of the brain stem is the *midbrain*. Below that, and plainly visible from the underview of the brain are the *pons* (ponz) and the *medulla oblongata* (meh-DUL-lah ob-long-GAH-tah). The pons connects the midbrain with the medulla, while the medulla connects the brain with the spinal cord through a large opening in the base of the skull (foramen magnum).
4. The *cerebellum* (ser-eh-BEL-um) is located immediately below the back part of the cerebral hemispheres and is connected with the cerebrum, brain stem, and spinal cord by means of the pons. The word *cerebellum* means "little brain."

Each of these divisions is described in greater detail later in this chapter.

## Coverings of the Brain and Spinal Cord

The *meninges* (men-IN-jez) are three layers of connective tissue that surround both the brain

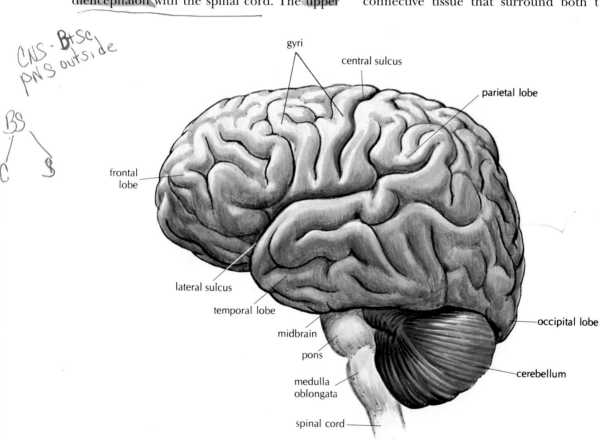

gyri
central sulcus
parietal lobe
frontal lobe
lateral sulcus
temporal lobe
midbrain
pons
medulla oblongata
spinal cord
occipital lobe
cerebellum

**Figure 9-1**

External surface of the brain showing the main parts and some of the lobes and sulci of the cerebrum.

and spinal cord to form a complete enclosure. The outermost of these membranes, the ***dura mater*** (DU-rah MA-ter), is the thickest and toughest of the meninges. Inside the skull, the dura mater splits in certain places to provide venous channels, called ***dural sinuses,*** for the drainage of blood coming from the brain tissue. The middle layer of the meninges is the ***arachnoid*** (ah-RAK-noyd). This membrane is loosely attached to the deepest of the meninges by weblike fibers allowing a space for the movement of cerebrospinal fluid between the two membranes. The innermost layer around the brain, the ***pia mater*** (PI-ah MA-ter), is attached to the nervous tissue of the brain and spinal cord and follows all the contours of these structures (Fig. 9-2). It is made of a delicate connective tissue in which there are many

blood vessels. The blood supply to the brain is carried, to a large extent, by the pia mater.

### Cerebrospinal Fluid

***Cerebrospinal*** (ser-e-bro-SPI-nal) ***fluid*** (CSF) is a clear liquid that circulates in and around the brain and spinal cord (Fig. 9-3). The function of CSF is to support neural tissue and to cushion shocks that would otherwise injure these delicate structures. This fluid also carries nutrients to the cells and transports waste products from the cells.

Cerebrospinal fluid flows freely through the brain and spinal cord and finally flows out into the subarachnoid space of the meninges. Much of the fluid is then returned to the blood through projections called *arachnoid villi* in the dural sinuses.

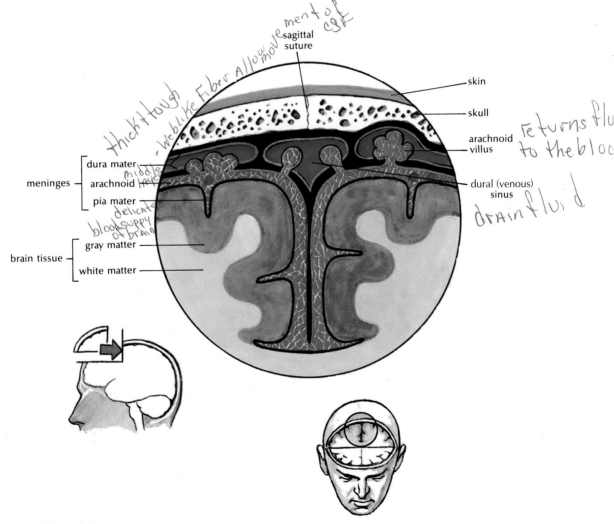

**Figure 9-2**

Frontal (coronal) section of the top of the head showing meninges and related parts.

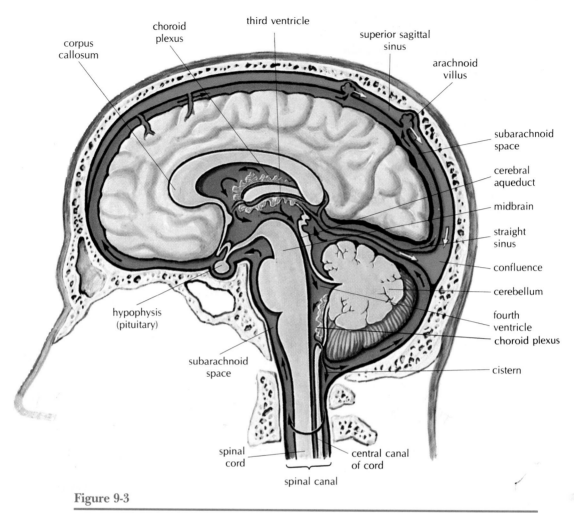

choroid plexus

third ventricle

superior sagittal sinus

arachnoid villus

corpus callosum

subarachnoid space

cerebral aqueduct

midbrain

straight sinus

confluence

cerebellum

fourth ventricle

choroid plexus

cistern

hypophysis (pituitary)

subarachnoid space

spinal cord

central canal of cord

spinal canal

**Figure 9-3**

Flow of CSF from choroid plexuses back to the blood in dural sinuses is shown by the black arrows; flow of blood is shown by the white arrows.

### THE VENTRICLES

Cerebrospinal fluid is formed in four spaces within the brain called *ventricles* (VEN-trih-klz). A vascular network in each ventricle, the *choroid* (KOR-oyd) *plexus,* forms CSF by filtration of the blood and by cellular secretion.

The four ventricles that produce CSF extend somewhat irregularly into the various parts of the brain (Fig. 9-4). The largest are the lateral ventricles in the two cerebral hemispheres. Their extensions into the lobes of the cerebrum are called *horns.* These paired ventricles communicate with a midline space, the third ventricle, by means of openings called *foramina* (fo-RAM-in-ah). The third ventricle is surrounded by the diencephalon. Continuing down from the third ventricle, a small canal, called the *cerebral aqueduct,* extends through the midbrain

into the fourth ventricle. The latter is continuous with the central canal of the spinal cord. In the roof of the fourth ventricle are three openings that allow the escape of CSF to the area that surrounds the brain and spinal cord.

## ▶ Divisions of the Brain

### The Cerebral Hemispheres

Each cerebral hemisphere is divided into four visible *lobes* named from the overlying cranial bones. These are the frontal, parietal, temporal, and occipital lobes (see Fig. 9-1). In addition, there is a small fifth lobe deep within each hemisphere that cannot be seen from the surface. Not much is known about this lobe, which is called the *insula.*

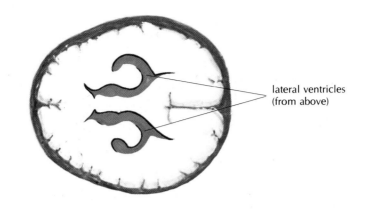

lateral ventricles
(from above)

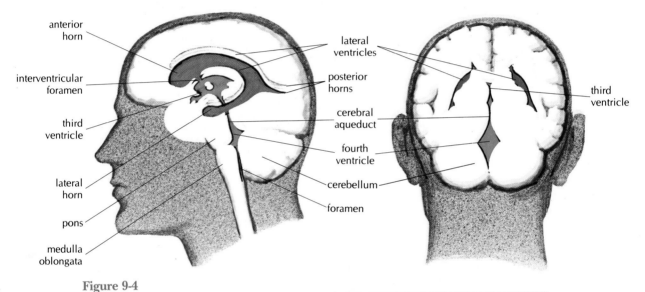

anterior horn

interventricular foramen

third ventricle

lateral horn

pons

medulla oblongata

lateral ventricles

posterior horns

cerebral aqueduct

fourth ventricle

cerebellum

foramen

third ventricle

**Figure 9-4**

Ventricles of the brain.

The outer nervous tissue of the cerebral hemispheres is gray matter that makes up the ***cerebral cortex.*** This thin layer of gray matter is the most highly evolved portion of the brain and is responsible for conscious thought, reasoning, and abstract mental functions. Specific functions are localized in the cortex of the different lobes, as described in greater detail below.

The cortex is arranged in folds forming elevated portions known as *gyri* (JI-ri), which are separated by shallow grooves called *sulci* (SUL-si) (see Fig. 9-1). Although there are many sulci, a few are especially important landmarks:

1. The ***central sulcus,*** which lies between the frontal and parietal lobes of each hemisphere at right angles to the longitudinal fissure

2. The ***lateral sulcus,*** which curves along the side of each hemisphere and separates the temporal lobe from the frontal and parietal lobes (see Fig. 9-1).

Internally, the cerebral hemispheres are made largely of white matter and a few islands of gray matter. The white matter consists of myelinated fibers that connect the cortical areas with each other and with other parts of the nervous system. An important band of white matter is the ***corpus callosum*** located at the bottom of the longitudinal fissure (see Fig. 9-3). This band acts as a bridge between the right and left hemispheres, permitting impulses to cross from one side of the brain to the other. The ***internal capsule*** is a crowded strip of white matter composed of many myelinated fibers (forming tracts). ***Basal*** *ganglia* are masses of gray matter located deep within each cerebral hemisphere. These groups of neurons help regulate body movement and facial expressions communicated from the cerebral cortex. The neurotransmitter *dopamine* (DO-pah-mene) is secreted by the neurons of the basal ganglia.

### FUNCTIONS OF THE CEREBRAL CORTEX

It is within the cerebral cortex, the layer of gray matter that forms the surface of each cerebral hemisphere, that impulses are received and analyzed. These form the basis of knowledge. The brain "stores" information, much of which can be recalled on demand by means of the phenomenon called *memory*. It is in the cerebral cortex that thought processes such as association, judgment, and discrimination take place. It is also from the cerebral cortex that conscious deliberation and voluntary actions emanate.

Although the various areas of the brain act in coordination to produce behavior, particular functions are localized in the cortex of each lobe. Some of these are described below:

1. The *frontal lobe,* which is relatively larger in the human being than in any other organism, lies in front of the central sulcus. This lobe contains the motor area, which directs movement (Fig. 9-5). Because of the way in which motor fibers cross to opposite sides in the CNS, the left side of the brain governs the right side of the body, and the right side of the brain governs the left side of the body. The frontal lobe also contains two areas important in speech (the speech centers are discussed below).

2. The *parietal lobe* occupies the upper part of each hemisphere and lies just behind the central sulcus. This lobe contains the *sensory area,* in which impulses from the skin, such as touch, pain, and temperature, are interpreted. The estimation of distances, sizes, and shapes also take place here.

3. The *temporal lobe* lies below the lateral sulcus and folds under the hemisphere on each side. This lobe contains the *auditory area* for receiving and interpreting impulses from the ear. The *olfactory area,* concerned with the sense of smell, is located in the medial part of the temporal lobe; it is stimulated by impulses arising from receptors in the nose.

4. The *occipital lobe* lies behind the parietal lobe and extends over the cerebellum. This lobe contains the *visual area* for interpreting impulses arising from the retina of the eye.

### COMMUNICATION AREAS

The ability to communicate by written and verbal means is an interesting example of the way in which areas of the cerebral cortex are interrelated (Fig. 9-6). The development and use of these areas are closely connected with the process of learning.

1. The *auditory areas* are located in the temporal lobe. In one of these areas sound impulses transmitted from the environment are detected, while in the surrounding area (auditory speech center) the sounds are interpreted and understood. The beginnings of language are learned by auditory means, so the auditory area for understanding sounds is very near the auditory receiving area of the cortex. Babies often seem to understand what is being said long before they do any talking themselves. It is usually several years before children learn to read or write words.

2. The *motor areas* for communication by speech and writing are located in front of the lowest part of the motor cortex in the frontal lobe. Because the lower part of the motor cortex controls the muscles of the head and neck, it seems logical to think of the motor speech center as an extension forward in this area. Control of the muscles of speech (in the tongue, the soft palate, and the larynx) is carried out here. Similarly, the written speech center is located in front of the cortical area that controls the muscles of the arm and hand. The ability to write words is usually one of the last phases in the development of learning words and their meanings.

3. The *visual areas* of the cortex in the occipital lobe are also involved in communication. Here, visual images of language are received. These visual impulses are then interpreted as words in the visual area that lies in front of the receiving cortex. The ability to read with understanding is also developed in this area. You might *see* writing in the Japanese language, for example, but this would involve only the visual receiving area in the occipital lobe unless you could also *read* the words.

There is a functional relation among areas of the brain. Many neurons must work together to enable a person to receive, interpret, and respond to verbal and written messages as well as to touch (tactile stimulus) and other sensory stimuli.

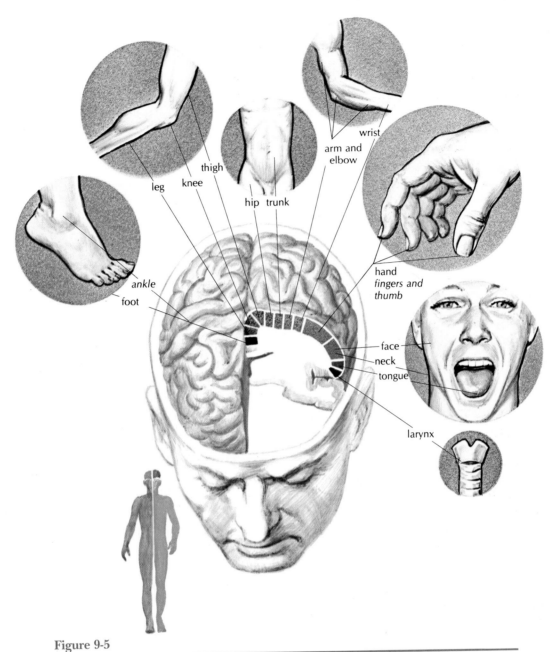

**Figure 9-5**

Motor areas of the cerebral cortex (frontal lobe). The parts of the body are drawn in proportion to the area of control.

### MEMORY AND THE LEARNING PROCESS

Memory is the mental faculty for recalling ideas. In the initial stage of the memory process, sensory signals (*e.g.,* visual, auditory) are retained for a very short time, perhaps only fractions of a second. Nevertheless, they can be used for further processing. *Short-term memory* refers to the retention of bits of information for a few seconds or maybe a few min-

utes, after which the information is lost unless reinforced. *Long-term memory* refers to the storage of information that can be recalled at a later time. There is a tendency for a memory to become more fixed the more often a person repeats the remembered experience; thus, short-term memory signals can lead to long-term memories. Furthermore, the more often a memory is recalled, the more indelible it

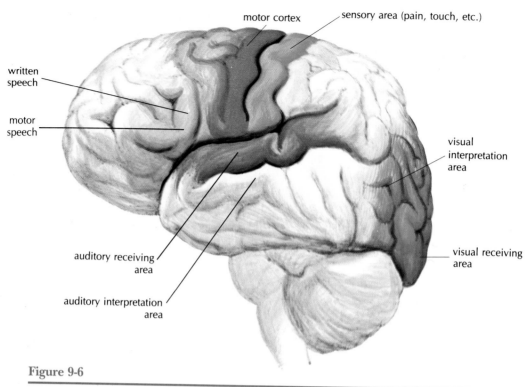

written speech

motor speech

motor cortex

sensory area (pain, touch, etc.)

visual interpretation area

visual receiving area

auditory receiving area

auditory interpretation area

**Figure 9-6**

Functional areas of the cerebral cortex.

becomes; such a memory can be so deeply fixed in the brain that it can be recalled immediately.

Careful anatomic studies have shown that tiny extensions called *fibrils* form at the synapses in the cerebral cortex so impulses can travel more easily from one neuron to another. The number of these fibrils increases with age. Physiologic studies show that rehearsal (repetition) of the same information again and again accelerates and potentiates the degree of transfer of short-term memory into long-term memory. A person who is wide awake memorizes far better than a person who is in a state of mental fatigue. It has also been noted that the brain is able to organize information so that new ideas are stored in the same areas in which similar ones have been stored before.

### The Diencephalon

The *diencephalon,* or interbrain, is located between the cerebral hemispheres and the brain stem. It can be seen by cutting into the central section of the brain. The *thalamus* (THAL-ah-mus) and the *hypothalamus* are included in the diencephalon (Fig. 9-7).

The two parts of the thalamus form the lateral walls of the third ventricle. Nearly all sensory impulses travel through the masses of gray matter that form the thalamus. The action of the thalamus is to sort out the impulses and direct them to particular areas of the cerebral cortex.

The hypothalamus is located in the midline area below the thalamus and forms the floor of the third ventricle. It contains cells that help control body temperature, water balance, sleep, appetite, and some emotions, such as fear and pleasure. Both the sympathetic and parasympathetic divisions of the autonomic nervous system are under the control of the hypothalamus, as is the pituitary gland. The hypothalamus thus influences the heart beat, the contraction and relaxation of the walls of blood vessels, hormone secretion, and other vital body functions.

### THE LIMBIC SYSTEM

Along the border between the cerebrum and the diencephalon is a region known as the *limbic system.* This system is involved in emotional states and behavior. It includes the *hippocampus* (shaped like a sea horse), located under the lateral ventricles, which

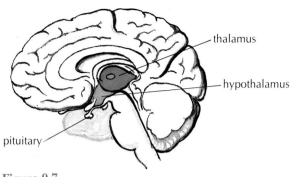

**Figure 9-7**

Diagram showing the relationship among the thalamus, hypothalamus, and pituitary (hypophysis). (Chaffee EE, Lytle IM: Basic Physiology and Anatomy, 4th ed, p. 211. Philadelphia, JB Lippincott, 1980)

functions in learning and the formation of long-term memory. It also includes regions that stimulate the *reticular formation,* a network that extends along the brain stem and governs wakefulness and sleep. The limbic system thus links the conscious functions of the cerebral cortex and the automatic functions of the brain stem.

### The Brain Stem

The brain stem is composed of the midbrain, the pons, and the medulla oblongata. These structures connect the cerebrum and diencephalon with the spinal cord.

### THE MIDBRAIN

The *midbrain,* located below the center of the cerebrum, forms the forward part of the brain stem. Four rounded masses of gray matter that are hidden by the cerebral hemispheres form the upper part of the midbrain; these four bodies act as relay centers for certain eye and ear reflexes. The white matter at the front of the midbrain conducts impulses between the higher centers of the cerebrum and the lower centers of the pons, medulla, cerebellum, and spinal cord. Cranial nerves III and IV originate from the midbrain.

### THE PONS

The *pons* lies between the midbrain and the medulla, in front of the cerebellum. It is composed largely of myelinated nerve fibers, which serve to connect the two halves of the cerebellum with the brain stem as well as with the cerebrum above and

the spinal cord below. The pons is an important connecting link between the cerebellum and the rest of the nervous system, and it contains nerve fibers that carry impulses to and from the centers located above and below it. Certain reflex (involuntary) actions, such as some of those regulating respiration, are integrated in the pons. Cranial nerves V through VIII originate from the pons.

### THE MEDULLA OBLONGATA

The *medulla oblongata* of the brain is located between the pons and the spinal cord. It appears white externally because, like the pons, it contains many myelinated nerve fibers. Internally, it contains collections of cell bodies (gray matter) called *nuclei,* or *centers.* Among these are vital centers such as the following:

1. The *respiratory center* controls the muscles of respiration in response to chemical and other stimuli.
2. The *cardiac center* helps regulate the rate and force of the heart beat.

## Sleep

Although everbody does it, there are still many mysteries about how, why, and when we sleep. Study of brain waves with the electroencephalograph has shown that sleep occurs in two phases. One of these, rapid eye movement (REM) sleep, is associated with dreaming.

As we descend from drowsiness into deep sleep, we progress through a non-rapid eye movement (NREM) phase. Although all the triggers for falling asleep are not known, a major factor is the amount of light received through the eyes. This signals the hypothalamus to regulate sleep–wake cycles. About 1 to 1½ hours after falling asleep, short periods of REM sleep with dreaming occur. These then alternate with NREM sleep at intervals of about 90 minutes. Extreme fatigue and sedatives decrease the amount of REM sleep in favor of deep NREM sleep. Also, infants have a much greater percentage of REM sleep; this declines with age.

Sleep rests the body, improves learning, and strengthens the immune system. There is also evidence that REM sleep helps the brain to sort memories.

3. The *vasomotor* (vas-o-MO-tor) *center* regulates the contraction of smooth muscle in the blood vessel walls and thus controls blood flow and blood pressure.

The ascending sensory fibers that carry messages through the spinal cord up to the brain travel through the medulla, as do descending motor fibers. These groups of fibers form tracts (bundles) and are grouped together according to function. The motor fibers from the motor cortex of the cerebral hemispheres extend down through the medulla, and most of them cross from one side to the other (decussate) while going through this part of the brain. It is in the medulla that the shifting of nerve fibers occurs that causes the right cerebral hemisphere to control muscles in the left side of the body and the upper portion of the cortex to control muscles in the lower portions of the body. The medulla is an important reflex center; here, certain neurons end and impulses are relayed to other neurons. The last four pairs of cranial nerves (IX through XII) are connected with the medulla.

### The Cerebellum

The *cerebellum* is made up of three parts: the middle portion (vermis) and two lateral hemispheres. Like the cerebral hemispheres, the cerebellum has an outer area of gray matter and an inner portion that is largely white matter. The functions of the cerebellum are:

1. To aid in the *coordination of voluntary muscles* so that they will function smoothly and in an orderly fashion. Disease of the cerebellum causes muscular jerkiness and tremors.
2. To aid in the *maintenance of balance* in standing, walking, and sitting, as well as during more strenuous activities. Messages from the internal ear and from sensory receptors in tendons and muscles aid the cerebellum.
3. To aid in the *maintenance of muscle tone* so that all muscle fibers are slightly tensed and ready to produce necessary changes in position as quickly as may be necessary.

## ▶ Brain Studies

### Imaging the Brain

A major tool for clinical study of the brain is the *CT* (computed tomography) *scan*, which provides multiple x-ray pictures taken from different angles simultaneously. By means of a computer, the information is organized and displayed as photographs of the bone, soft tissue, and cavities of the brain (Fig. 9-8). Anatomic lesions such as tumors or scar tissue accumulations are readily seen.

*MRI* (magnetic resonance imaging) gives even clearer pictures of the brain without the use of dyes or x-rays. The method is based on computerized interpretation of the movements of atomic nuclei following their exposure to radio waves within a powerful magnetic field. Although the method is more expensive and takes longer than CT imaging, it gives more views of the brain and may reveal tumors, scar tissue, and hemorrhaging not shown by CT.

With *PET* (positron emission tomography) one can actually visualize the brain in action. With this method, a radioactively labeled substance, glucose for example, is followed as it moves through the brain. As tasks are performed, regions of the cortex that are involved become "hot."

### The Electroencephalograph

The interactions of the billions of nerve cells in the brain give rise to measurable electric currents. These may be recorded by an instrument called the *electroencephalograph* (e-lek-tro-en-SEF-ah-lo-graf). The recorded tracings or brain waves produce an electroencephalogram (EEG).

## ▶ Cranial Nerves

### Location of the Cranial Nerves

There are 12 pairs of cranial nerves (henceforth, when a cranial nerve is identified, a pair is meant). They are numbered according to their connection with the brain, beginning at the front and proceeding back (Fig. 9-9). The first 9 pairs and the 12th pair supply structures in the head.

### General Functions of the Cranial Nerves

From a functional point of view, we may think of the kinds of messages the cranial nerves handle as belonging to one of four categories:

1. *Special sensory impulses,* such as those for smell, taste, vision, and hearing

**Figure 9-8**

CT scanner. (Photograph courtesy of Philips Medical Systems)

2. *General sensory impulses,* such as those for pain, touch, temperature, deep muscle sense, pressure, and vibrations

3. *Somatic motor impulses* resulting in voluntary control of skeletal muscles

4. *Visceral motor impulses* producing involuntary control of glands and involuntary muscles (cardiac muscle and smooth muscle). These motor pathways are part of the autonomic nervous system, parasympathetic division.

### Names and Functions of the Cranial Nerves

The 12 cranial nerves are always numbered according to the traditional Roman style. A few of the cranial nerves—I, II, and VIII—contain only sensory fibers; some—III, IV, VI, XI, and XII—contain all or mostly motor fibers. The remainder—V, VII, IX, and X—contain both sensory and motor fibers; they are known as *mixed nerves.* All 12 nerves are listed below:

I. The *olfactory nerve* carries smell impulses from receptors in the nasal mucosa to the brain.

II. The *optic nerve* carries visual impulses from the eye to the brain.

III. The *oculomotor nerve* is concerned with the contraction of most of the eye muscles.

IV. The *trochlear* (TROK-le-ar) *nerve* supplies one eyeball muscle.

V. The *trigeminal* (tri-JEM-in-al) *nerve* is the great sensory nerve of the face and head. It has three branches that transport general sense impulses (*e.g.,* pain, touch, temperature) from the eye, the upper jaw, and the lower jaw. The third branch is joined by motor fibers to the muscles of mastication (chewing).

VI. The *abducens* (ab-DU-senz) *nerve* is another nerve sending controlling impulses to an eyeball muscle.

VII. The *facial nerve* is largely motor. The muscles of facial expression are all supplied by branches from the facial nerve. This nerve also includes special sensory fibers for taste (anterior two thirds of the tongue), and it contains secretory fibers to the smaller salivary glands (the submandibular and sublingual) and to the lacrimal (tear) gland.

VIII. The *vestibulocochlear* (ves-tib-u-lo-KOK-le-ar) *nerve* contains special sensory fibers for hearing as well as those for balance from the semicircular canals of the internal ear. This nerve is also called the *auditory* or *acoustic nerve.*

IX. The *glossopharyngeal* (glos-o-fah-RIN-ge-al) *nerve* contains general sensory fibers from the back of the tongue and the pharynx (throat).

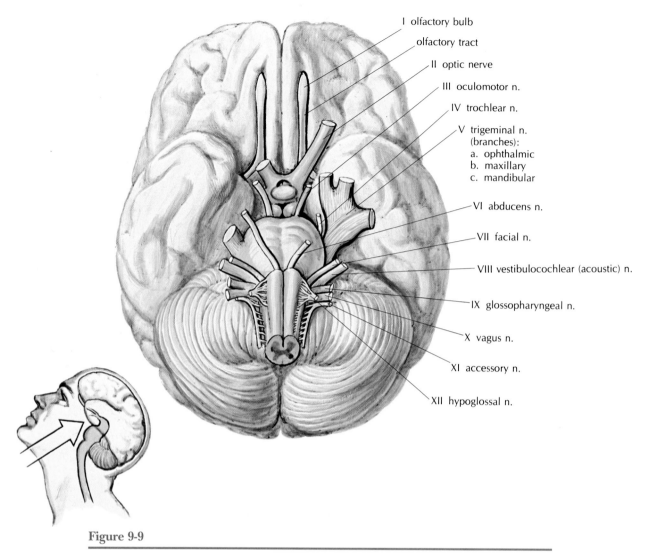

I olfactory bulb
olfactory tract
II optic nerve
III oculomotor n.
IV trochlear n.
V trigeminal n.
(branches):
a. ophthalmic
b. maxillary
c. mandibular
VI abducens n.
VII facial n.
VIII vestibulocochlear (acoustic) n.
IX glossopharyngeal n.
X vagus n.
XI accessory n.
XII hypoglossal n.

**Figure 9-9**

Base of the brain showing cranial nerves.

This nerve also contains sensory fibers for taste from the posterior third of the tongue, secretory fibers that supply the largest salivary gland (parotid), and motor nerve fibers to control the swallowing muscles in the pharynx.

X. The *vagus* (VA-gus) *nerve* is the longest cranial nerve. (Its name means "wanderer.") It supplies most of the organs in the thoracic and abdominal cavities. This nerve also contains motor fibers to the larynx (voice box) and pharynx, and to glands that produce digestive juices and other secretions.

XI. The ***accessory nerve*** (formerly called the *spinal accessory nerve*) is a motor nerve with two branches. One branch controls two muscles of the neck, the trapezius and sternocleidomastoid; the other supplies muscles of the larynx.

XII. The ***hypoglossal nerve,*** the last of the 12 cranial nerves, carries impulses controlling the muscles of the tongue.

## SUMMARY

**I. Protective structures of the nervous system**
   **A.** Meninges—coverings of brain and spinal cord
      **1.** Dura mater—tough outermost layer
      **2.** Arachnoid—weblike middle layer
      **3.** Pia mater—vascular innermost layer
   **B.** Cerebrospinal fluid (CSF)
      **1.** Cushions and protects
      **2.** Circulates around and within brain and spinal cord
      **3.** Ventricles—four spaces within brain where CSF is produced
      **4.** Choroid plexus—vascular network in ventricle that produces CSF

**II. Divisions of brain**
   **A.** Cerebrum—largest part of brain
      **1.** Right and left hemispheres
         **a.** Lobes—frontal, parietal, temporal, occipital, insula
         **b.** Cortex—outer layer of gray matter
   **B.** Diencephalon—area between cerebral hemispheres and brain stem
      **1.** Thalamus—directs sensory impulses to cortex
      **2.** Hypothalamus—maintains homeostasis, controls pituitary

   **3.** Limbic system
      **a.** Contains parts of cerebrum and diencephalon
      **b.** Controls emotion and behavior
   **C.** Brain stem
      **1.** Midbrain—involved in eye and ear reflexes
      **2.** Pons—connecting link for other divisions
      **3.** Medulla oblongata
         **a.** Connects with spinal cord
         **b.** Contains vital centers for respiration, heart rate, vasomotor activity
   **D.** Cerebellum—regulates coordination, balance, muscle tone

**III. Brain studies**
   **A.** Imaging methods
      **1.** CT—computed tomography
      **2.** MRI—magnetic resonance imaging
      **3.** PET—Positron emission tomography
   **B.** Electroencephalogram (EEG)

**IV. Cranial nerves**—12 pairs
   **A.** Sensory (I, II, VIII)
   **B.** Motor (III, IV, VI, XI, XII)
   **C.** Mixed (V, VII, IX, X)

## QUESTIONS FOR STUDY AND REVIEW

1. Name and locate the main parts of the brain, and briefly describe the main functions of each.
2. Name the covering of the brain and the spinal cord. Name and describe its three layers.
3. What is the purpose of the cerebrospinal fluid? Where and how is cerebrospinal fluid formed?
4. Name the four surface lobes of the cerebral hemispheres and describe functions of the cortex in each.
5. Name and describe the speech centers.
6. What are the differences between short-term and long-term memory, and how is each attained?
7. Describe the thalamus; where is it located? What are its functions?
8. What activities does the hypothalamus regulate?
9. What is the limbic system and what are its functions?
10. Name and locate three divisions of the brain stem.
11. Name four general functions of the cranial nerves.
12. Name and describe the functions of the 12 cranial nerves.

# The Sensory System

## Behavioral Objectives

After careful study of this chapter, you should be able to:

- Describe the function of the sensory system
- Differentiate between the special and general senses and give examples of each
- Describe the structure of the eye
- Define *refraction* and list the refractive media of the eye
- Differentiate between the rods and the cones of the eye
- Describe the three divisions of the ear
- Describe the receptors for hearing and for equilibrium with respect to location and function

Memmler, RL, Cohen, BJ, Wood, DL. *STRUCTURE AND FUNCTION OF THE HUMAN BODY*, 6/e,
© 1996 Lippincott-Raven Publishers

## ▶ Senses and Sensory Mechanisms

The sensory system serves to protect the individual by detecting changes in the environment. An environmental change becomes a *stimulus* when it initiates a nerve impulse, which then travels to the CNS by way of a sensory (afferent) neuron. Many stimuli arrive from the external environment and are detected at or near the surface of the body. Others, such as the stimuli from the viscera, originate internally and help to maintain homeostasis. The part of the nervous system that detects a stimulus is the *receptor*. This may be:

1. The free dendrite of a sensory neuron, such as the receptors for pain
2. A modified ending, or *end-organ*, on the dendrite of an afferent neuron, such as those for touch and temperature
3. A special cell associated with an afferent neuron, such as the rods and cones of the retina of the eye and the receptors in the other special sense organs.

Regardless of the type of stimulus or the type of receptor involved, a stimulus becomes a sensation—something we experience—only when the nerve impulse it generates is interpreted by a specialized area of the cerebral cortex to which it travels.

One way of classifying the senses is according to the distribution of the receptors. A *special sense* is localized in a special sense organ; a *general sense* is widely distributed throughout the body.

1. **Special senses**
   a. *Vision* from receptors in the eye
   b. *Hearing* from receptors in the internal ear
   c. *Equilibrium* from receptors in the internal ear
   d. *Taste* from the tongue receptors
   e. *Smell* from receptors in the upper nasal cavities
2. **General senses**
   a. *Pressure*, *heat*, *cold*, *pain*, and *touch* from the skin and internal organs
   b. Sense of *position* from the muscles, tendons, and joints.

## ▶ The Eye

### Protection of the Eyeball and Its Parts

In the embryo, the eye develops as an outpocketing of the brain. The eye is a delicate organ, and nature has carefully protected it by means of the following structures:

1. The skull bones form the walls of the eye orbit (cavity) and serve to protect more than half of the dorsal part of the eyeball.
2. The lids and eyelashes aid in protecting the eye anteriorly.
3. Tears lubricate the eye, wash away small foreign objects that enter the eye, and contain an enzyme that protects against infection.
4. A sac lined with an epithelial membrane separates the front of the eye from the eyeball proper and aids in the destruction of some of the pathogenic bacteria that may enter from the outside.

### Coats of the Eyeball

The eyeball has three separate coats, or tunics (Fig. 10-1). The outermost tunic, called the *sclera* (SKLE-rah), is made of tough connective tissue. It is commonly referred to as the *white of the eye*. The second tunic of the eyeball is the *choroid* (KO-royd). Composed of a delicate network of connective tissue interlaced with many blood vessels, this coat

---

## Mechanoreceptors

Sensory receptors can be classified according to the stimulus to which they respond. Thermoreceptors, such as those in the skin, detect change in temperature; photoreceptors in the retina of the eye respond to light; and chemoreceptors, such as those for taste and smell, detect chemicals.

Mechanoreceptors respond to movement, such as stretch, pressure, or vibration. The receptors that provide awareness of body position are examples, as are those that monitor stretching and tone of muscle fibers. Pressure and touch receptors in the skin are also mechanoreceptors.

In the ear, the receptors for hearing and equilibrium are mechanoreceptors. Both function by means of ciliated cells that respond to movement. In the vestibular apparatus, movements of the head trigger impulses that help to regulate balance. In the cochlea, sound waves cause movement of cilia and generate the impulses involved in hearing.

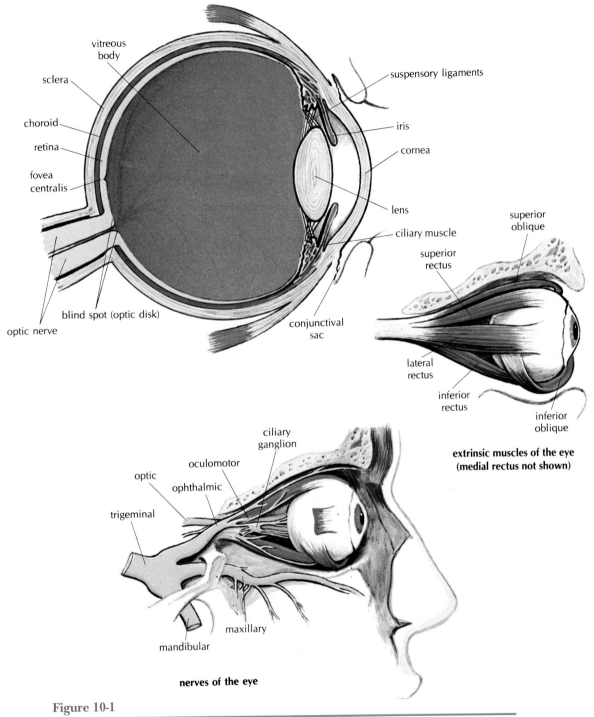

**Figure 10-1**

The eye.

contains much dark brown pigment. The choroid may be compared to the dull black lining of a camera in that it prevents incoming light rays from scattering and reflecting off the inner surface of the eye. The innermost tunic, called the **retina**

(RET-ih-nah), includes ten layers of nerve cells, including the cells commonly called **rods** and **cones** (Fig. 10-2). These are the receptors for the sense of vision. The rods are highly sensitive to light and thus function in dim light but do not provide a very sharp

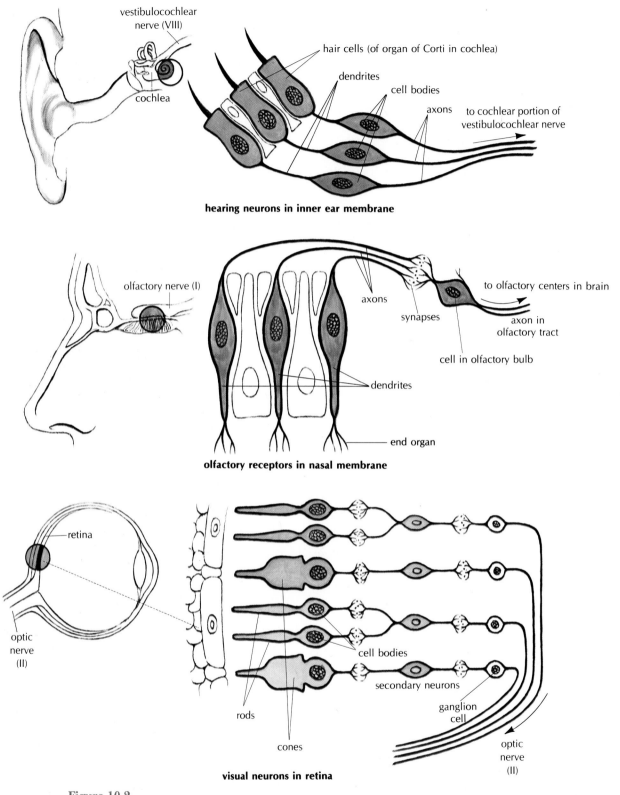

**Figure 10-2**

Diagram of neurons for receiving impulses from the special sense organs.

image. The cones function in bright light and are sensitive to color. When you enter a darkened room, such as a movie theater, you cannot see for a short period of time. It is during this time that the rods are beginning to function, a change described as *dark adaptation.*

As far as is known, there are three types of cones, each sensitive to red, green, or blue light. Persons who completely lack cones are totally color blind; those who lack one type of cone are partially color blind. Color blindness is an inherited condition that occurs almost exclusively in males.

The rods and cones function by means of pigments that are sensitive to light. Manufacture of these pigments requires vitamin A, so a person who lacks vitamin A in the diet may have difficulty seeing in dim light, that is, may have *night blindness.*

### Pathway of Light Rays

Light rays pass through a series of transparent, colorless eye parts. On the way they undergo a process of bending known as *refraction.* This refraction of the light rays makes it possible for light from a very large area to be focused on a very small surface, the retina, where the receptors are located. The following are, in order from outside in, the transparent refracting parts, or *media,* of the eye:

1. The *cornea* (KOR-ne-ah) is a forward continuation of the sclera, but it is transparent and colorless, whereas the rest of the sclera is opaque and white.
2. The *aqueous* (A-kwe-us) *humor,* a watery fluid that fills much of the eyeball in front of the lens, helps maintain the slight forward curve of the cornea.
3. The *lens,* technically called the *crystalline lens,* is a clear, circular structure made of a firm, elastic material.
4. The *vitreous* (VIT-re-us) *body* is a soft jellylike substance that fills the entire space behind the lens and keeps the eyeball in its spherical shape.

The cornea is referred to frequently as the *window* of the eye. It bulges forward slightly and is the most important refracting structure. Injuries caused by foreign objects or by infection may result in scar formation in the cornea, leaving an area of opacity through which light rays cannot pass. If such an injury involves the central area in front of the pupil (the hole in the center of the colored part of the eye), blindness may result. Eye banks store corneas obtained from donors, and corneal transplantation is a fairly common procedure.

The next light-bending medium is the aqueous humor, followed by the crystalline lens. The lens has two bulging surfaces, so it may be best described as biconvex. Because the lens is elastic, its thickness can be adjusted to focus light for near or distance vision.

The last of these transparent refracting parts of the eye is the vitreous body. Like the aqueous humor, it is important in maintaining the shape of the eyeball as well as in aiding in refraction.

### Muscles of the Eye

The muscles inside the eyeball are *intrinsic* (in-TRIN-sik) *muscles;* those attached to bones of the eye orbit as well as to the sclera are *extrinsic* (eks-TRIN-sik) *muscles.*

The intrinsic muscles are found in two circular structures, as follows:

1. The *iris,* the colored or pigmented part of the eye, is composed of two types of muscles. The size of the central opening of the iris, called the *pupil,* is governed by the action of these two sets of muscles, one of which is arranged in a circular fashion, while the other extends in a radial manner like the spokes of a wheel.
2. The *ciliary body* is shaped somewhat like a flattened ring with a hole the size of the outer edge of the iris. This muscle alters the shape of the lens during the process of accommodation (described below).

The purpose of the iris is to regulate the amount of light entering the eye. If a strong light is flashed in the eye, the circular muscle fibers of the iris, which form a sphincter, contract and thus reduce the size of the pupil. In contrast, if the light is very dim, the radial involuntary iris muscles, which are attached at the outer edge, contract; the opening is thereby pulled outward and thus enlarged. This pupillary enlargement is known as *dilation* (di-LA-shun).

The pupil changes size, too, according to whether one is looking at a near object or a distant one. Viewing a near object causes the pupil to become smaller; viewing a distant object causes it to enlarge.

The muscle of the ciliary body is similar in direction and method of action to the radial muscle of the iris. When the ciliary muscle contracts, it draws forward and removes the tension on the *suspensory liga-*

**ments,** which hold the lens in place (see Fig. 10-1). The elastic lens then recoils and becomes thicker in much the same way that a rubber band thickens when the pull on it is released. When the ciliary body relaxes, the lens becomes flattened. These actions change the refractive ability of the lens.

The process of **accommodation** involves coordinated eye changes to enable one to focus on near objects. The ciliary body contracts, thereby thickening the lens, and the circular muscle fibers of the iris contract to decrease the size of the pupillary opening. In young persons the lens is elastic, and therefore its thickness can be readily adjusted according to the need for near or distance vision. With aging, the lens loses its elasticity and therefore its ability to adjust to near vision by thickening, making it difficult to focus clearly on close objects. This condition is called **presbyopia** (pres-be-O-pe-ah), which literally means "old eye."

The six extrinsic muscles connected with each eye are ribbonlike and extend forward from the apex of the orbit behind the eyeball (see Fig. 10-1). One end of each muscle is attached to a bone of the skull, while the other end is attached to the sclera. These muscles pull on the eyeball in a coordinated fashion that causes the two eyes to move together to center on one visual field. Another muscle located within the orbit is attached to the upper eyelid. When this muscle contracts, it keeps the eye open; as the muscle becomes weaker with age, the eyelids may droop and interfere with vision.

### Nerve Supply to the Eye

Two sensory nerves supply the eye (see Fig. 10-1):

1. The **optic nerve** (cranial nerve II) carries visual impulses from the retinal rods and cones to the brain.
2. The **ophthalmic** (of-THAL-mik) **branch of the trigeminal nerve** (cranial nerve V) carries impulses of pain, touch, and temperature from the eye and surrounding parts to the brain.

The optic nerve arises from the retina a little toward the medial or nasal side of the eye. Visual impulses are transmitted from the retina, ultimately to the cortex of the occipital lobe. There are no rods and cones in the retina near the area of the optic nerve fibers, so this circular white area is a blind spot, known as the **optic disk.** Near the optic disk there is a tiny depressed area in the retina, called the *fovea centralis* (FO-ve-ah sen-TRA-lis), which contains a high concentration of cones and is the point of most acute vision.

There are three nerves that carry motor impulses to the muscles of the eyeball. The largest is the oculomotor nerve (cranial nerve III), which supplies voluntary and involuntary motor impulses to all the muscles but two. The other two nerves, the trochlear (cranial nerve IV) and the abducens (cranial nerve VI), each supply one voluntary muscle.

### The Conjunctiva and the Lacrimal Apparatus

The **conjunctiva** (kon-junk-TI-vah) is a membrane that lines the eyelid and covers the anterior part of the sclera. As the conjunctiva extends from the eyelid to the front of the eyeball, recesses, or sacs, are formed. Tears, produced by the **lacrimal** (LAK-rih-mal) **gland,** serve to keep the conjunctiva moist. As tears flow across the eye from the lacrimal gland, located in the upper lateral part of the orbit, the fluid carries away small particles that have entered the conjunctival sacs. The tears are then carried into ducts near the nasal corner of the eye where they drain into the nose by way of the **nasolacrimal** (na-zo-LAK-rih-mal) **duct** (Fig. 10-3). Any excess of tears causes a "runny nose"; a greater overproduction of them results in the spilling of tears onto the cheeks.

With age there is often a thinning and drying of the conjunctiva, resulting in inflammation and enlarged blood vessels. The lacrimal gland produces less secretion, but those tears that are produced may overflow to the cheek because of plugging of the nasolacrimal ducts, which would normally carry the tears away from the eye.

## ▶ The Ear

The ear is a sense organ for both hearing and equilibrium (Fig. 10-4). It may be divided into three main sections:

1. The **external ear** includes the outer projection and a canal.
2. The **middle ear** is an air space containing three small bones.
3. The **internal ear** is the most complex because it contains the sensory receptors for hearing and equilibrium.

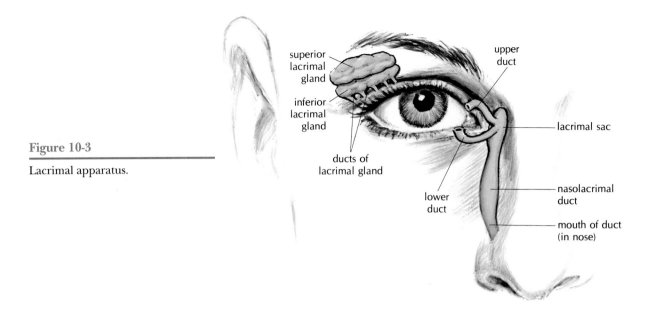

**Figure 10-3**

Lacrimal apparatus.

*Labels on figure:*
superior lacrimal gland
inferior lacrimal gland
ducts of lacrimal gland
upper duct
lacrimal sac
lower duct
nasolacrimal duct
mouth of duct (in nose)

## The External Ear

The projecting part of the ear is known as the *pinna* (PIN-nah), or the *auricle* (AW-rih-kl). Its purpose is to direct sound waves into the ear, but it is probably of little importance in the human. Then follows an opening, the *external auditory canal,* or *meatus* (me-A-tus), which extends medially for about 2.5 cm or more, depending on which wall of the canal is measured. The skin lining this tube is very thin, and in the first part of the canal contains many *ceruminous* (seh-RU-mih-nus) *glands.* The *cerumen* (seh-RU-men), or wax, may become dried and impacted in the canal and must then be removed.

At the end of the auditory canal is the *tympanic* (tim-PAN-ik) *membrane,* or eardrum, which serves as a boundary between the external auditory canal and the middle ear cavity. It vibrates freely as sound waves enter the ear. The *eustachian* (u-STA-shun) *tube* (auditory tube) connects the middle ear cavity and the throat, or *pharynx* (FAR-inks). This tube opens to allow pressure to equalize on the two sides of the tympanic membrane.

## The Middle Ear

The middle ear cavity is a small, flattened space that contains air and three small bones, or *ossicles* (OS-ih-klz). Air enters the cavity from the pharynx through the eustachian tube. The mucous membrane of the pharynx is continuous through the eustachian tube into the middle ear cavity, and

infection may travel along the membrane, causing middle ear disease. At the back of the middle ear cavity is an opening into the mastoid air cells, which are spaces inside the mastoid process of the temporal bone.

The three ossicles are joined in such a way that they amplify the sound waves received by the tympanic membrane and then transmit the sounds to the fluid in the internal ear. The handle-like part of the first bone, or *malleus* (MAL-e-us), is attached to the tympanic membrane, while the head-like portion is connected with the second bone, called the *incus* (ING-kus). The innermost of the ossicles is shaped somewhat like a stirrup and is called the *stapes* (STA-peze). It is connected with the membrane of the *oval window,* which in turn vibrates and transmits these waves to the fluid of the internal ear.

## The Internal Ear

The most complicated and important part of the ear is the internal portion, which consists of three separate spaces hollowed out inside the temporal bone (Fig. 10-5). This part of the ear, called the *bony labyrinth* (LAB-ih-rinth), consists of three divisions. One is the *vestibule,* next to the oval window. The second and third divisions are the *cochlea* (KOK-le-ah) and the *semicircular canals.* All three divisions contain a fluid called *perilymph* (PER-e-limf). The cochlea is a bony tube, shaped like a snail shell and

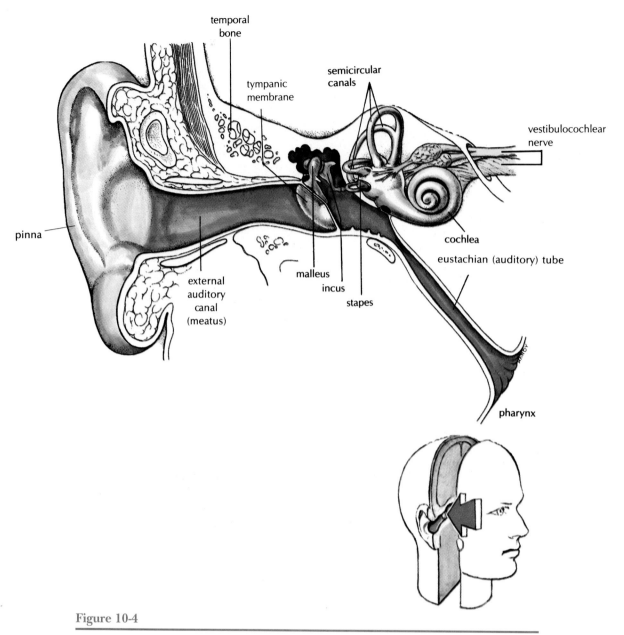

**Figure 10-4**

The ear, showing the external, middle, and internal subdivisions. (Adapted from Chaffee EE, Lytle IM: Basic Physiology and Anatomy, 4th ed, p. 227. Philadelphia, JB Lippincott, 1980)

located toward the front; the semicircular canals are three projecting bony tubes toward the back. In the fluid of the bony semicircular canals are the ***membranous*** (MEM-brah-nus) ***canals,*** which contain another fluid called ***endolymph*** (EN-do-limf). A ***membranous cochlea,*** situated in the perilymph of the bony cochlea, also is filled with endolymph.

The organ of hearing, called the ***organ of Corti,*** consists of receptors connected with nerve fibers in the ***cochlear nerve*** (a part of the eighth cranial nerve). This organ is located inside the membranous cochlea, or ***cochlear duct.*** The sound waves enter the external auditory canal and cause the tympanic membrane to vibrate. These vibrations are amplified by the ossicles and transmitted by them to the perilymph. They then are conducted by the perilymph through the membrane to the endolymph. The waves of the endolymph stimulate the

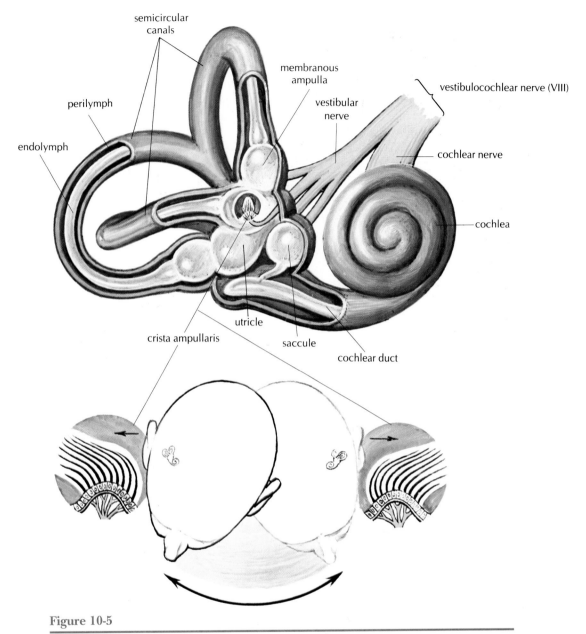

**Figure 10-5**

The internal ear, including a section showing the crista ampullaris, where sensory receptors for balance are located.

tiny, hairlike cilia on the receptor cells, setting up nerve impulses that are then conducted to the brain.

Other sensory receptors in the internal ear include those related to equilibrium, some of which are located at the base of the semicircular canals (see Fig. 10-5). These canals are connected with two small sacs in the vestibule, one of which contains sensory cells for obtaining information on the posi-tion of the head. Receptors for the sense of equilib-rium are also ciliated cells. As the head moves, a shift in the position of the cilia generates a nerve impulse. Nerve fibers from these sacs and from the canals form the ***vestibular*** (ves-TIB-u-lar) ***nerve,*** which joins the cochlear nerve to form the vestibu-locochlear nerve, the eighth cranial nerve (formerly called the *auditory* or *acoustic nerve).*

# ▶ **Other Special Sense Organs**

### Sense of Taste

The sense of taste involves receptors in the tongue and two different nerves that carry taste impulses to the brain (Fig. 10-6). The taste receptors, known as **taste buds,** are located along the edges of small depressed areas called **fissures.** Taste buds are stimulated only if the substance to be tasted is in solution. Receptors for four basic tastes are localized in different regions, forming a "taste map" of the tongue:

1. **Sweet** tastes are most acutely experienced at the tip of the tongue (hence the popularity of lollipops and ice cream cones).
2. **Sour** tastes are most effectively detected by the taste buds located at the sides of the tongue.
3. **Salty** tastes are most acute at the anterior sides of the tongue.
4. **Bitter** tastes are detected at the back part of the tongue.

The nerves of taste include the facial and the glossopharyngeal cranial nerves (VII and IX). The

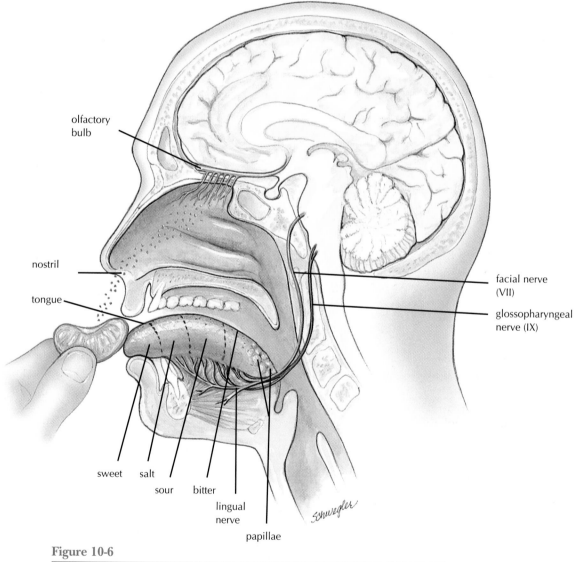

**Figure 10-6**

Organs of taste and smell.

interpretation of taste impulses probably is accomplished by the lower front portion of the brain, although there may be no sharply separate taste, or *gustatory* (GUS-tah-to-re), *center.*

### Sense of Smell

The receptors for smell are located in the *olfactory* (ol-FAK-to-re) *epithelium* of the upper part of the nasal cavity. Again, the chemicals detected must be in solution in the fluids that line the nose. Because these receptors are high in the nasal cavity, one must "sniff" to bring odors upward in the nose. The impulses from the receptors for smell are carried by the olfactory nerve (I), which leads directly to the olfactory center in the brain. The interpretation of smell is closely related to the sense of taste, but a greater variety of dissolved chemicals can be detected by smell than by taste. The smell of foods is just as important in stimulating appetite and the flow of digestive juices as is the sense of taste.

The olfactory receptors deteriorate with age, with the result that food becomes less appealing. It is important when presenting food to elderly persons that the food look inviting so as to stimulate their appetites.

## ❱ General Senses

Unlike the *special* sensory receptors, which are limited to a relatively small area, the *general* sensory receptors are scattered throughout the body. These include receptors for pressure, heat, cold, pain, touch, position, and balance (Fig. 10-7).

### Sense of Pressure

It has been found that even when the skin is anesthetized, there is still consciousness of pressure. These sensory end-organs for deep sensibility are located in the subcutaneous and deeper tissues. They are sometimes referred to as *receptors for deep touch.*

### Sense of Temperature

There are separate receptors for heat and cold and they vary in distribution. A warm object stimulates only the heat receptors, and a cool object affects only the cold receptors. As in the case of other sensory receptors, continued stimulation results in *adaptation;* that is, the receptors adjust themselves in such a way that the sensation becomes less acute if the original stimulus is continued. For example, if you immerse your hand in very warm water, it may be uncomfortable; however, if you leave your hand there, very soon the water will feel less hot (even if it has not cooled appreciably). Receptors for warmth, cold, and light pressure adapt rapidly. In contrast, those for pain adapt slowly; this is nature's way of being certain that the warnings of the pain sense are heeded.

### Sense of Touch

The touch receptors are small, rounded bodies called *tactile* (TAK-til) *corpuscles.* They are found mostly in the dermis and are especially numerous and close together in the tips of the fingers and the toes. The lips and the tip of the tongue also contain many of these receptors and are very sensitive to touch. Other areas, such as the back of the hand and the back of the neck, have fewer receptors and are less sensitive to touch.

### Sense of Pain

Pain is the most important protective sense. The receptors for pain are very widely distributed. They are found in the skin, muscles, and joints, and to a lesser extent in most internal organs (including the blood vessels and viscera). Pain receptors are not enclosed in capsules as are the sensory end-organs; rather, they are merely branchings of the nerve fiber, called *free nerve endings.*

There are two pathways for the transmission of pain to the CNS, one for acute, sharp pain and the other for slow, chronic pain. Thus, a single strong stimulus produces the immediate sharp pain followed in a second or so by the slow, diffuse, burning pain that increases in severity with the passage of time.

### Sense of Position

Receptors located in muscles, tendons, and joints relay impulses that aid in judging one's position and changes in the locations of body parts in relation to each other. They also inform the brain of the amount of muscle contraction and tendon tension. These rather widespread end-organs, known as *proprioceptors* (pro-pre-o-SEP-tors), are aided in this function by the equilibrium receptors of the internal ear. Information received by these receptors is

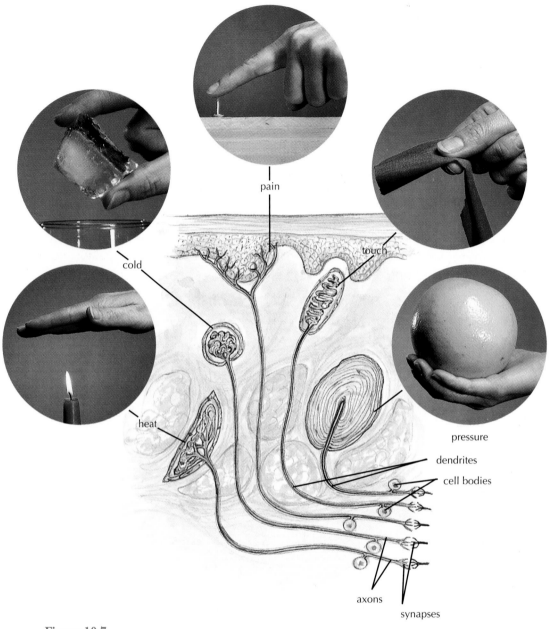

**Figure 10-7**

Diagram showing the superficial receptors and the deeper cell bodies and synapses, suggesting the continuity of sensory pathways into the CNS.

needed for the coordination of muscles and is important in such activities as walking, running, and many more complicated skills, such as playing a musical instrument. These muscle receptors also play an important part in maintaining muscle tone and good posture, as well as allowing for the adjustment of the muscles for the particular kind of work to be done. The nerve fibers that carry impulses from these receptors enter the spinal cord and ascend to the brain in the posterior part of the cord. The cerebellum is a main coordinating center for these impulses.

# SUMMARY

I. **Sensory system**—protects by detecting changes (stimuli) in the environment
   A. Special senses—vision, hearing, equilibrium, taste, smell
   B. General senses—pain, touch, temperature, pressure, position

II. **Eye**
   A. Protective structures—eyelids, bony orbit, eyelashes, tears (lacrimal apparatus), conjunctiva
   B. Coats (tunics)
      1. Sclera—white of the eye
         a. Cornea—anterior
      2. Choroid—pigmented; contains blood vessels
      3. Retina—nervous tissue
         a. Rods—detect white and black; function in dim light
         b. Cones—detect color; function in bright light
   C. Refraction
      1. Bending of light rays
      2. Refracting parts (media)—cornea, aqueous humor, crystalline lens, vitreous body
   D. Muscles
      1. Extrinsic muscles—six move each eyeball
      2. Intrinsic muscles
         a. Iris—colored ring around pupil; regulates the amount of light entering the eye
         b. Ciliary body—regulates the thickness of the lens for accommodation
   E. Nerve supply
      1. Optic nerve (II)—carries impulses from retina to brain
      2. Ophthalmic branch of trigeminal (V)—sensory nerve
      3. Oculomotor (III), trochlear (IV), abducens (VI) nerves—move eyeball

III. **Ear**
   A. Divisions
      1. External ear—pinna, auditory canal (meatus), tympanic membrane (eardrum)
      2. Middle ear—ossicles (malleus, incus, stapes)
         a. Eustachian tube—connects middle ear with pharynx to equalize pressure
      3. Internal ear—bony labyrinth
         a. Cochlea—contains receptors for hearing (organ of Corti)
         b. Vestibule—contains receptors for equilibrium
         c. Semicircular canals—contain receptors for equilibrium
   B. Nerve—vestibulocochlear (auditory) nerve (VIII)

IV. **Other special senses**
   A. Taste—gustatory sense
      1. Receptors—taste buds
      2. Tastes—sweet, sour, salty, bitter
      3. Nerves—facial (VII) and glossopharyngeal (IX)
   B. Smell—olfactory sense
      1. Receptors—in upper part of nasal cavity
      2. Nerve—olfactory nerve (I)

V. **General senses**
   A. Pressure
   B. Temperature
   C. Touch
   D. Pain—receptors are free nerve endings
   E. Position—receptors are proprioceptors in muscles, tendons, joints

## QUESTIONS FOR STUDY AND REVIEW

1. What is the function of the sensory system?
2. Define *sensory receptor* and give several examples.
3. List the five special senses.
4. List six general senses.
5. Name the main parts of the eye.
6. Trace the path of a light ray from the outside of the eye to the retina.
7. List the extrinsic muscles of the eye. What is their function?
8. List the intrinsic muscles of the eye. What is their function?
9. What is near accommodation? What structures of the eye are necessary for near accommodation?
10. List five protective devices for the eye.
11. List five cranial nerves associated with the eye and give the function of each.
12. Describe some changes that may occur in the eye with age.
13. Describe the structures that sound waves pass through in traveling through the ear to the receptors for hearing.
14. What cranial nerve carries impulses from the ear? Name the two branches.
15. Name the four kinds of taste. Where are the taste receptors?
16. Describe the olfactory apparatus.
17. What does *adaptation* mean with respect to the senses? How does sensory adaptation differ among the senses?
18. Where are the receptors for the senses of position and balance (equilibrium) located?
19. Differentiate between the terms in each of the following pairs:
    a. *special sense* and *general sense*
    b. *rods* and *cones*
    c. *endolymph* and *perilymph*
    d. *gustatory* and *olfactory*

# The Endocrine System: Glands and Hormones

## Behavioral Objectives

After careful study of this chapter, you should be able to:

- Compare the effects of the nervous system and the endocrine system in controlling the body
- Describe the functions of hormones
- Explain how hormones are regulated
- Identify the glands of the endocrine system on a diagram
- List the hormones produced by each endocrine gland and describe the effects of each on the body
- Describe how the hypothalamus controls the anterior and posterior pituitary
- List tissues other than the endocrine glands that produce hormones
- Explain how the endocrine system responds to stress

Memmler, RL., Cohen, BJ, Wood, DL. *STRUCTURE AND FUNCTION OF THE HUMAN BODY*, 6/e,
© 1996 Lippincott-Raven Publishers

The nervous system and the endocrine system are the two main controlling and coordinating systems of the body. The nervous system controls such rapid activity as muscle movement and intestinal activity by means of electric and chemical stimuli. The effects of the endocrine system occur more slowly and over a longer period. The glands of the endocrine system produce chemical messengers, or *hormones*, which have widespread effects. The two systems, however, are closely related. The activity of the pituitary gland, which in turn regulates other glands, is controlled by the nervous system. The connections between the nervous system and the endocrine system enable endocrine function to adjust to the demands of a changing environment.

## ▶ Functions of Hormones

Hormones are the chemical messengers that have specific regulatory effects on certain other cells or organs in the body. Originally, the word *hormone* applied to the secretions of the endocrine glands only. The term now includes various substances produced in the body that have regulatory actions, either locally or at a distance from where they are produced. Many body tissues produce substances that have strong effects in regulating the local environment.

Hormones are released into tissue fluid and the bloodstream and are carried to all parts of the body. They regulate growth, metabolism, reproduction, and other body processes. Some affect many tissues, for example, thyroid hormone and insulin. Others affect only specific tissues. One pituitary hormone, thyroid-stimulating hormone (TSH), acts only on the thyroid gland; another (ACTH) stimulates only the outer portion of the adrenal gland. The specific tissue acted upon by each hormone is the *target tissue*. The cells that make up these tissues have *receptors* in the cell membrane or within the cytoplasm to which the hormone attaches. Once attached to the cell, the hormone affects metabolism by regulating the manufacture of proteins.

Some hormones are effective for relatively long periods (thyroid hormones act for about 2 weeks). Others act rapidly and are removed rapidly. The level of antidiuretic hormone (ADH) in the circulation can vary from low to high in 15 minutes.

Chemically, hormones fall into several categories. The principal ones are:

1. **Amino acid compounds.** Most of the hormones are proteins or related compounds made of amino

acids. All hormones except those of the adrenal cortex and the sex glands fall into this category.
2. **Steroids** are hormones derived from lipids. They are produced by the adrenal cortex and the sex glands.

Substances known for other chemical effects act as hormones under certain circumstances—for example, neurotransmitters.

## Regulation of Hormones

The amount of each hormone that is secreted is normally kept within a specific range. Negative feedback is the method used to regulate these levels. The endocrine gland tends to oversecrete its hormone, exerting more effect on the target tissue. When the target tissue becomes too active, there is a negative effect on the endocrine gland which decreases its secretory action. The release of hormones may fall into a rhythmic pattern. Hormones of the adrenal cortex follow a 24-hour cycle related to a person's sleeping pattern, with the level of secretion greatest just before arising and least at bedtime. Hormones of the female menstrual cycle follow a monthly pattern.

The remainder of this chapter deals with the main hormone-secreting organs. Refer to Figure 11-1 to locate each of the organs as you study it.

## ▶ The Endocrine Glands and Their Hormones

### The Pituitary

The *pituitary* (pih-TU-ih-tar-e), or *hypophysis* (hi-POF-ih-sis), is a small gland about the size of a cherry. It is located in a saddle-like depression of the sphenoid bone just behind the point at which the optic nerves cross. It is surrounded by bone except where it connects with the brain by a stalk called the *infundibulum* (in-fun-DIB-u-lum). The gland is divided into two parts, the *anterior lobe* and the *posterior lobe*. The hormones released from each lobe are shown in Figure 11-2.

The pituitary is often called the *master gland* because it releases hormones that affect the working of other glands, such as the thyroid, gonads, and adrenal glands. (Hormones that stimulate other glands may be recognized by the ending *-tropin*, as in *thyrotropin*, which means "acting on the thyroid gland.") However, the pituitary itself is controlled by

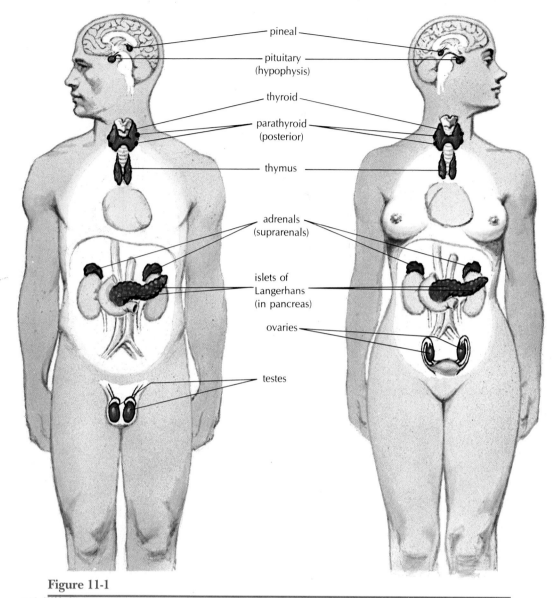

**Figure 11-1**

The main hormone-secreting organs.

the **hypothalamus** of the brain to which it is connected by the infundibulum or stalk (see Fig. 11-2).

The hormones produced in the anterior pituitary are not released from the gland until chemical messengers called **releasing hormones** arrive from the hypothalamus. These releasing hormones are sent to the anterior pituitary by way of a special type of circulatory pathway called a **portal system.** By this circulatory "detour," some of the blood that leaves the hypothalamus travels to capillaries in the anterior pituitary before returning to the heart. As the blood circulates through the capillaries, it delivers the hor-

mones that stimulate the release of anterior pituitary secretions.

The two hormones of the posterior pituitary (ADH and oxytocin) are actually produced in the hypothalamus and stored in the posterior pituitary. Their release is controlled by nerve impulses that travel over pathways (tracts) between the hypothalamus and the posterior pituitary.

### HORMONES OF THE ANTERIOR LOBE

1. **Growth hormone (GH),** or **somatotropin** (so-mah-to-TRO-pin), acts directly on most body tissues, pro-

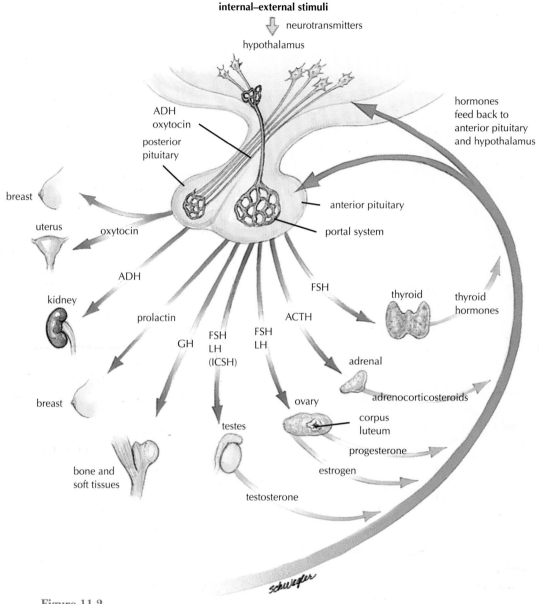

**Figure 11-2**

Pituitary gland and its relations with the brain and target tissues. Hypothalamic releasing hormones influence the anterior pituitary through a portal system. Tropic hormones from the anterior pituitary affect the working of various other glands. The hypothalamus communicates with the posterior pituitary through tracts.

moting protein deposits that are essential for growth. Produced throughout life, GH causes increase in size and height to occur in youth, before the closure of the epiphyses of long bones.

2. ***Thyroid-stimulating hormone (TSH),*** or ***thyrotropin*** (thi-ro-TRO-pin) stimulates the thyroid gland to produce thyroid hormones.

3. ***Adrenocorticotropic*** (ad-re-no-kor-te-ko-TRO-pik) ***hormone (ACTH)*** stimulates the cortex of the adrenal glands.

4. ***Prolactin*** (pro-LAK-tin) ***(PRL)*** stimulates the production of milk in the female.

***Gonadotropins*** (gon-ah-do-TRO-pinz), acting on the gonads, regulate the growth, development, and

## Growth Hormone

 Growth hormone, also called *somatotropin*, is produced by the anterior pituitary gland. Its release is regulated by growth hormone–releasing hormone (GH–RH) and growth hormone–inhibiting hormone (GH–IH), also called *somatostatin*, from the hypothalamus.

Although growth hormone mainly affects the development of bones and muscles, it has a general effect on most other tissues as well, stimulating an increase in protein synthesis. It also stimulates the liver to release fatty acids for energy when the level of blood glucose drops. In this capacity, it is released during times of stress. The production of growth hormone is controlled by its concentration in the blood, by the level of other hormones, including those from the adrenal, thyroid, and sex glands, and by the level of nutrients in the blood.

Insufficient growth hormone results in pituitary dwarfism. This condition can be prevented by administering growth hormone to children who lack it. Growth hormone is now produced by genetic engineering. The gene for the hormone is introduced into harmless bacteria, which then produce large quantities in laboratory cultures.

functioning of the reproductive systems in both males and females. The two gonadotropins are the following:

5. *Follicle-stimulating hormone (FSH)* stimulates the development of eggs in the ovaries and sperm cells in the testes.
6. *Luteinizing* (LU-te-in-i-zing) *hormone (LH)* causes ovulation in females and sex hormone secretion in both males and females; in males the hormone is called *interstitial cell–stimulating hormone (ICSH)*.

### HORMONES OF THE POSTERIOR LOBE

1. *Antidiuretic* (an-ti-di-u-RET-ik) *hormone (ADH)* promotes the reabsorption of water from the kidney tubules and thus decreases the excretion of water. Large amounts of this hormone cause contraction of the smooth muscle of blood vessels and raise blood pressure.
2. *Oxytocin* (ok-se-TO-sin) causes contraction of the muscle of the uterus and milk ejection from the

breasts. Under certain circumstances, commercial preparations of this hormone are administered during or after childbirth to cause the uterus to contract.

### The Thyroid Gland

The largest of the endocrine glands is the *thyroid*, which is located in the neck (Fig. 11-3). It has two oval parts called the *lateral lobes,* one on either side of the larynx (voice box). A narrow band called the *isthmus* (IS-mus) connects these two lobes. The entire gland is enclosed by a connective tissue capsule. The principal hormone produced by this gland is *thyroxine* (thi-ROK-sin). The hormone *calcitonin* (kal-sih-TO-nin), produced in the thyroid and active in calcium metabolism, will be discussed later in connection with the parathyroid gland.

The main function of thyroid hormones is to increase the rate of metabolism of most body cells. In particular, these hormones increase energy metabolism and protein metabolism. Thyroid hormones are necessary along with growth hormones for normal growth to occur. For the thyroid gland to produce hormones, there must be an adequate supply of iodine in the blood.

### The Parathyroid Glands

Behind the thyroid gland, and embedded in its capsule, are four tiny epithelial bodies called the *parathyroid glands* (Fig. 11-4). The secretion of these glands, *parathyroid hormone (PTH)*, is one of three substances that regulate calcium metabolism. The other two are *calcitonin* and *hydroxycholecalciferol* (hi-drok-se-ko-le-kal-SIF-eh-rol). PTH promotes the release of calcium from bone tissue, thus increasing the amount of calcium circulating in the bloodstream. Calcitonin, which is produced in the thyroid gland, acts to lower the amount of calcium circulating in the blood. Hydroxycholecalciferol is produced from vitamin D after first being modified by the liver and then the kidney. It increases absorption of calcium by the intestine to raise blood calcium levels.

### The Adrenal Glands

The *adrenals,* or *suprarenals,* are two small glands located above the kidneys. Each adrenal gland has two parts that act as separate glands. The inner area is called the *medulla,* and the outer portion is called the *cortex.*

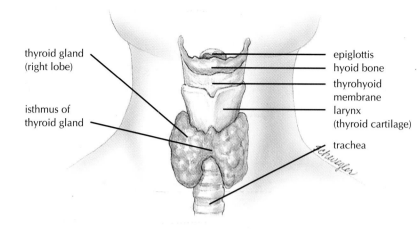

**Figure 11-3**

Thyroid gland (anterior view) in relation to the larynx and trachea.

## HORMONES FROM THE ADRENAL MEDULLA

The hormones of the adrenal medulla are released in response to stimulation by the sympathetic nervous system. The principal hormone produced by the medulla is one we already have learned about because it acts as a neurotransmitter. It is *epinephrine,* also called *adrenaline.* Another hormone, *norepinephrine,* is closely related chemically and is similar in its actions. These are referred to as the *fight-or-flight hormones* because of their effects during emergency situations. Some of these effects are as follows:

1. Stimulation of the involuntary muscle in the walls of the arterioles, causing these muscles to contract and blood pressure to rise accordingly
2. Conversion of the glycogen stored in the liver into glucose. The glucose is poured into the blood and brought to the voluntary muscles, permitting them to do an extraordinary amount of work.
3. Increase in the heart rate
4. Increase in the metabolic rate of body cells
5. Dilation of the bronchioles, through relaxation of the smooth muscle of their walls.

## HORMONES FROM THE ADRENAL CORTEX

There are three main groups of hormones secreted by the adrenal cortex:

1. *Glucocorticoids* (glu-ko-KOR-tih-koyds) maintain the carbohydrate reserve of the body by controlling the conversion of amino acids into sugar instead of protein. These hormones are produced in larger than normal amounts in times of stress and so aid the body in responding to unfavorable conditions. They have the ability to suppress the inflammatory response and are often administered as medication for this purpose. A major hormone of this group is *cortisol,* which is also called *hydrocortisone.*
2. *Mineralocorticoids* (min-er-al-o-KOR-tih-koyds) are important in the regulation of electrolyte balance. They control the reabsorption of sodium

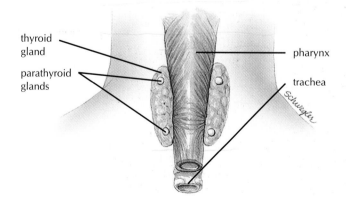

**Figure 11-4**

Posterior view of the thyroid gland showing the parathyroid glands embedded in its surface.

and the secretion of potassium by the kidney tubules. The major hormone of this group is **aldosterone** (al-DOS-ter-one).

3. **Sex hormones** are normally secreted, but in small amounts; their effects in the body are slight.

## The Islets of the Pancreas

Scattered throughout the **pancreas** are small groups of specialized cells called **islets** (I-lets), also known as the **islets of Langerhans** (LAHNG-er-hanz). A microscopic view of the pancreas showing the islet cells is presented in Figure 11-5. These cells make up the endocrine portion of the pancreas and function independently from the exocrine part of the pancreas, which produces digestive juices and releases them through ducts. The most important hormone secreted by these islets is **insulin.**

Insulin is active in the transport of glucose across the cell membrane. Once inside the cell, glucose can be used for energy. Insulin also increases the transport of amino acids into the cells and improves their use in the manufacture of proteins. When blood sugar and insulin decrease, most body cells switch to using fat for basic energy needs. Brain cells are an exception, because they use only glucose for energy and do not require insulin for glucose transport and metabolism. Insulin increases the rate at which the liver changes excess sugar into fatty acids, which can then be stored in fat cells. Through its actions, insulin has the effect of lowering the level of sugar in the blood.

A second hormone produced by the islet cells is **glucagon** (GLU-kah-gon), which works with insulin to regulate blood sugar levels. Glucagon causes the liver to release stored glucose into the bloodstream. Glucagon also increases the rate at which glucose is made from proteins in the liver. In these two ways, glucagon increases blood sugar.

If for some reason the pancreatic islets fail to produce enough insulin, sugar is not oxidized ("burned") in the tissues for transformation into energy. Instead, the sugar remains in the blood and then must be excreted in the urine. This condition, called **diabetes mellitus** (di-ah-BE-teze mel-LI-tus), is the most common endocrine disorder.

## The Sex Glands

The sex glands, the ovaries of the female and the testes of the male, not only produce the sex cells but also are important endocrine organs. The hormones produced by these organs are needed in the development of the sexual characteristics, which usually appear in the early teens, and for the maintenance of the reproductive organs once full development has been attained. Those features that typify a male or female other than the structures directly concerned with reproduction are termed **secondary sex characteristics.** They include a deep voice and facial and body hair in males; and wider hips and a greater ratio of fat to muscle in females.

The main male sex hormone or **androgen** (AN-dro-jen) produced by the male sex glands is **testosterone** (tes-TOS-ter-one).

In the female, the hormones that most nearly parallel testosterone in their actions are the **estrogens** (ES-tro-jens). Estrogens contribute to the development of the female secondary sex characteristics and stimulate the development of the mammary glands, the onset of menstruation, and the development and functioning of the reproductive organs.

The other hormone produced by the female sex glands, called **progesterone** (pro-JES-ter-one), assists

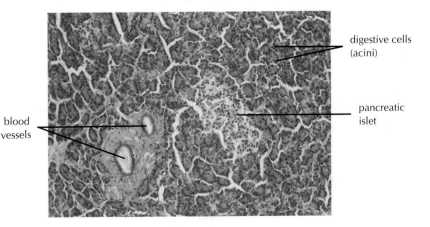

**Figure 11-5**

Microscopic view of pancreatic cells. Light staining islet cells are seen among the cell clusters (acini) that produce digestive juices. (Courtesy of Dana Morse Bittus and B.J. Cohen)

blood vessels

digestive cells (acini)

pancreatic islet

in the normal development of pregnancy. All the sex hormones are discussed in more detail in Chapter 20. Table 11-1 gives a summary of the major endocrine glands and the hormones they secrete.

## ▶ Other Hormone-Producing Tissues

The *thymus gland* is a mass of lymphoid tissue that lies in the upper part of the chest above the heart. This gland is important in the development of immunity. Its hormone, *thymosin* (THI-mo-sin), assists in the maturation of certain white blood cells called *T-lymphocytes* (T cells) after they have left the thymus gland and taken up residence in lymph nodes throughout the body.

The *pineal* (PIN-e-al) *body* is a small, flattened, cone-shaped structure located posterior to the midbrain and connected to the roof of the third ventricle. The pineal produces the hormone, *melatonin* (mel-ah-TO-nin), during the dark period of each day. Little hormone is produced during daylight hours. This pattern of hormone secretion influences the regulation of sleep/wake cycles. Melatonin also seems to delay the onset of puberty.

Cells in the lining of the stomach and small intestine secrete hormones that stimulate the production of digestive juices and help regulate the process of digestion.

In response to a decreased supply of oxygen, the kidneys produce a hormone called *erythropoietin* (e-rith-ro-POY-eh-tin), which stimulates red blood cell production in the bone marrow.

The atria of the heart produce a substance called *atrial natriuretic* (na-tre-u-RET-ik) *peptide (ANP)* in response to increased filling of the atria. ANP increases loss of salt by the kidneys and lowers blood pressure.

The *placenta* (plah-SEN-tah) produces several hormones during pregnancy. These cause changes in the uterine lining and later in pregnancy help prepare the breasts for lactation. Pregnancy tests are based on the presence of placental hormones.

### Prostaglandins

*Prostaglandins* (pros-tah-GLAN-dins) are a group of local hormones made by most body tissues. They are produced, act, and are rapidly inactivated in or close to the sites of origin. A bewildering array of functions has been ascribed to these substances. Some prostaglandins cause constriction of blood vessels, of bronchial tubes, and of the intestine, whereas others cause dilation of these same structures. Prostaglandins are active in promoting inflammation; certain anti-inflammatory drugs, such as aspirin, act by blocking the production of pro- staglandins. Some of the prostaglandins have been used to induce labor or abortion and have been recommended as possible contraceptive agents. Overproduction of prostaglandins by the uterine lining (endometrium) can cause painful cramps of the muscle of the uterus. Treatment with drugs that are prostaglandin inhibitors has been successful in some cases. Much has been written about these substances, and extensive research on them continues.

## ▶ Hormones and Stress

Stress in the form of physical injury, disease, emotional anxiety, and even pleasure calls forth a specific response from the body that involves both the nervous system and the endocrine system. The nervous system response, the "fight-or-flight" response, is mediated by parts of the brain, especially the hypothalamus, and by the autonomic nervous system. The hypothalamus also triggers the release of ACTH from the anterior pituitary. The hormones released from the adrenal cortex as a result of stimulation by ACTH raise the blood sugar level, inhibit inflammation, decrease the immune response, and limit the release of histamine. Growth hormone, thyroid hormones, sex hormones, and insulin may also be released. These substances help the body meet stressful situations, but unchecked they are harmful to the body and may lead to such stress-related disorders as high blood pressure, heart disease, ulcers, back pain, and headaches.

Although no one would enjoy a life totally free of stress in the form of stimulation and challenge, unmanaged stress, or "distress," has negative effects on the body. For this reason, techniques such as biofeedback and meditation to control stress are useful. Simply setting priorities, getting adequate periods of relaxation, and getting regular physical exercise are important in maintaining total health.

## TABLE 11-1
## The Major Endocrine Glands and Their Hormones

| Gland | Hormone | Principal Functions |
|---|---|---|
| Anterior pituitary | GH (growth hormone) | Promotes growth of all body tissues |
| | TSH (thyroid-stimulating hormone) | Stimulates thyroid gland to produce thyroid hormones |
| | ACTH (adrenocorticotropic hormone) | Stimulates adrenal cortex to produce cortical hormones; aids in protecting body in stress situations (injury, pain) |
| | PRL (prolactin) | Stimulates secretion of milk by mammary glands |
| | FSH (follicle-stimulating hormone) | Stimulates growth and hormone activity of ovarian follicles; stimulates growth of testes; promotes development of sperm cells |
| | LH (luteinizing hormone); ICSH (interstitial cell–stimulating hormone) in males | Causes development of corpus luteum at site of ruptured ovarian follicle in female; stimulates secretion of testosterone in male |
| Posterior pituitary | ADH (antidiuretic hormone; vasopressin) | Promotes reabsorption of water in kidney tubules; stimulates smooth muscle tissue of blood vessels to constrict |
| | Oxytocin | Causes contraction of muscle of uterus; causes ejection of milk from mammary glands |
| Thyroid | Thyroid hormone (thyroxine and triiodothyronine) | Increases metabolic rate, influencing both physical and mental activities; required for normal growth |
| | Calcitonin | Decreases calcium level in blood |
| Parathyroids | Parathyroid hormone | Regulates exchange of calcium between blood and bones; increases calcium level in blood |
| Adrenal medulla | Epinephrine and norephinephrine | Increases blood pressure and heart rate; activates cells influenced by sympathetic nervous system plus many not affected by sympathetic nerves |
| Adrenal cortex | Cortisol (95% of glucocorticoids) | Aids in metabolism of carbohydrates, proteins, and fats; active during stress |
| | Aldosterone (95% of mineralocorticoids) | Aids in regulating electrolytes and water balance |
| | Sex hormones | May influence secondary sexual characteristics |
| Pancreatic islets | Insulin | Aids transport of glucose into cells; required for cellular metabolism of foods, especially glucose; decreases blood sugar levels |
| | Glucagon | Stimulates liver to release glucose, thereby increasing blood sugar levels |
| Testes | Testosterone | Stimulates growth and development of sexual organs (testes, penis, others) plus development of secondary sexual characteristics such as hair growth on body and face and deepening of voice; stimulates maturation of sperm cells |
| Ovaries | Estrogens (*e.g.,* estradiol) | Stimulates growth of primary sexual organs (uterus, tubes, etc.) and development of secondary sexual organs such as breasts, plus changes in pelvis to ovoid, broader shape |
| | Progesterone | Stimulates development of secretory parts of mammary glands; prepares uterine lining for implantation of fertilized ovum; aids in maintaining pregnancy |

## SUMMARY

I. **Hormones**
   A. Functions
      1. Affect certain other cells or organs—target tissue
      2. Carried by bloodstream
      3. Widespread effects on growth, metabolism, reproduction
      4. Local effects
   B. Main types
      1. Amino acid compounds—proteins and related compounds
      2. Steroids
         a. Derived from lipids
         b. Produced by adrenal cortex and sex glands
   C. Regulation—negative feedback
II. **Endocrine glands**
   A. Pituitary
      1. Regulated by hypothalamus
         a. Anterior pituitary—releasing hormones sent through portal system
         b. Posterior pituitary—stores hormones; released by nervous stimulation
      2. Anterior lobe hormones
         a. Growth hormone (GH)
         b. Thyroid-stimulating hormone (TSH)
         c. Adrenocorticotropic hormone (ACTH)
         d. Prolactin (PRL)
         e. Follicle-stimulating hormone (FSH)
         f. Luteinizing hormone (LH)
      3. Posterior lobe hormones
         a. Antidiuretic hormone (ADH)
         b. Oxytocin
   B. Thyroid hormones
      1. Thyroxine—influences cell metabolism
      2. Calcitonin—decreases blood calcium levels
   C. Parathyroids—secrete parathyroid hormone (PTH), which increases blood calcium levels
   D. Adrenals
      1. Hormones of adrenal medulla (inner region)
         a. Epinephrine and norepinephrine—act as neurotransmitters
      2. Hormones of adrenal cortex (outer region)
         a. Glucocorticoids—released during stress; example, cortisol
         b. Mineralocorticoids—regulate water and electrolyte balance; example, aldosterone
         c. Sex hormones—produced in small amounts
   E. Islet cells of pancreas—secrete hormones
      1. Insulin
         a. Lowers blood glucose
         b. Lack causes diabetes mellitus
      2. Glucagon
         a. Raises blood glucose
   F. Sex glands—needed for reproduction and development of secondary sex characteristics
      1. Testes—secrete testosterone
      2. Ovaries—secrete estrogen and progesterone
III. **Other hormone-producing tissues**
   A. Thymus—secretes thymosin, which aids in development of T-lymphocytes
   B. Pineal—secretes melatonin
      1. Regulates sexual development and sleep/wake cycles
      2. Controlled by environmental light
   C. Stomach and small intestine—secrete hormones that regulate digestion
   D. Kidneys—secrete erythropoietin, which increases production of red blood cells
   E. Atria of heart—ANP causes loss of sodium by kidney and lowers blood pressure
   F. Placenta—secretes hormones that maintain pregnancy and prepare breasts for lactation
   G. Other—cells throughout body produce prostaglandins, which have varied effects

## QUESTIONS FOR STUDY AND REVIEW

1. Compare the actions of the nervous system and the endocrine system.
2. Define *hormone*. Describe some general functions of hormones.
3. Give three examples of amino acid hormones; of steroid hormones.
4. Define *target tissue*.
5. Name the two divisions of the pituitary gland. Name the hormones released from each division and describe the effects of each.
6. What type of system connects the anterior pituitary with the hypothalamus? What is carried to the pituitary by this system?
7. Where is the thyroid gland located? What is its main hormone, and what does it do?
8. What is the main purpose of PTH?
9. Name the two divisions of the adrenal glands and describe the effects of the hormones from each.
10. What are the main purposes of insulin in the body? Name and describe the condition characterized by insufficient production of insulin.
11. Name the male and female sex hormones and briefly describe what each does.
12. Name the hormone produced by the thymus gland; by the pineal body. What are the effects of each?
13. Name five organs other than the endocrine glands that secrete hormones.
14. What are some of the various functions that have been ascribed to prostaglandins?
15. List several hormones that are released during stress.

# UNIT **IV**

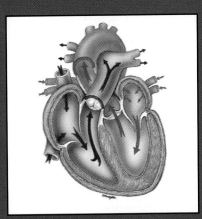

# Circulation and Body Defense

The chapters of this unit will discuss the blood, the heart, the blood vessels, and the lymphatic system as well as body defenses. It is the purpose of this unit to emphasize the importance of transportation and immune systems in maintaining normal body functions.

# CHAPTER 12

# The Blood

Memmler, RL, Cohen, BJ, Wood, DL. *STRUCTURE AND FUNCTION OF THE HUMAN BODY*, 6/e,
© 1996 Lippincott-Raven Publishers

## Selected Key Terms

The following terms are defined in the Glossary:

**agglutination**

**antiserum**

**centrifuge**

**coagulation**

**erythrocyte**

**fibrin**

**hematocrit**

**hemoglobin**

**hemostasis**

**leukocyte**

**plasma**

**platelet (thrombocyte)**

**serum**

## Behavioral Objectives

After careful study of this chapter, you should be able to:

- List the functions of the blood
- List the main ingredients in plasma
- Name the three types of formed elements in the blood and give the function of each
- Describe five types of leukocytes
- Describe the formation of blood cells
- Cite three steps in hemostasis
- Briefly describe the steps in blood clotting
- Define *blood type* and explain the relation between blood type and transfusions
- Describe the tests used to study blood

Blood is classified as a connective tissue because nearly half of it is made up of cells. Blood cells share many characteristics of origination and development with other connective tissues. However, blood differs from other connective tissues in that its cells are not fixed in position; instead, they move freely in the liquid portion of the blood, the *plasma.*

Blood is a viscous (thick) fluid that varies in color from bright scarlet to dark red, depending on how much oxygen it is carrying. The quantity of circulating blood differs with the size of the person; the average adult male, weighing 70 kg (154 pounds), has about 5 liters (5.2 quarts) of blood. This volume accounts for about 8% of total body weight.

The blood is carried through a closed system of vessels pumped by the heart. The circulating blood is of fundamental importance in maintaining the internal environment in a constant state (homeostasis).

## ▶ Functions of the Blood

The circulating blood serves the body in three ways: transportation, regulation, and protection.

### Transportation

1. Oxygen from inhaled air diffuses into the blood through the thin lung membranes and is carried to all the tissues of the body. Carbon dioxide, a waste product of cell metabolism, is carried from the tissues to the lungs, where it is breathed out.
2. The blood transports nutrients and other needed substances, such as electrolytes (salts) and vitamins, to the cells. These materials may enter the blood from the digestive system or may be released into the blood from body stores.
3. The blood transports the waste products from the cells to the sites from which they are released. The kidney removes excess water, electrolytes, and urea (from protein metabolism) and maintains the acid–base balance of the blood. The liver removes bile pigments and drugs.
4. The blood carries hormones from their sites of origin to the organs they affect.

### Regulation

1. Buffers in the blood help keep the pH of body fluids at about 7.4.
2. The blood serves to regulate the amount of fluid in the tissues by means of substances (mainly proteins) that maintain the proper osmotic pressure.

3. The blood transports heat that is generated in the muscles to other parts of the body, thus aiding in the regulation of body temperature.

### Protection

1. The blood carries the cells that are among the body's defenders against pathogens. It also contains substances (antibodies) that are concerned with immunity to disease.
2. The blood contains factors that protect against blood loss from the site of an injury.

## ▶ Blood Constituents

The blood is composed of two prime elements: as already mentioned, the liquid element is called *plasma;* the cells and fragments of cells are called *formed elements* or *corpuscles* (KOR-pus-ls) (Fig. 12-1). The formed elements are classified as follows:

1. *Erythrocytes* (eh-RITH-ro-sites), from *erythro,* meaning "red," are the red blood cells, which transport oxygen.
2. *Leukocytes* (LU-ko-sites), from *leuko,* meaning "white," are the several types of white blood cells, which protect against infection.
3. *Platelets,* also called *thrombocytes* (THROM-bo-sites), are cell fragments that participate in blood clotting.

### Blood Plasma

More than half of the total volume of blood is plasma. The plasma itself is 90% water. Many different substances, dissolved or suspended in the water, make up the other 10%. The plasma content varies somewhat, because substances are removed and added as the blood circulates to and from the tissues. However, the body tends to maintain a fairly constant level of these substances. For example, the level of glucose, a simple sugar, is maintained at a remarkably constant level of about one tenth of one percent (0.1%) in solution.

After water, the next largest percentage of material in the plasma is *protein.* Proteins are the principal constituents of cytoplasm and are essential to the growth and the rebuilding of body tissues. The plasma proteins include the following:

1. *Albumin* (al-BU-min), the most abundant protein in plasma, is important for maintaining the

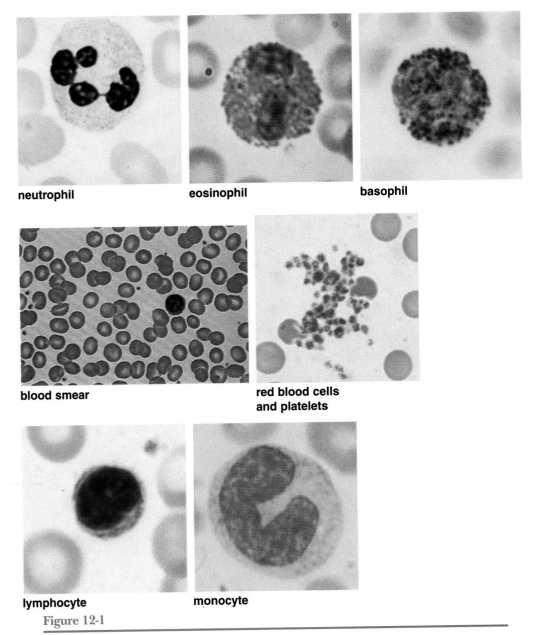

**neutrophil**   **eosinophil**   **basophil**

**blood smear**

**red blood cells
and platelets**

**lymphocyte**   **monocyte**

Figure 12-1

Normal blood smear and close-up view of individual blood cells.

osmotic pressure of the blood. This protein is manufactured in the liver.

2. Blood *clotting factors* are also manufactured in the liver.
3. *Antibodies* combat infection.
4. A system of enzymes made of several proteins, collectively known as *complement,* helps antibodies in their fight against pathogens (see Chap. 15).

Nutrients are also found in the plasma. The principal carbohydrate found in the plasma is *glucose,* which is absorbed by the capillaries of the intestine following digestion. Glucose is stored mainly in the liver as glycogen and is released as needed to supply energy.

*Amino acids,* the products of protein digestion, are also found in the plasma. These are also absorbed into the blood through the intestinal capillaries.

*Lipids* constitute a small percentage of blood plasma. Lipids include fats. They may be stored as fat for reserve energy or carried to the cells as a source of energy.

The *electrolytes* in the plasma appear primarily as chloride, carbonate, or phosphate salts of sodium, potassium, calcium, and magnesium. These salts have a variety of functions, including the formation of bone (calcium and phosphorus), the production of hormones by certain glands (iodine for the production of thyroid hormone), the transportation of oxygen (iron), and the maintenance of the acid–base balance (sodium and potassium carbonates and phosphates). Small amounts of other elements also help maintain homeostasis.

Many other materials, such as vitamins, hormones, waste products and drugs, are transported in the plasma.

### The Formed Elements
#### ERYTHROCYTES
*Erythrocytes,* the red cells, are tiny, disk-shaped bodies with a central area that is thinner than the edges (see Fig. 12-1). They are different from other cells in that the mature form found in the circulating blood is lacking a nucleus and most of the other organelles commonly found in cells. As red cells mature, these components are lost to provide more space for the cells to carry oxygen. This vital gas is bound in the red cells to *hemoglobin* (he-mo-GLO-bin), a protein that contains iron. Hemoglobin combined with oxygen gives the blood its characteristic red color. The more oxygen carried by the hemoglobin, the brighter is the red color of the blood. Therefore, the blood that goes from the lungs to the tissues is a bright red because it carries a great supply of oxygen; in contrast, the blood that returns to the lungs is a much darker red, because it has given up much of its oxygen to the tissues. Hemoglobin that has given up its oxygen is able to carry hydrogen ions; in this way, hemoglobin acts as a buffer and plays an important role in acid–base balance (see Chap. 19). The red cells also carry a small amount of carbon dioxide from the tissues to the lungs for elimination by exhalation.

The ability of hemoglobin to carry oxygen can be blocked by carbon monoxide. This harmful gas combines with hemoglobin to form a stable compound that can severely restrict the ability of the erythrocytes to carry oxygen. Carbon monoxide is a by-product of the incomplete burning of fuels such as gasoline and other petroleum products, coal, wood, and other carbon-containing materials. It also occurs in cigarette smoke and auto exhaust.

## Hemoglobin

The hemoglobin molecule is a protein made of four chains of amino acids (the "globin" part of the molecule), each of which holds an iron-containing "heme" unit. It is the heme portion that binds oxygen.

Hemoglobin allows the blood to carry much more oxygen than it could carry simply dissolved in the plasma. Hemoglobin picks up oxygen in the lungs and releases it in the body tissues. When cells are active, they need more oxygen. At the same time, they generate heat and acidity. These changing conditions promote the release of oxygen from hemoglobin.

Hemoglobin is produced in the liver and in the red bone marrow. It is constantly broken down as red blood cells die and disintegrate. Some of its components are recycled, but we still need protein, iron, and vitamin $B_{12}$ in our diet to maintain supplies.

The erythrocytes are by far the most numerous of the corpuscles, averaging from 4.5 to 5 million per cubic millimeter ($mm^3$) of blood. (A cubic millimeter is the same as a microliter [µL], one millionth of a liter, another way of expressing the concentration of blood cells). Because mature red cells have no nucleus and cannot divide, they must be replaced constantly. The production of red cells is stimulated by the hormone *erythropoietin* (eh-rith-ro-POY-eh-tin). This hormone is released from the kidney in response to a decrease in oxygen supply.

#### LEUKOCYTES
The *leukocytes,* or white blood cells, are very different from the erythrocytes in appearance, quantity, and function (see Fig. 12-1). The cells themselves are round, but they contain nuclei of varying shapes and sizes. Occurring at a concentration of 5,000 to 10,000 per cubic millimeter of blood, leukocytes are outnumbered by red cells by about 700 to 1. Whereas the red cells have a definite color, the leukocytes tend to be colorless.

The different types of white cells are identified by their size, the shape of the nucleus, and the appearance of granules in the cytoplasm when the cells are stained, usually with Wright's blood stain. *Gran-*

ulocytes (GRAN-u-lo-sites) include *neutrophils* (NU-tro-fils), which show lavender granules; *eosinophils* (e-o-SIN-o-fils), which have beadlike, bright pink granules; and *basophils* (BA-so-fils), which have large, dark blue granules that often obscure the nucleus. The neutrophils, which are active in fighting infections, are the most numerous of the white cells, constituting up to 60% of all leukocytes. The eosinophils and basophils make up a very small percentage of the white cells, but increase in number during allergic reactions.

Because the nuclei of the neutrophils are of various shapes, they are also called *polymorphs* (meaning "many forms") or simply *polys*. Other nicknames are *PMNs*, an abbreviation of *poly*morpho*nuclear neutrophils, and *segs*. Before reaching full maturity and becoming segmented, the nucleus of the neutrophil looks like a thick, curved band. An increase in the number of these *band cells* (also called *stab* or *staff cells*) is a sign of infection and active production of neutrophils.

The *agranulocytes,* so named because they lack easily visible granules, are the *lymphocytes and monocytes.* The ratio of the different types of leukocytes is often a valuable clue in arriving at a medical diagnosis. The relative percentages and the functions of the various leukocytes are given in Table 12-1.

The most important function of leukocytes is to destroy pathogens. Whenever pathogens enter the tissues, as through a wound, certain white blood cells (neutrophils and monocytes) are attracted to that area. They leave the blood vessels and proceed by *ameboid* (ah-ME-boyd) or ameba-like motion to the area of infection. There they engulf the invaders by the process of *phagocytosis* (fag-o-si-TO-sis) (Fig. 12-2). If the pathogens are extremely virulent (strong) or numerous, they may destroy the leukocytes. A collection of dead and living bacteria, together with dead and living leukocytes, forms *pus.*

Leukocytes may enter the tissues, undergo transformation, and become active in many different ways. Monocytes mature to become *macrophages* (MAK-ro-faj-ez) or fuse together to become giant cells, highly active in disposing of invaders or foreign material. Some lymphocytes become *plasma cells,* active in the production of circulating antibodies. The activities of the various white cells are further discussed in Chapter 15.

### PLATELETS

Of all the formed elements, the blood *platelets* (thrombocytes) are the smallest (see Fig. 12-1). These tiny structures are not cells in themselves, but fragments of cells. The number of platelets in the circulating blood has been estimated to range from 150,000 to 450,000 per cubic millimeter. Platelets are essential to blood *coagulation* (clotting). When,

**TABLE 12-1**
**Leukocytes (White Blood Cells)**

| Type of Cell | Relative Percentage (Adult) | Function |
|---|---|---|
| **Granulocytes** | | |
| Neutrophils (NU-tro-fils) | 54%–62% | Phagocytosis |
| Eosinophils (e-o-SIN-o-fils) | 1%–3% | Allergic reactions; defense against parasites |
| Basophils (BA-so-fils) | less than 1% | Allergic reactions; inflammatory reactions |
| **Agranulocytes** | | |
| Lymphocytes (LIM-fo-sites) | 25%–38% | Immunity (T cells and B cells) |
| Monocytes (MON-o-sites) | 3%–7% | Phagocytosis |

leukocyte    capillary    bacteria

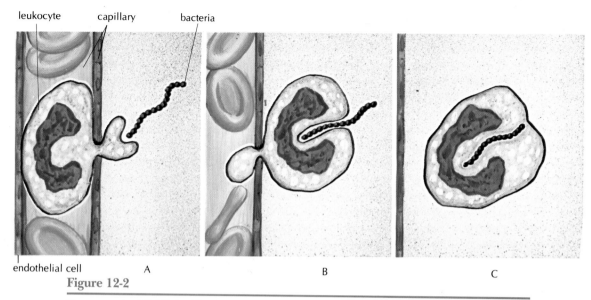

endothelial cell          A                              B                              C

**Figure 12-2**

Phagocytosis. (**A**) White blood cell squeezes through a capillary wall in the region of an infection. (**B,C**) White cell engulfs the bacteria.

as a result of injury, blood comes in contact with any tissue other than the lining of the blood vessels, the platelets stick together and form a plug that seals the wound. They then release chemicals that take part in a series of reactions that eventually results in the formation of a clot. The last step in these reactions is the conversion of a plasma protein called *fibrinogen* (fi-BRIN-o-jen) into solid threads of *fibrin,* which form the clot.

### Origin of the Formed Elements

The red blood cells (erythrocytes), the platelets (thrombocytes), and most of the white blood cells (leukocytes) are formed in the red marrow, which is located in the ends of long bones and in the inner mass of all other bones.

The ancestors of blood cells are called *stem cells.* Each kind of stem cell develops into one of the blood cell types found within the red marrow. One group of white cells, the lymphocytes, develops to maturity in lymphoid tissue and can multiply in this tissue as well (see Chap. 15).

As already mentioned, when an invader enters the tissues, the leukocytes in the blood are attracted to the area. In addition, the bone marrow goes into emergency production, with the result that the number of leukocytes in the blood is enormously increased. Detection in a blood examination of an abnormally large number of white cells may thus be an indication of an infection.

The platelets are believed to originate in the red marrow as fragments of certain giant cells called *megakaryocytes* (meg-ah-KAR-e-o-sites), which are formed in the red marrow. Platelets do not have nuclei or DNA, but they do contain active enzymes and mitochondria.

The life spans of the different types of blood cells vary considerably. For example, after leaving the bone marrow, erythrocytes circulate in the bloodstream for approximately 120 days. Leukocytes may appear in the circulating blood for only 6 to 8 hours. However, they may then enter the tissues, where they survive for longer periods—days, months, or even years. Blood platelets have a life span of about 10 days. In comparison with other tissue cells, most of those in the blood are very short lived. Thus, the need for constant replacement of blood cells means that normal activity of the red bone marrow is absolutely essential to life.

## ❱ Hemostasis

*Hemostasis* (he-mo-STA-sis) is the process that prevents the loss of blood from the circulation when a blood vessel is ruptured by an injury. Events in hemostasis include the following:

1. ***Contraction*** of the smooth muscles in the wall of the blood vessel. This reduces the flow of blood and loss from the defect in the vessel wall. The term for this reduction in the diameter of a vessel is *vasoconstriction.*
2. Formation of a ***platelet plug.*** Activated platelets become sticky and adhere to the defect to form a temporary plug.
3. Formation of a ***blood clot.***

### Blood Clotting

The many substances necessary for blood clotting, or coagulation, are normally inactive in the bloodstream. A balance is maintained between compounds that promote clotting, known as ***procoagulants,*** and those that prevent clotting, known as ***anticoagulants.*** In addition, there are chemicals in the circulation that act to dissolve clots. Under normal conditions the substances that prevent clotting prevail. However, when an injury occurs, the procoagulants are activated and a clot is formed.

Basically, the clotting process consists of the following essential steps (Fig. 12-3):

1. The injured tissues release ***thromboplastin*** (throm-bo-PLAS-tin), a substance that triggers the clotting mechanism.
2. Thromboplastin reacts with certain protein factors and calcium ions to form ***prothrombin*** activator, which in turn reacts with calcium ions to convert the prothrombin to ***thrombin.***
3. Thrombin, in turn, converts soluble fibrinogen into insoluble fibrin. ***Fibrin*** forms a network of threads that entraps red blood cells and platelets to form a clot.

When blood is removed from the body it will automatically clot as it comes into contact with some surface other than the lining of a blood vessel. The fluid that remains after clotting has occurred is called ***serum*** (pl. *sera*). Serum contains all the components of blood plasma *except* the clotting factors.

Several methods used to measure the body's ability to coagulate blood are described later in this chapter.

## ▶ Blood Typing and Transfusions

### Blood Groups

If for some reason the amount of blood in the body is severely reduced, through ***hemorrhage*** (HEM-eh-rij) (excessive bleeding) or disease, the body cells suffer from lack of oxygen and nutrients. The obvious measure to take in such an emergency is to administer blood from another person into the veins of the patient, a procedure called ***transfusion.***

Care must be taken in transferring blood from one person to another, however, because the patient's plasma may contain substances, called ***antibodies,*** or *agglutinins,* that can cause the red cells of the donor's blood to rupture and release their hemoglobin. Such cells are said to be ***hemolyzed*** (HE-mo-lized), and the resulting condition can be very dangerous.

These reactions are determined by certain proteins, called ***antigens*** (AN-ti-gens), or *agglutinogens,* on the red cell membranes. There are many types of these proteins, but only two groups are particularly likely to cause a transfusion reaction, the so-called *A* and *B antigens* and the *Rh factor.* Four blood types involving the A and B antigens have been recognized: A, B, AB, and O. These letters indicate the type of antigen present on the red cells. If only the A antigen is present on the red cells, the person is type A; if only the B antigen is present, he or she is type B. Type AB red cells have both antigens, and type O have neither. It is these antigens on the donor's red cells that react with the antibodies in the patient's plasma and cause a transfusion reaction.

Blood sera containing antibodies to the A and B antigens are used to test for blood type. Blood serum containing antibodies that can agglutinate and destroy red cells with A antigen on the surface is called ***anti-A serum;*** blood serum containing antibodies that can destroy red cells with B antigen on the surface is called ***anti-B serum.*** When combined in the laboratory, each antiserum will cause the corresponding red

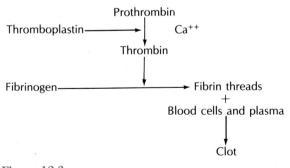

**Figure 12-3**

Formation of a blood clot.

cells to clump together in a process known as **_agglu-_**
**_tination_** (ah-glu-tih-NA-shun). The blood's pattern of
agglutination, when mixed _separately_ with these two
sera, reveals its blood type (Fig. 12-4). Type A reacts
with A antiserum only; type B reacts with B antiserum
only. Type AB agglutinates with both, and type O agglutinates with neither A nor B.

Of course no one has antibodies to his or her
own blood type antigens or their own cells would be
destroyed by their plasma. Each person does, however, develop antibodies that react with the AB antigens he or she is lacking.

An individual's blood type is determined by
heredity, and the percentage of persons with each

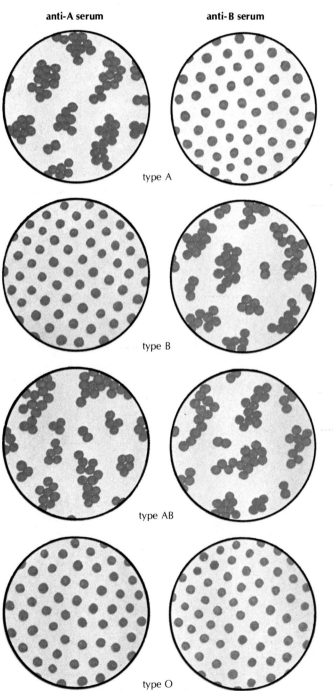

**Figure 12-4**

Blood typing. Red cells in type A blood are agglutinated (clumped) by anti-A serum; those in type B
blood are agglutinated by anti-B serum. Type AB
blood cells are agglutinated by both sera, and type
O blood is not agglutinated by either serum.

of the different blood types varies in different populations. About 45% of the white population of the United States is type O. In an emergency, type O blood can be given to any ABO type because the cells lack both A and B antigens and will not react with either A or B antibodies. Conversely, the blood of type AB individuals contains no antibodies to agglutinate red cells and they can therefore receive blood from any type of ABO donor (Table 12-2).

Usually a person can safely give blood to any person with the same blood type. However, because of other factors that may be present in the blood, determination of blood type must be accompanied by additional tests (cross matching) for compatibility before a transfusion is given.

### The Rh Factor

More than 85% of the population of the United States has another red cell antigen group called the *Rh factor,* named for *Rh*esus monkeys in which it was first found. Rh is also known as the *D antigen.* Such individuals are said to be *Rh positive.* Those persons who lack this protein are said to be *Rh negative.* If Rh-positive blood is given to an Rh-negative person, he or she may produce antibodies to the "foreign" Rh antigens. The blood of this "Rh-sensitized" person will then destroy any Rh-positive cells received in a later transfusion.

Rh incompatability is a potential problem in certain pregnancies. A mother who is Rh negative may develop antibodies to the Rh protein of an Rh-positive fetus (the fetus having inherited this factor from the father). The response is evoked by red cells from the fetus that enter the mother's circulation during childbirth. During a subsequent pregnancy with an Rh-positive fetus, some of the anti-Rh antibodies may pass from the mother's blood into the blood of her fetus and cause destruction of the fetus's red cells. Should this occur, the newborn can sometimes be saved by a transfusion during which much of the blood is replaced.

Red cell destruction may be prevented by administration of immune globulin $Rh_o(D)$, or Rho-GAM, to the mother during pregnancy and shortly after delivery. These preformed antibodies clear the mother's circulation of Rh antigens and prevent the activation of her immune system.

## ▶ Blood Studies

Many kinds of studies may be made of the blood, and some of these have become a standard part of a routine physical examination. Machines that are able to perform several tests at the same time have largely replaced manual procedures, particularly in large institutions.

### The Hematocrit

The *hematocrit* (he-MAT-o-krit), the volume percentage of red blood cells in whole blood, is determined by spinning of a blood sample in a high-speed centrifuge for 3 to 5 minutes. In this way the cellular elements are separated out from the plasma.

The hematocrit is expressed as the volume of packed red cells per unit volume (100 mL) of whole blood. For example, if a laboratory report states "hematocrit, 38%," that means that there are 38 mL red cells per 100 mL whole blood. In other words, 38% of the whole blood is red cells. For adult men, the normal range is 42 mL to 54 mL per 100 mL blood, whereas for adult women the range is slightly

**TABLE 12-2**
**The ABO Blood Group System**

| Blood Type | Red Blood Cell Antigen | Reacts with Antiserum | Plasma Antibodies | Can Take From | Can Donate to |
|---|---|---|---|---|---|
| A | A | Anti-A | Anti-B | A, O | A, AB |
| B | B | Anti-B | Anti-A | B, O | B, AB |
| AB | A, B | Anti-A, anti-B | None | AB, A, B, O | AB |
| O | None | None | Anti-A, anti-B | O | O, A, B, AB |

lower, 36 mL to 46 mL per 100 mL blood. These normal ranges, like all normal ranges for humans, may vary depending on the method used and the interpretation of the results by an individual laboratory. Hematocrit values much below or much above these figures point to an abnormality requiring further study.

### Hemoglobin Tests

A sufficient amount of hemoglobin in red cells is required for adequate delivery of oxygen to the tissues. To measure its level, the hemoglobin is released from the red cells and the color of the blood is compared with a known color scale. Hemoglobin is expressed in grams per 100 mL (dL) whole blood. Normal hemoglobin concentrations for adult males range from 14 to 17 g per 100 mL blood. Values for adult females are in a somewhat lower range, at 12 to 15 g per 100 mL blood.

### Blood Cell Counts

Most laboratories use automated methods for obtaining the data for blood counts. However, visual counts using a *hemocytometer* (he-mo-si-TOM-eh-ter) are still done for reference and quality control in all automated laboratories (Fig. 12-5).

The normal count for red blood cells varies from 4.5 to 5.5 million cells per cubic millimeter ($\mu$L) of blood. The leukocyte count varies from 5000 to 10,000 cells per cubic millimeter of blood.

It is difficult to count platelets directly because they are so small. More accurate counts can be obtained with automated methods. The normal platelet count ranges from 150,000 to 450,000 per cubic millimeter of blood, but counts may fall to 100,000 or less without causing serious bleeding problems. If a count is very low, a platelet transfusion may be given.

### The Blood Slide (Smear)

In addition to the above tests, the *complete blood count* (*CBC*) includes the examination of a stained blood slide. In this procedure a drop of blood is spread very thinly and evenly over a glass slide and a special stain (Wright's) is applied to differentiate the otherwise colorless white cells. The slide is then studied under the microscope. The red cells are examined for abnormalities in size, color, or shape and for variations in the percentage of immature forms. The number of platelets is estimated. Parasites such as the malarial organism and others may

be found. In addition, a *differential white count* is done. This is an estimation of the percentage of each type of white blood cell in the smear. Because each type of white blood cell has a specific function, changes in their proportions can be a valuable diagnostic aid.

### Blood Chemistry Tests

Batteries of tests on blood serum are often done by machine. One machine, the Sequential Multiple Analyzer (SMA), provides for the running of some 20 tests per minute. Tests for electrolytes, such as sodium, potassium, chloride, and bicarbonate, as well as for blood glucose, blood urea nitrogen (BUN), and *creatinine* (kre-AT-in-in), another nitrogen waste product, may be performed at the same time.

Other tests check for enzymes. Increased levels of *CPK* (creatine phosphokinase), *LDH* (lactic dehydrogenase), and other enzymes indicate tissue damage, such as that which may occur in heart disease. An excess of *alkaline phosphatase* (FOS-fah-tase) could indicate a liver disorder or cancer involving bone.

Blood can be tested for amounts of lipids such as cholesterol and triglycerides (fats) or for amounts of plasma proteins. Many of these tests help in evaluating disorders that may involve various vital organs. For example, the presence of more than the normal amount of glucose (sugar) dissolved in the blood, a condition called *hyperglycemia* (hi-per-gli-SE-me-ah), is found most frequently in persons with unregulated diabetes. Sometimes several evaluations of sugar content are done following the administration of a known amount of glucose. This procedure is called the *glucose tolerance test* and usually is given along with another test that determines the amount of sugar in the urine. This combination of tests can indicate faulty cell metabolism. The list of blood chemistry tests is extensive and is constantly increasing. We may now obtain values for various hormones, vitamins, antibodies, and toxic or therapeutic drug levels.

### Coagulation Studies

Nature prevents the excessive loss of blood from small vessels by the formation of a clot. Preceding surgery and under some other circumstances, it is important to know that the time required for coagulation to take place is not long. Because clotting is a rather complex process involving many reactants,

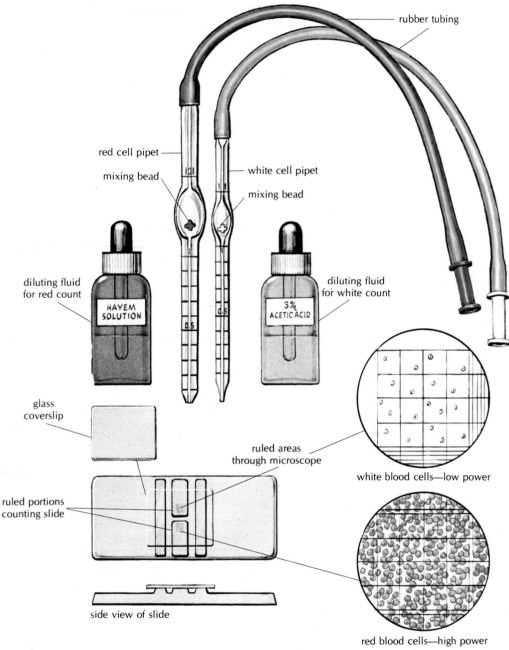

rubber tubing

red cell pipet

mixing bead

white cell pipet

mixing bead

diluting fluid
for red count

HAYEM
SOLUTION

3%
ACETIC ACID

diluting fluid
for white count

glass
coverslip

ruled areas
through microscope

white blood cells—low power

ruled portions
counting slide

side view of slide

red blood cells—high power

**Figure 12-5**

Parts of a hemocytometer.

a delay may be due to a number of different causes, including lack of certain hormone-like substances, calcium salts, or vitamin K.

Each of the various clotting factors has been designated by a Roman numeral ranging from I to XIII. Factor I is fibrinogen, factor II is prothrombin, factor III is thromboplastin, and factor IV is assigned to

calcium ions. The amounts of all these factors may be determined and evaluated on a percentage basis, aiding in the diagnosis and treatment of some bleeding disorders.

Additional tests for coagulation include tests for bleeding time, clotting time, capillary strength, and platelet function.

# SUMMARY

I. **Functions of blood**
  A. Transportation—of oxygen, carbon dioxide, nutrients, minerals, vitamins, waste, hormones
  B. Regulation—of pH, fluid balance, body temperature
  C. Protection—against foreign organisms, blood loss

II. **Blood constituents**
  A. Plasma—liquid component
    1. Water—main ingredient
    2. Proteins—albumin, clotting factors, antibodies, complement
    3. Nutrients—carbohydrates, lipids, amino acids
    4. Minerals
    5. Waste products
    6. Hormones and other materials
  B. Formed elements
    1. Erythrocytes—red blood cells
    2. Leukocytes—white blood cells
      a. Granulocytes — neutrophils (polymorphs, segs), eosinophils, basophils
      b. Agranulocytes—lymphocytes, monocytes
    3. Platelets (thrombocytes)
  C. Origin of formed elements—produced in red bone marrow from stem cells

III. **Hemostasis**—prevention of blood loss
  A. Contraction of blood vessels
  B. Formation of platelet plug
  C. Formation of blood clot

  1. Factors
    a. Procoagulants—promote clotting
    b. Anticoagulants—prevent clotting
  2. Main steps in blood clotting
    a. Thromboplastin converts prothrombin to thrombin
    b. Thrombin converts fibrinogen to solid threads of fibrin
    c. Threads form clot
  3. Serum—fluid that remains after blood has clotted

IV. **Blood typing**
  A. AB antigens—result in A, B, AB, and O blood types
  B. Rh antigens—positive or negative
  C. Tested by mixing with antisera to different antigens
  D. Incompatible transfusions will cause destruction of donor red cells

V. **Blood studies**
  A. Hematocrit—measures percentage of packed red cells in whole blood
  B. Hemoglobin tests—color test, electrophoresis
  C. Blood cell counts
  D. Blood slide (smear)
  E. Blood chemistry tests—electrolytes, waste products, enzymes, glucose, hormones
  F. Coagulation studies—clotting factor assays, bleeding time, clotting time, capillary strength, platelet function

## QUESTIONS FOR STUDY AND REVIEW

1. How does the color of blood vary with the amount of oxygenation?
2. Name the three main purposes of blood.
3. Name the two main fractions of blood.
4. Name four main ingredients of blood plasma. What are their purposes?
5. Name and describe the three main types of formed elements in blood.
6. Name and give the functions of the five types of leukocytes.
7. What is the main function of erythrocytes? leukocytes? platelets? Where do they originate?
8. Name the four blood types in the ABO system. What determines the different types?
9. What is the Rh factor? What proportion of the population of the United States possesses this factor? In what situations is this factor of medical importance? Why?
10. Define *hemostasis*. Describe the three main steps in hemostasis.
11. Describe the three basic steps involved in the clotting process.
12. What does the hematocrit measure? Cite normal hematocrit ranges for males and females.
13. What can be learned by studying the blood smear? What is determined by a differential white count?
14. What are some evaluations made by blood chemistry tests?

# CHAPTER 13

# The Heart

## Behavioral Objectives

After careful study of this chapter, you should be able to:

- Describe the three layers of the heart
- Compare the functions of the right heart and left heart
- Name the four chambers of the heart
- Name the valves at the entrance and exit of each ventricle
- Briefly describe blood circulation through the myocardium
- Explain the effects of the autonomic nervous system on the heart rate
- Briefly describe the cardiac cycle
- Name the components of the heart's conduction system
- List and define several terms that describe different heart rates
- Explain what produces the two main heart sounds
- Briefly describe five instruments used in the study of the heart

Memmler, RL, Cohen, BJ, Wood, DL. *STRUCTURE AND FUNCTION OF THE HUMAN BODY*, 6/e,
© 1996 Lippincott-Raven Publishers

# Circulation and the Heart

The next two chapters investigate the manner in which the blood delivers oxygen and nutrients to the cells and carries away the waste products of cell metabolism. This continuous one-way movement of the blood is known as its *circulation.* The prime mover that propels blood throughout the body is the *heart.* We shall have a look at the heart before going into the subject of the blood vessels in detail.

The heart is a muscular pump that drives the blood through the blood vessels. Slightly bigger than a fist, this organ is located between the lungs in the center and a bit to the left of the midline of the body. The strokes (contractions) of this pump average about 72 per minute and are carried on unceasingly for the whole of a lifetime.

The importance of the heart has been recognized for centuries. The fact that its rate of beating is affected by the emotions may be responsible for the very frequent references to the heart in song and poetry. However, the vital functions of the heart and its disorders are of more practical importance to us.

# Structure of the Heart

## Layers of the Heart Wall

The heart is a hollow organ, the walls of which are formed of three different layers. Just as a warm coat might have a smooth lining, a thick and bulky interlining, and an outer layer of a third fabric, so the heart wall has three tissue layers (Fig. 13-1).

1. The *endocardium* (en-do-KAR-de-um) is a very thin smooth layer of cells that resembles squamous epithelium. This membrane lines the interior of the heart. The valves of the heart are formed by reinforced folds of this material.
2. The *myocardium* (mi-o-KAR-de-um), the muscle of the heart, is the thickest layer and is responsible for pumping blood through the vessels. The unique structure of cardiac muscle is described in more detail below.
3. The *epicardium* (ep-ih-KAR-de-um) forms the thin, outermost layer of the heart wall and is continuous with the serous lining of the fibrous sac that encloses the heart. These membranes together make up the *pericardium* (per-ih-KAR-de-um). The serous lining of the pericardial sac is

separated from the epicardium on the heart surface by a thin film of fluid.

## SPECIAL FEATURES OF THE MYOCARDIUM

Cardiac muscle cells are lightly striated (striped) and have specialized partitions between the cells that appear faintly under the light microscope (Fig. 13-2). These *intercalated* (in-TER-cah-la-ted) *disks* are actually modified cell membranes that allow for rapid transfer of electric impulses between the cells. The adjective *intercalated* means "inserted between."

Another feature of cardiac muscle tissue is the branching of the muscle fibers (cells). These fibers are interwoven so that the stimulation that causes the contraction of one fiber results in the contraction of a whole group. These structural features play an important role in the working of the heart muscle.

## Two Hearts and a Partition

Health professionals often refer to the *right heart* and the *left heart.* This is because the human heart is really a double pump (Fig. 13-3). The right side pumps blood low in oxygen to the lungs through the *pulmonary circuit.* The left side pumps oxygenated blood to the remainder of the body through the *systemic circuit.* The two sides are completely separated from each other by a partition called the *septum.* The upper part of this partition is called the *interatrial* (in-ter-A-tre-al) *septum,* while the larger, lower portion is called the *interventricular* (in-ter-ven-TRIK-u-lar) *septum.* The septum, like the heart wall, consists largely of myocardium.

## Four Chambers

On either side of the heart are two chambers, one a receiving chamber (atrium) and the other a pumping chamber (ventricle):

1. The *right atrium* is a thin-walled chamber that receives the blood returning from the body tissues. This blood, which is low in oxygen, is carried in *veins,* the blood vessels leading *to* the heart from the body tissues.
2. The *right ventricle* pumps the venous blood received from the right atrium into the lungs.
3. The *left atrium* receives blood high in oxygen content as it returns from the lungs.
4. The *left ventricle,* which is the chamber with the thickest walls, pumps oxygenated blood to all

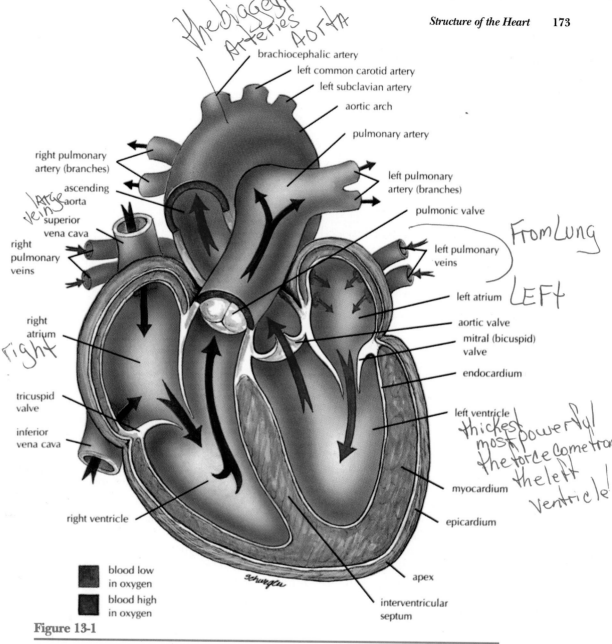

The biggest Arteries AORTA

Large Veins

From Lung LEFT

Right

thickest most powerful the force come from the left ventricle!

**Figure 13-1**

Heart and great vessels.

parts of the body. This blood goes through the *arteries,* the vessels that take blood *from* the heart to the tissues.

### Four Valves

One-way valves that direct the flow of blood through the heart are located at the entrance and the exit of each ventricle. The entrance valves are the *atrioventricular* (a-tre-o-ven-TRIK-u-lar) *valves;* the exit valves are the *semilunar* (sem-e-LU-

nar) *valves.* (So named because each flap of these valves resembles a half-moon.) Each valve has a specific name, as follows:

1. The *right atrioventricular (AV) valve* is also known as the *tricuspid* (tri-KUS-pid) *valve,* because it has three cusps, or flaps, that open and close. When this valve is open, blood flows freely from the right atrium into the right ventricle. However, when the right ventricle begins to contract, the valve closes so that blood cannot return to

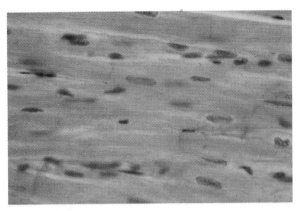

**Figure 13-2**

Cardiac muscle tissue. Note the central nuclei, branching of fibers, and intercalated disks. (Courtesy of Nancy C. Maguire, Thomas Jefferson University)

the right atrium; this ensures forward flow into the pulmonary artery.

2. The *left atrioventricular (AV) valve* is the bicuspid valve, but it is usually referred to as the *mitral* (MI-tral) *valve.* It has two rather heavy cusps that permit blood to flow freely from the left atrium into the left ventricle. However, the cusps close when the left ventricle begins to contract; this prevents blood from returning to the left atrium and ensures the forward flow of blood into the *aorta* (a-OR-tah). Both the tricuspid and mitral valves are attached by means of thin fibrous threads to muscles in the walls of the ventricles (see Fig. 13-7). The function of these threads, called the *chordae tendineae* (KOR-de ten-DIN-e-e), is to stabilize the valve flaps when the ventricles contract so that the force of the blood will not push them up into the atria. In this manner they help to prevent a backflow of blood when the heart beats.

3. The *pulmonic* (pul-MON-ik) *valve*, also called the *pulmonary valve,* is a semilunar valve located between the right ventricle and the pulmonary artery that leads to the lungs. As soon as the right ventricle has finished emptying itself, the valve closes to prevent blood on its way to the lungs from returning to the ventricle.

4. The *aortic* (a-OR-tik) *valve* is a semilunar valve located between the left ventricle and the aorta. Following contraction of the left ventricle, the aortic valve closes to prevent the flow of blood back from the aorta to the ventricle.

The appearance of the heart valves in the closed position is illustrated in Figure 13-4.

### Blood Supply to the Myocardium

Only the endocardium comes into contact with the blood that flows through the heart chambers. Therefore, the myocardium must have its own blood vessels to provide oxygen and nourishment and to remove waste products. The arteries that supply blood to the muscle of the heart are called the *right* and *left coronary arteries* (Fig. 13-5). These arteries, which are the first branches of the aorta, arise just above the aortic semilunar valve (see Fig. 13-4). They receive blood when the heart relaxes and branch to all regions of the heart muscle. After passing through capillaries in the myocardium, blood drains into the cardiac veins and finally into the *coronary sinus* for return to the right atrium.

## ◗ Function of the Heart

### The Work of the Heart

Although the right and left sides of the heart are separated from each other, they work together. The blood is squeezed through the chambers by a contraction of heart muscle beginning in the thin-walled upper chambers, the atria, followed by a contraction of the thick muscle of the lower chambers, the ventricles. This active phase is called *systole* (SIS-to-le), and in each case it is followed by a resting period known as *diastole* (di-AS-to-le). The contraction of the walls of the atria is completed at the time the contraction of the ventricles begins. Thus, the resting phase (diastole) begins in the atria at the same time as the contraction (systole) begins in the ventricles. After the ventricles have emptied, both chambers are relaxed for a short period as they fill with blood. Then another beat begins with contraction of the atria followed by contraction of the ventricles. This sequence of heart relaxation and contraction is called the *cardiac cycle.* Each cycle takes an average of 0.8 seconds (Fig. 13-6).

A unique property of heart muscle is its ability to adjust the strength of contraction to the amount of blood received. When the heart chamber is filled and the wall stretched (within limits), the contraction is strong. As less blood enters the heart, the contraction becomes less forceful. Thus, as more blood enters the heart, as occurs during exercise, the mus-

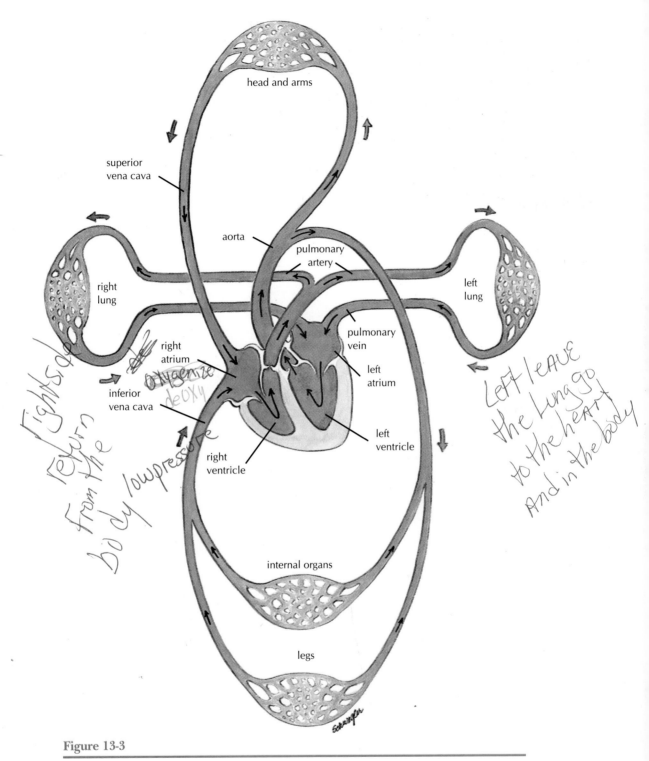

**Figure 13-3**

The heart is a double pump. The pulmonary circuit carries blood to the lungs to be oxygenated; the systemic circuit carries blood to all other parts of the body.

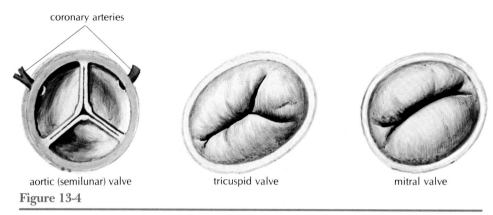

coronary arteries

aortic (semilunar) valve   tricuspid valve   mitral valve

**Figure 13-4**

Valves of the heart, seen from above, in the closed position.

cle contracts with greater strength to push the larger volume of blood out into the blood vessels.

The volume of blood pumped by each ventricle in 1 minute is termed the ***cardiac output.*** It is the product of the ***stroke volume***—the volume of blood ejected from the ventricle with each beat, and the ***heart rate***—the number of times the heart beats per minute.

## The Conduction System of the Heart

The cardiac cycle is regulated by specialized areas in the heart wall that form the conduction system of the heart (Fig. 13-7). Two of these areas are tissue masses called ***nodes;*** the third is a group of fibers called the ***atrioventricular bundle.*** The ***sinoatrial (SA) node,*** which is located in the upper wall of the right atrium, initiates the heartbeat and for this

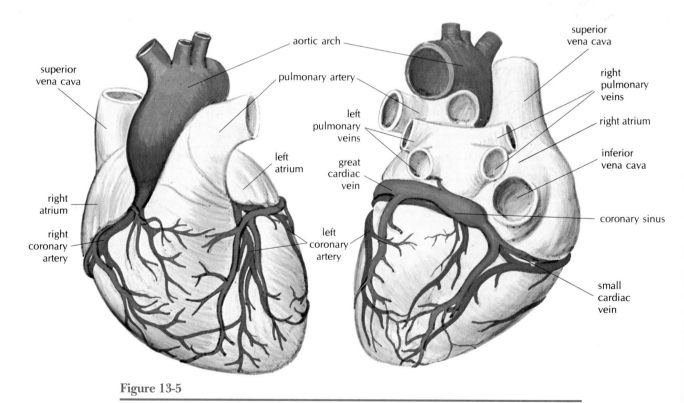

**Figure 13-5**

Coronary arteries and cardiac veins. (*Left*) Anterior view. (*Right*) Posterior view.

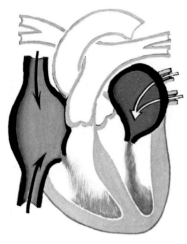

**Diastole**
Atria fill with blood which begins to flow into ventricles as soon as their walls relax.

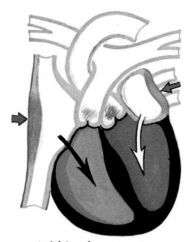

**Atrial Systole**
Contraction of atria pumps blood into the ventricles.

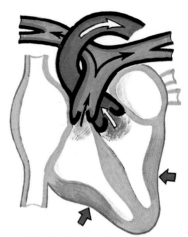

**Ventricular Systole**
Contraction of ventricles pumps blood into aorta and pulmonary arteries.

**Figure 13-6**

Pumping cycle of the heart.

reason is called the *pacemaker.* The second node, located in the interatrial septum at the bottom of the right atrium, is called the ***atrioventricular (AV) node.*** The ***atrioventricular bundle,*** also known as the ***bundle of His,*** is located at the top of the interventricular septum; it has branches that extend to all parts of the ventricular walls. Fibers travel first down both sides of the interventricular septum in groups called the ***right*** and ***left bundle branches.*** Smaller ***Purkinje*** (pur-KIN-je) ***fibers*** then travel in a branching network throughout the myocardium of the ventricles. The special membranes between the cells (intercalated disks) allow the rapid flow of impulses throughout the heart muscle. The order in which these impulses travel is as follows:

1. The sinoatrial node generates the electric impulse that begins the heartbeat.
2. The excitation wave travels throughout the muscle of each atrium, causing it to contract.
3. The atrioventricular node is stimulated. The relatively slower conduction through this node allows time for the atria to contract and complete the filling of the ventricles.
4. The excitation wave travels rapidly through the bundle of His and then throughout the ventricular walls by means of the bundle branches and Purkinje fibers. The entire musculature of the ventricles contracts practically at once.

As a safety measure, a region of the conduction system other than the sinoatrial node can generate a heartbeat if the sinoatrial node fails, but it does so at a slower rate. A normal heart rhythm originating at the SA node is termed a ***sinus rhythm.***

## Cardiac Reserve

Like many other organs, the heart has great reserves of strength. The ***cardiac reserve*** is a measure of how many times more than average the heart can produce when needed. Based on a heart rate of 75 beats/minute and a stroke volume of 70 mL/beat, the average cardiac output for an adult at rest is about 5L/minute. This means that *at rest* the heart pumps the equivalent of the total blood volume each minute.

During mild exercise this volume might double and even double again during strenuous exercise. For most people the cardiac reserve is 4 to 5 times the resting output. In athletes exercising vigorously the ratio may reach 6 to 7 times. In contrast, those with heart disease may have little or no cardiac reserve. They may be fine at rest but quickly become short of breath or fatigued when exercising.

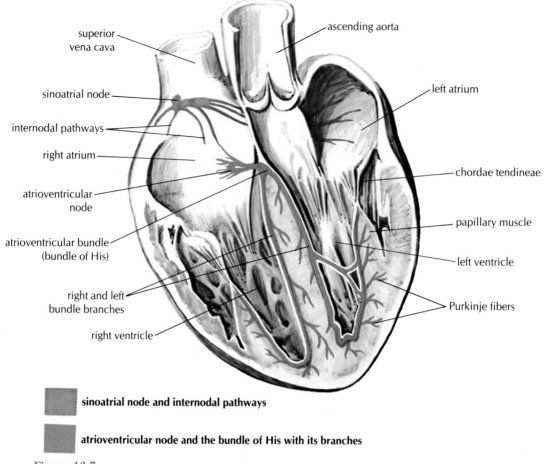

superior vena cava

ascending aorta

sinoatrial node

left atrium

internodal pathways

right atrium

chordae tendineae

atrioventricular node

papillary muscle

atrioventricular bundle (bundle of His)

left ventricle

right and left bundle branches

Purkinje fibers

right ventricle

**sinoatrial node and internodal pathways**

**atrioventricular node and the bundle of His with its branches**

**Figure 13-7**

Conduction system of the heart.

## Control of the Heart Rate

Although the fundamental beat of the heart originates within the heart itself, the heart rate can be influenced by the nervous system and by other factors in the internal environment. The autonomic nervous system (ANS) plays a major role in modifying the heart rate according to need (Fig. 13-8). Stimulation from the sympathetic nervous system increases the heart rate. During a fight-or-flight response, the sympathetic nerves can boost the cardiac output two to three times the resting value by increasing the rate and force of heart contractions. Stimulation from the parasympathetic nervous system decreases the heart rate to restore homeostasis. The parasympathetic nerve that supplies the heart is the vagus nerve (cranial nerve X). It slows the heart rate by acting on the SA and AV nodes. These influences allow the heart to meet changing needs rapidly. The heart rate is also affected by such factors as hormones, ions, and drugs in the blood.

## Heart Rates

1. *Bradycardia* (brad-e-KAR-de-ah) is a relatively slow heart rate of less than 60 beats/minute. During rest and sleep, the heart may beat less than 60 beats/minute but usually does not fall below 50 beats/minute.
2. *Tachycardia* (tak-e-KAR-de-ah) refers to a heart rate over 100 beats/minute.
3. *Sinus arrhythmia* (ah-RITH-me-ah) is a regular variation in heart rate due to changes in the rate and depth of breathing. It is a normal phenomenon.
4. *Premature beat,* also called *extrasystole,* is a beat that comes before the expected normal beat. These may occur in normal persons initiated by caf-

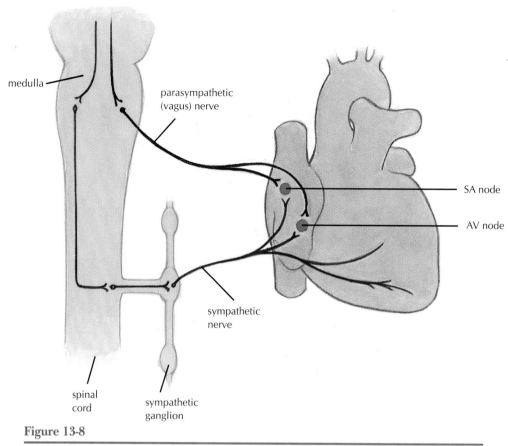

medulla

parasympathetic (vagus) nerve

SA node

AV node

sympathetic nerve

spinal cord

sympathetic ganglion

**Figure 13-8**

Nervous stimulation of the heart.

feine, nicotine, or psychologic stresses. They are also common in persons with heart disease.

### Heart Sounds and Murmurs

The normal heart sounds are usually described by the syllables "lubb" and "dupp." The first is a longer, lower-pitched sound that occurs at the start of ventricular systole. It is probably caused by a combination of things, mainly closure of the atrioventricular valves. The second, or "dupp," sound is shorter and sharper. It occurs at the beginning of ventricular relaxation and is due in large part to sudden closure of the semilunar valves. An abnormal sound is called a *murmur* and is usually due to faulty action of a valve. For example, if a valve fails to close tightly and blood leaks back, a murmur is heard. Another condition giving rise to an abnormal sound is the narrowing (stenosis) of a valve opening. The many conditions that can cause abnormal heart sounds include congenital defects, disease, and physiologic variations. A murmur due to rapid filling of the ventricles is called a *functional* or *flow murmur;*

such a murmur is not abnormal. To differentiate, an abnormal sound caused by any structural change in the heart or the vessels connected with the heart is called an *organic murmur.*

### The Heart in the Elderly

As a result of accumulated damage, the heart may become less efficient with age. On the average, by the age of 70 there has been a decrease in cardiac output of about 35%. Because of a decrease in the reserve strength of the heart, elderly persons are often limited in their ability to respond to emergencies, infections, blood loss, or stress.

## ▶ Instruments Used in Heart Studies

The *stethoscope* (STETH-o-skope) is a relatively simple instrument used for conveying sounds from within the patient's body to the ear of the examiner.

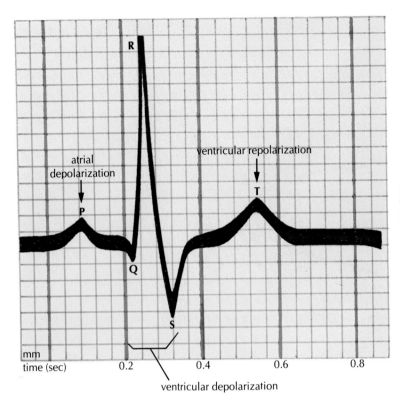

atrial
depolarization

ventricular repolarization

R

P

Q

S

T

mm

time (sec)    0.2      0.4      0.6      0.8

ventricular depolarization

**Figure 13-9**

Normal EKG showing one cardiac cycle.

Experienced listeners can gain much information using this device.

The *electrocardiograph* (*EKG* or *ECG*) is used for making records of the changes in electric currents produced by the contracting heart muscle. It may thus reveal certain myocardial injuries. Electric activity is picked up by electrodes placed on the surface of the skin and appears as *waves* on the EKG tracing. The P wave represents the activity of the atria; the QRS and T waves represent the activity of the ventricles (Fig. 13-9). Changes in the waves and the intervals between them are used to diagnose heart damage and arrhythmias.

The *fluoroscope* (flu-OR-o-scope), an instrument for examining deep structures with x-rays, may be used to reveal heart action and to establish the sizes and relations of some of the thoracic organs. It may be used in conjunction with *catheterization* (kath-eh-ter-i-ZA-shun) of the heart. In right heart catheterization, an extremely thin tube (catheter) is passed through the veins of the right arm or right groin and then into the right side of heart. During the procedure, the fluoroscope is used for observation of the route taken by the catheter. The tube is passed all the way through the pulmonic valve into the large lung arteries. Samples of blood are obtained along the way and removed for testing, pressure readings being taken meanwhile.

In left heart catheterization, a catheter is passed through an artery in the left groin or arm to the heart. Dye can then be injected into the coronary arteries. The tube may also be passed through the aortic valve into the left ventricle for further studies.

*Ultrasound* consists of sound waves generated at a frequency above the range of sensitivity of the human ear. In *echocardiography* (ek-o-kar-de-OG-rah-fe), also known as *ultrasound cardiography*, high-frequency sound waves are sent to the heart from a small instrument on the surface of the chest. The ultrasound waves bounce off the heart and are recorded as they return, showing the heart in action. Movement of the echoes is traced on an electronic instrument called an *oscilloscope* and recorded on film. (The same principle is employed by submarines to detect ships.) The method is safe and painless, and it does not use x-rays. It provides information on the sizes and shapes of heart structures, on cardiac function, and on possible heart defects.

# SUMMARY

## I. Structure of the heart

A. Layers
1. Endocardium—thin inner layer
2. Myocardium—thick muscle layer
   a. Lightly striated, intercalated disks, branching of fibers
3. Epicardium—thin outer layer

B. Pericardium—membrane-lined sac that encloses the heart

C. Chambers
1. Atria—left and right receiving chambers
2. Ventricles—left and right pumping chambers
3. Septa—partitions between chambers

D. Valves—prevent backflow of blood
1. Tricuspid—right atrioventricular valve
2. Mitral (bicuspid)—left atrioventricular valve
3. Pulmonic (semilunar) valve—at entrance to pulmonary artery
4. Aortic (semilunar) valve—at entrance to aorta

E. Blood supply to myocardium
1. Coronary arteries—first branches of aorta; fill when heart relaxes
2. Coronary sinus—collects venous blood from heart and empties into right atrium

## II. Function of the heart

A. Cardiac cycle
1. Diastole—relaxation phase
2. Systole—contraction phase

B. Cardiac output—volume pumped by each ventricle per minute
1. Stroke volume—amount pumped with each beat
2. Heart rate—number of beats per minute

C. Conduction system
1. Sinoatrial node (pacemaker)—at top of right atrium
2. Atrioventricular node—between atria and ventricles
3. Atrioventricular bundle (bundle of His)—at top of interventricular septum
   a. Bundle branches—right and left, on either side of septum
   b. Purkinje fibers—branch through myocardium of ventricles

D. Control of heart rate
1. Autonomic nervous system
   a. Sympathetic system—speeds heart rate
   b. Parsympathetic system—slows heart rate via vagus nerve
2. Others—hormones, ions, drugs

E. Heart rates
1. Bradycardia—slower rate than normal; less than 60 beats/minute
2. Tachycardia—faster rate than normal; more than 100 beats/minute
3. Extrasystole—premature beat

F. Heart sounds
1. Normal
   a. "Lubb"—occurs at closing of atrioventricular valves
   b. "Dupp"—occurs at closing of semilunar valves
2. Abnormal—murmur

## III. Instruments used in heart studies

A. Stethoscope—used to listen to heart sounds

B. Electrocardiograph (EKG, ECG)—records electric activity as waves

C. Fluoroscope—examines deep tissue with x-rays

D. Catheter—thin tube inserted into heart for blood samples, pressure readings, etc.

E. Echocardiograph—uses ultrasound to record picture of heart in action

## QUESTIONS FOR STUDY AND REVIEW

1. What are the three layers of the heart wall?
2. Describe the characteristics of heart muscle tissue.
3. Name the sac around the heart.
4. What is a partition in the heart called? Name two.
5. Name the chambers of the heart and tell what each does.
6. Name the valves of the heart and explain the purpose of each valve.
7. Why does the myocardium need its own blood supply? Name the arteries that supply blood to the heart.
8. Explain the contraction pattern of the cardiac cycle.
9. How does the heart's ability to contract differ from that of other muscles? What is required to maintain an effective rate of heartbeat?
10. Define *cardiac output*. What determines cardiac output?
11. What are the parts of the heart's conduction system called and where are these structures located? Outline the order in which the excitation waves travel.
12. Compare the effects of the sympathetic and parasympathetic nervous systems on the working of the heart.
13. What two syllables are used to indicate normal heart sounds, and at what time in the heart cycle can they be heard?
14. What is a fluoroscope and what is its purpose in heart studies?
15. Of what value is heart catheterization and how is it carried out?
16. What is an electrocardiograph and what is its purpose?
17. How does echocardiography differ from electrocardiography?
18. Differentiate between the terms in each of the following pairs:
    a. *pulmonary* and *systemic circuit*
    b. *interatrial* and *interventricular*
    c. *systole* and *diastole*
    d. *stroke volume* and *heart rate*
    e. *tachycardia* and *bradycardia*
    f. *functional murmur* and *organic murmur*

# CHAPTER 14

# Blood Vessels and Blood Circulation

## Behavioral Objectives

After careful study of this chapter, you should be able to:

- Differentiate among the three main types of vessels in the body with regard to structure and function
- Compare the locations and functions of the pulmonary and systemic circuits
- Name the four sections of the aorta
- Name the main branches of the aorta
- Name the main vessels that drain into the superior and inferior venae cavae
- Define *venous sinus* and give four examples
- Describe the structure and function of the hepatic portal system
- Explain how materials are exchanged across the capillary wall
- Describe the factors that regulate blood flow
- Define *pulse* and list factors that affect the pulse rate
- List several factors that affect blood pressure
- Explain how blood pressure is commonly measured

Memmler, RL, Cohen, BJ, Wood, DL. *STRUCTURE AND FUNCTION OF THE HUMAN BODY, 6/e,*
© 1996 Lippincott-Raven Publishers

The vascular system will be easier to understand if you refer to the appropriate illustrations in this chapter as the vessels are described. If this information is added to what you already know about the blood and the heart, a picture of the circulatory system as a whole will emerge.

## ▶ Blood Vessels

### Functional Classification

The blood vessels, together with the four chambers of the heart, form a closed system for the flow of blood; only if there is an injury to some part of the wall of this system does any blood escape. On the basis of function, blood vessels may be divided into three groups:

1. *Arteries* carry blood from the ventricles (pumping chambers) of the heart out to the capillaries in the tissues. The smallest arteries are called *arterioles* (ar-TE-re-olz).
2. *Veins* drain capillaries in the tissues and return the blood to the heart. The smallest veins are the *venules* (VEN-ulz).
3. *Capillaries* allow for exchanges between the blood and body cells, or between the blood and air in the lung tissues. The capillaries connect the arterioles and venules.

All the vessels together may be subdivided into two groups or circuits: pulmonary and systemic. The vessels in these two circuits are shown in Figure 14-1; the anatomic relation of the circuits to the heart is shown in Figure 13-3.

1. The *pulmonary circuit* eliminates carbon dioxide from the blood and replenishes its supply of oxygen. The pulmonary vessels that carry blood to and from the lungs include:
   a. the pulmonary artery and its branches that carry blood from the right ventricle to the lungs
   b. the capillaries in the lungs through which gases are exchanged
   c. the pulmonary veins that carry blood back to the left atrium

The pulmonary vessels differ from those in the systemic circuit in that the pulmonary arteries carry blood that is *low* in oxygen and the pulmonary veins carry blood that is *high* in oxygen.

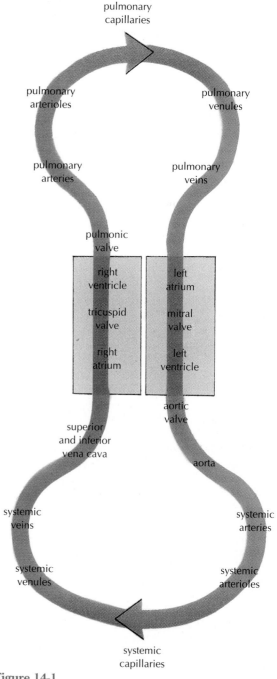

**Figure 14-1**

Blood vessels form a closed system for the flow of blood. Blood high in oxygen (oxygenated) is shown in red; blood low in oxygen (deoxygenated) is shown in blue. Changes in oxygen content occur as blood flows through capillaries.

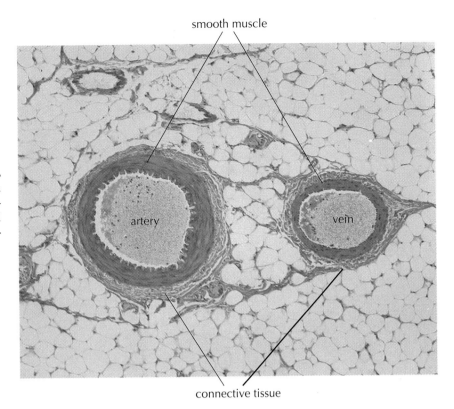

smooth muscle

artery

vein

connective tissue

**Figure 14-3**

Cross section of an artery and a vein as seen through a microscope. (Cormack DH: Essential Histology, plate 11–1. Philadelphia, JB Lippincott, 1993)

4. The *abdominal aorta* is the longest section of the aorta, spanning the abdominal cavity.

The thoracic and abdominal aorta together make up the descending aorta.

### Branches of the Ascending Aorta

The first, or ascending, part of the aorta has two branches near the heart, called the *left* and *right coronary arteries,* that supply the heart muscle. These form a crown around the base of the heart and give off branches to all parts of the myocardium.

### Branches of the Aortic Arch

The arch of the aorta, located immediately beyond the ascending aorta, gives off three large branches.

1. The *brachiocephalic* (brak-e-o-seh-FAL-ik) *trunk* is a short artery formerly called the *innominate.* Its name means that it supplies the head and the arm. After extending upward somewhat less than 5 cm (2 inches), it divides into the *right subclavian* (sub-KLA-ve-an) *artery,* which supplies the right upper extremity (arm), and the *right common carotid* (kah-ROT-id) *artery,* which supplies the right side of the head and the neck.

2. The *left common carotid artery* extends upward from the highest part of the aortic arch. It supplies the left side of the neck and the head.

3. The *left subclavian artery* extends under the left collar bone (clavicle) and supplies the left upper extremity. This is the last branch of the aortic arch.

### Branches of the Thoracic Aorta

The thoracic aorta supplies branches to the chest wall, to the *esophagus* (e-SOF-ah-gus), and to the bronchi (the subdivisions of the trachea) and their treelike subdivisions in the lungs. There are usually 9 to 10 pairs of *intercostal* (in-ter-KOS-tal) *arteries* that extend between the ribs, sending branches to the muscles and other structures of the chest wall.

### Branches of the Abdominal Aorta

As in the case of the thoracic aorta, there are unpaired branches extending forward and paired

*(Text continues on page 190)*

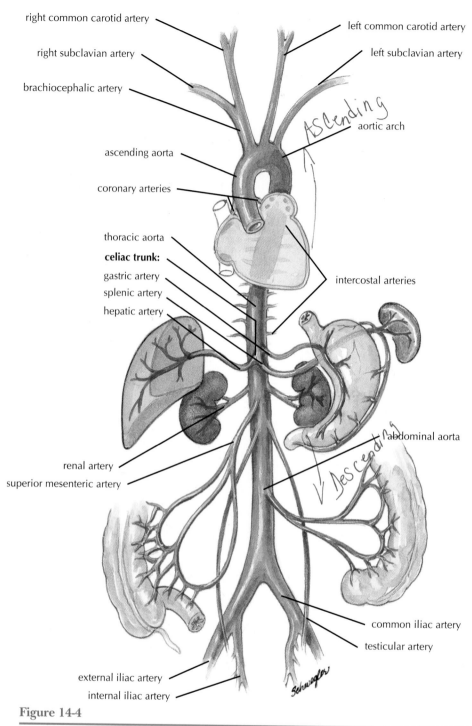

right common carotid artery

right subclavian artery

brachiocephalic artery

ascending aorta

coronary arteries

thoracic aorta

**celiac trunk:**

gastric artery

splenic artery

hepatic artery

renal artery

superior mesenteric artery

external iliac artery

internal iliac artery

left common carotid artery

left subclavian artery

*ASCending*

aortic arch

intercostal arteries

*DeSCending*

abdominal aorta

common iliac artery

testicular artery

**Figure 14-4**

Aorta and its branches.

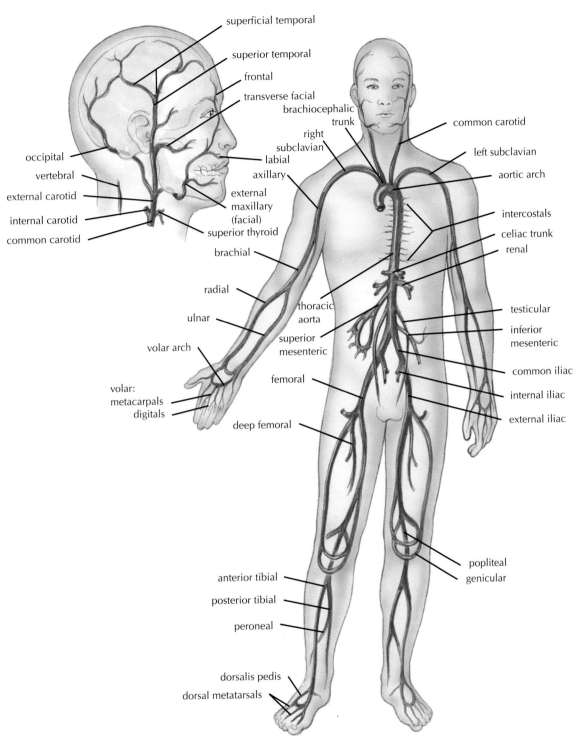

**Figure 14-5**

Principal systemic arteries.

arteries extending toward the side. The unpaired vessels are large arteries that supply the abdominal viscera. The most important of these visceral branches are listed below:

1. The *celiac* (SE-le-ak) *trunk* is a short artery about 1.25 cm (1/2 inch) long that subdivides into three branches: the *left gastric artery* goes to the stomach, the *splenic* (SPLEN-ik) *artery* goes to the spleen, and the *hepatic* (heh-PAT-ik) *artery* carries oxygenated blood to the liver.
2. The *superior mesenteric* (mes-en-TER-ik) *artery,* the largest of these branches, carries blood to most of the small intestine and to the first half of the large intestine.
3. The much smaller *inferior mesenteric artery,* located below the superior mesenteric and near the end of the abdominal aorta, supplies the second half of the large intestine.

The paired lateral branches of the abdominal aorta include the following right and left vessels:

1. The *phrenic* (FREN-ik) *arteries* supply the diaphragm.
2. The *suprarenal* (su-prah-RE-nal) *arteries* supply the adrenal (suprarenal) glands.
3. The *renal* (RE-nal) *arteries,* the largest in this group, carry blood to the kidneys.
4. The *ovarian arteries* in the female and *testicular* (tes-TIK-u-lar) *arteries* in the male (formerly called the *spermatic arteries),* supply the sex glands.
5. Four pairs of *lumbar* (LUM-bar) *arteries* extend into the musculature of the abdominal wall.

### Iliac Arteries and Their Subdivisions

The abdominal aorta finally divides into two *common iliac* (IL-e-ak) *arteries.* Both of these vessels, about 5 cm (2 inches) long, extend into the pelvis, where each one subdivides into an *internal* and an *external iliac artery.* The internal iliac vessels then send branches to the pelvic organs, including the urinary bladder, the rectum, and some of the reproductive organs. Each external iliac artery continues into the thigh as the *femoral* (FEM-or-al) *artery.* This vessel gives off branches in the thigh and then becomes the *popliteal* (pop-LIT-e-al) *artery,* which subdivides below the knee. The subdivisions include the *tibial artery* and the *dorsalis pedis* (dor-SA-lis PE-dis), which supply the leg and the foot.

### Other Subdivisions of Systemic Arteries

Just as the larger branches of a tree give off limbs of varying sizes, so the arterial tree has a multitude of subdivisions. Hundreds of names might be included, but we shall mention only a few. For example, each common carotid artery gives off branches to the thyroid gland and other structures in the neck before dividing into the *external* and *internal carotid arteries,* which supply parts of the head. The hand receives blood from the subclavian artery, which becomes the *axillary* (AK-sil-ar-e) in the axilla (armpit). The longest part of this vessel, the *brachial* (BRA-ke-al) *artery,* is in the arm proper. It subdivides into two branches near the elbow: the *radial artery,* which continues down the thumb side of the forearm and wrist, and the *ulnar artery,* which extends along the medial or little finger side into the hand.

### Anastomoses

A communication between two vessels is called an *anastomosis* (ah-nas-to-MO-sis). By means of arterial anastomoses, blood reaches vital organs by more than one route. Some examples of such unions of end arteries are described below:

1. The *circle of Willis* (Fig. 14-6) receives blood from the two internal carotid arteries and from the *basilar* (BAS-il-ar) *artery,* which is formed by the union of two vertebral arteries. This arterial circle lies just under the center of the brain and sends branches to the cerebrum and other parts of the brain.
2. The *volar* (VO-lar) *arch* is formed by the union of the radial and ulnar arteries in the hand. It sends branches to the hand and the fingers.
3. The *mesenteric arches* are made of communications between branches of the vessels that supply blood to the intestinal tract.
4. *Arterial arches* are formed by the union of branches of the tibial arteries in the foot. Similar anastomoses are found in other parts of the body.

Arteriovenous anastomoses are blood shunts found in a few areas, including the external ears, the hands, and the feet. Vessels with muscular walls connect arteries directly with veins and thus bypass the capillaries (Fig. 14-7). This provides a more rapid flow and a greater volume of blood to these areas than elsewhere, thus protecting these exposed parts from freezing in cold weather.

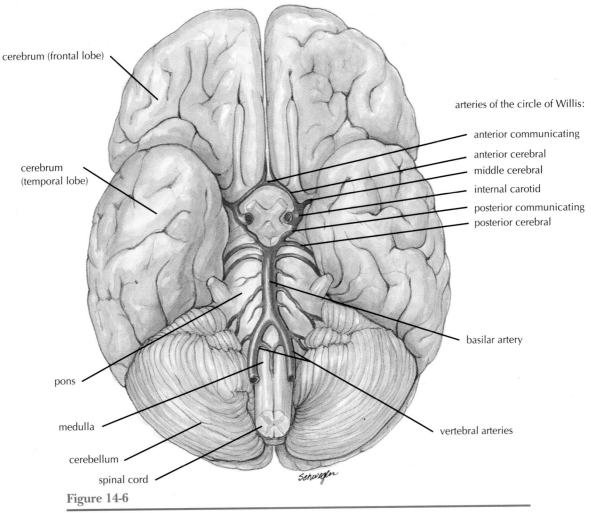

cerebrum (frontal lobe)

arteries of the circle of Willis:

anterior communicating
anterior cerebral
middle cerebral
internal carotid
posterior communicating
posterior cerebral

cerebrum
(temporal lobe)

basilar artery

pons

medulla

vertebral arteries

cerebellum

spinal cord

Schwegn

**Figure 14-6**

Arteries that supply the brain, showing the arteries that make up the circle of Willis.

## ▶ Names of Systemic Veins

### Superficial Veins

Whereas most arteries are located in protected and rather deep areas of the body, many veins are found near the surface (Fig. 14-8). The most important of these superficial veins are in the extremities. These include the following:

1. The veins on the back of the hand and at the front of the elbow. Those at the elbow are often used for removing blood samples for test purposes, as well as for intravenous injections. The largest of this group of veins are the *cephalic* (seh-FAL-ik), the *basilic* (bah-SIL-ik), and the *median cubital* (KU-bih-tal) *veins.*

2. The *saphenous* (sah-FE-nus) *veins* of the lower extremities, which are the longest veins of the body. The great saphenous vein begins in the foot and extends up the medial side of the leg, the knee, and the thigh. It finally empties into the femoral vein near the groin.

### Deep Veins

The deep veins tend to parallel arteries and usually have the same names as the corresponding arteries. Examples of these include the *femoral* and the *iliac* vessels of the lower part of the body and the *brachial, axillary,* and *subclavian* vessels of the upper extremities. However, exceptions are found in the veins of the head and the neck. The *jugular* (JUG-u-

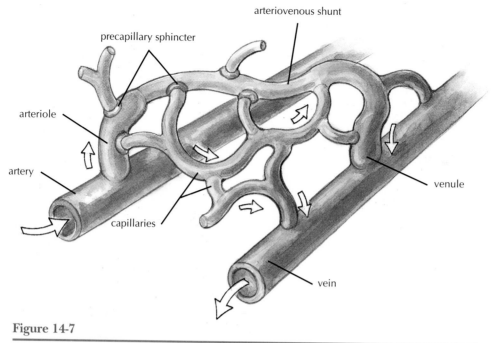

**Figure 14-7**

Capillary network showing an arteriovenous shunt (anastomosis).

lar) *veins* drain the areas supplied by the carotid arteries. Two *brachiocephalic* (innominate) *veins* are formed, one on each side, by the union of the subclavian and the jugular veins. (Remember, there is only *one* brachiocephalic artery.)

### Superior Vena Cava

The veins of the head, neck, upper extremities, and chest all drain into the *superior vena cava* (VE-nah KA-vah), which goes to the heart. It is formed by the union of the right and left brachiocephalic veins, which drain the head, neck, and upper extremities. The *azygos* (AZ-ih-gos) *vein* drains the veins of the chest wall and empties into the superior vena cava just before the latter empties into the heart (see Fig. 14-8).

### Inferior Vena Cava

The *inferior vena cava,* which is much longer than the superior vena cava, returns the blood from the parts of the body below the diaphragm. It begins in the lower abdomen with the union of the two common iliac veins. It then ascends along the back wall of the abdomen, through a groove in the posterior part of the liver, through the diaphragm, and finally through the lower thorax to empty into the right atrium of the heart.

Drainage into the inferior vena cava is more complicated than drainage into the superior vena cava. The large veins below the diaphragm may be divided into two groups:

1. The right and left veins that drain paired parts and organs. They include the *iliac veins* from near the groin, four pairs of *lumbar veins* from the dorsal part of the trunk and from the spinal cord, the *testicular veins* from the testes of the male and the *ovarian veins* from the ovaries of the female, the *renal* and *suprarenal veins* from the kidneys and adrenal glands near the kidneys, and finally the large *hepatic veins* from the liver. For the most part, these vessels empty directly into the inferior vena cava. The left testicular in the male and the left ovarian in the female empty into the left renal vein, which then takes this blood to the inferior vena cava; these veins thus constitute exceptions to the rule that the paired veins empty directly into the vena cava.

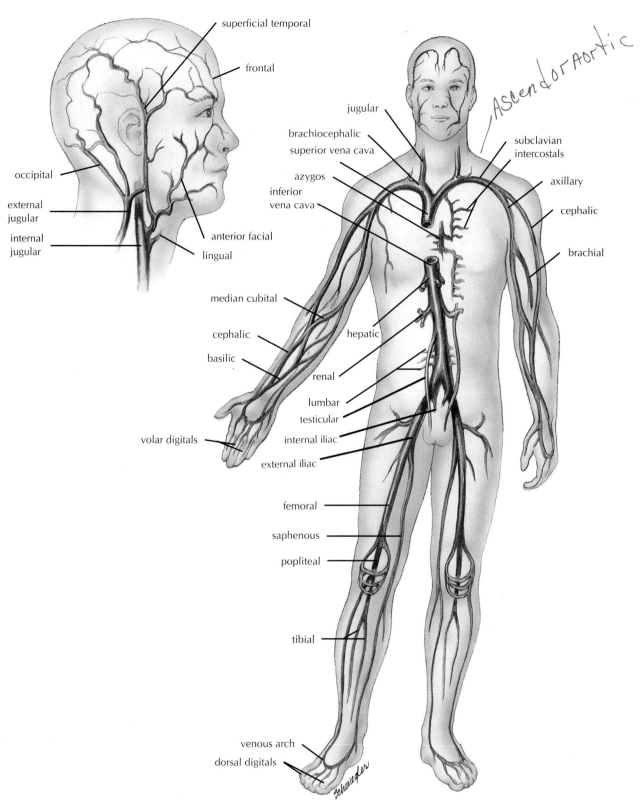

superficial temporal

frontal

occipital

external jugular

internal jugular

anterior facial

lingual

jugular

brachiocephalic

superior vena cava

azygos

inferior vena cava

median cubital

cephalic

basilic

volar digitals

hepatic

renal

lumbar

testicular

internal iliac

external iliac

femoral

saphenous

popliteal

tibial

venous arch

dorsal digitals

subclavian

intercostals

axillary

cephalic

brachial

*Ascend or Aortic*

Schweqler

**Figure 14-8**

Principal systemic veins.

2. Unpaired veins that come from the spleen and from parts of the digestive tract (stomach and intestine) empty into a vein called the *hepatic portal vein.* Unlike other lower veins, which empty into the inferior vena cava, the hepatic portal vein is part of a special system that enables blood to circulate through the liver before returning to the heart.

### Venous Sinuses

The word *sinus* means "space" or "hollow." A *venous sinus* is a large channel that drains deoxygenated blood but does not have the usual tubular structure of the veins. An important example of a venous sinus is the *coronary sinus,* which receives most of the blood from the veins of the heart wall (see Fig. 13-5). It lies between the left atrium and left ventricle on the posterior surface of the heart, and it empties directly into the right atrium along with the two venae cavae.

Other important venous sinuses are the *cranial venous sinuses,* which are located inside the skull and drain the veins that come from all over the brain (Fig. 14-9). The largest of the cranial venous sinuses are described below.

1. The two *cavernous sinuses,* situated behind the eyeballs, serve to drain the *ophthalmic* (of-THAL-mik) *veins* of the eyes.
2. The *superior sagittal* (SAJ-ih-tal) *sinus* is a single long space located in the midline above the brain and in the fissure between the two hemispheres of the cerebrum. It ends in an enlargement called the *confluence* (KON-flu-ens) *of sinuses.*
3. The two *transverse sinuses,* also called the *lateral sinuses,* are large spaces between the layers of the dura mater (the outermost membrane around the brain). They begin posteriorly, in the region of the confluence of sinuses, and then extend toward either side. As each sinus extends around the inside of the skull, it receives blood draining those parts not already drained by the superior sagittal and other sinuses that join the back portions of the transverse sinuses. This means that nearly all the blood that comes from the veins of the brain eventually empties into one or the other of the transverse sinuses. On either side the sinus extends far enough forward to empty into an internal jugular vein, which then passes through a hole in the skull to continue downward in the neck.

### Hepatic Portal System

Almost always, when blood leaves a capillary bed it flows directly back to the heart. In a portal system, however, blood circulates through a second capillary bed, usually in a second organ, before it returns to the heart. A portal system is a kind of detour in the pathway of venous return that can transport materials directly from one organ to another. The largest portal system in the body is the *hepatic portal system,* which carries blood from the abdominal organs to the liver (Fig. 14-10). In a similar fashion, a small, local portal system is located in the brain to carry substances from the hypothalamus to the pituitary (see Chap. 11). The hepatic portal system includes the veins that drain blood from capillaries in the spleen, stomach, pancreas, and intestine. Instead of emptying their blood directly into the inferior vena cava, they deliver it by way of the hepatic portal vein to the liver. The largest tributary of the portal vein is the *superior mesenteric vein.* It is joined by the *splenic vein* just under the liver. Other tributaries of the portal circulation are the *gastric, pancreatic,* and *inferior mesenteric veins.*

On entering the liver, the portal vein divides and subdivides into ever smaller branches. Eventually, the portal blood flows into a vast network of sinus-like vessels called *sinusoids* (SI-nus-oyds). These enlarged capillary channels allow liver cells close contact with the blood coming from the abdominal organs. (Similar blood channels are found in the spleen and endocrine glands, including the thyroid and adrenals.) After leaving the sinusoids, blood is finally collected by the hepatic veins, which empty into the inferior vena cava.

The purpose of the hepatic portal system of veins is to transport blood from the digestive organs and the spleen to the liver sinusoids so the liver cells can carry out their functions. For example, when food is digested, most of the end products are absorbed from the small intestine into the bloodstream and transported to the liver by the portal system. In the liver, these nutrients are processed, stored, and released as needed into the general circulation.

## ▶ The Physiology of Circulation

### How Capillaries Work

In a general way, the circulating blood might be compared to a train that travels around the country, loading and unloading freight in each of the

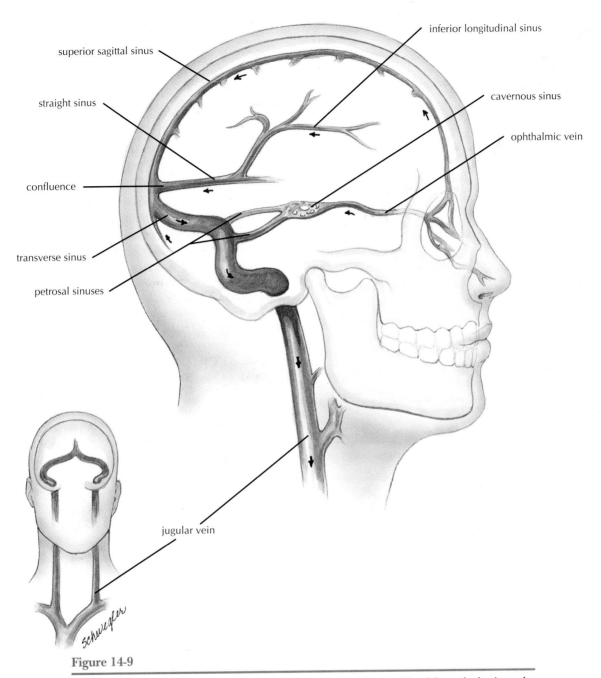

superior sagittal sinus

inferior longitudinal sinus

straight sinus

cavernous sinus

ophthalmic vein

confluence

transverse sinus

petrosal sinuses

jugular vein

**Figure 14-9**

Cranial venous sinuses. The paired transverse sinuses, which carry blood from the brain to the jugular veins, are shown in the inset.

cities it serves. For example, as blood flows through capillaries surrounding the air sacs in the lungs, it picks up oxygen and unloads carbon dioxide. Later, when this oxygenated blood is pumped to capillaries in other parts of the body, it unloads the oxygen and picks up carbon dioxide (Fig. 14-11) and other substances resulting from cellular activities. The microscopic capillaries are of fundamental importance in these activities. It is only through and between the cells of these thin-walled vessels that the aforementioned exchanges can take place.

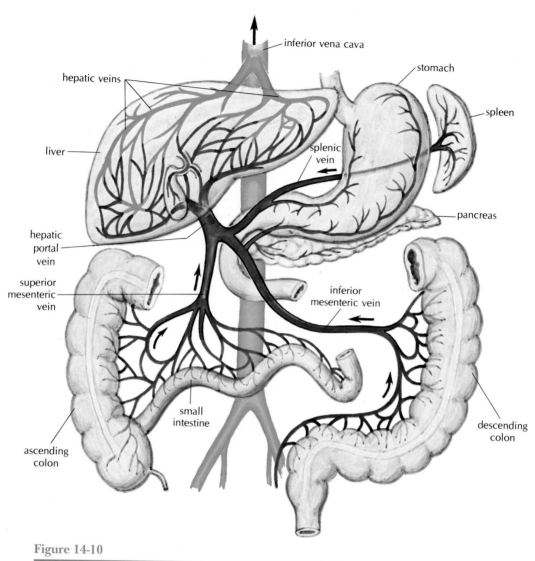

**Figure 14-10**

Hepatic portal circulation.

All living cells are immersed in a slightly salty liquid called *tissue fluid*. Looking again at Figure 14-11, one can see how this fluid serves as "middleman" between the capillary membrane and the neighboring cells. As water, oxygen, and other materials necessary for cellular activity pass through the capillary walls, they enter the tissue fluid. Then these substances make their way by diffusion to the cells. At the same time, carbon dioxide and other end products of cell metabolism come from the cells and move in the opposite direction. These substances enter the capillary and are carried away in the bloodstream, to reach other organs or to be eliminated from the body.

Diffusion is the main process by which substances move between the cells and the capillary blood. A secondary force is the pressure of the blood as it flows through the capillaries. This force acts to filter, or "push," water and dissolved materials out of the capillary into the tissue fluid. Fluid is drawn back into the capillary by osmotic pressure, the "pulling force" of substances dissolved and suspended in the blood. Osmotic pressure is maintained by plasma proteins (mainly albumin), which are too large to go through the capillary wall. These processes result in the constant exchange of fluids across the capillary wall.

The movement of blood through the capillaries is relatively slow due to the much larger size of the cross-sectional area of the capillaries compared with that of the larger vessels from which capillaries

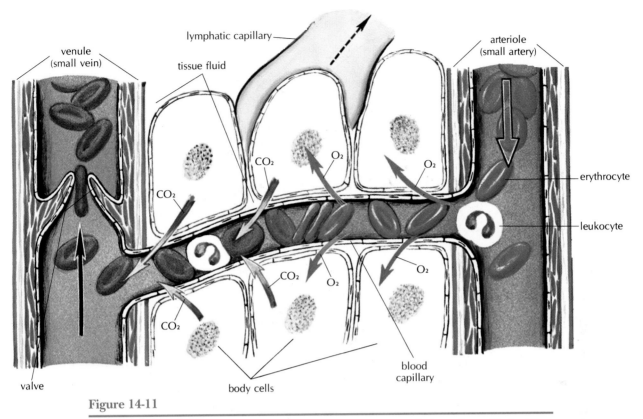

**Figure 14-11**

Diagram showing the connection between the small blood vessels through capillaries. Note the lymphatic capillary, which aids in tissue drainage.

branch. This slow progress through the capillaries allows time for exchanges to occur.

### Dynamics of Blood Flow

The flow of blood is carefully regulated to supply the needs of the tissues without unnecessary burden on the heart. Some organs, such as the brain, liver, and kidneys, require large quantities of blood even at rest. The requirements of some tissues, such as those of skeletal muscles and digestive organs, increase greatly during periods of activity. (The blood flow in muscle can increase 25 times during exercise.) The volume of blood flowing to a particular organ can be regulated by changing the size of the blood vessels supplying that organ.

#### VASODILATION AND VASOCONSTRICTION

An increase in the diameter of a blood vessel is called *vasodilation.* This change allows for the delivery of more blood to an area. *Vasoconstriction* is a decrease in the diameter of a blood vessel, causing a decrease in blood flow. These *vasomotor activities* result from the contraction or relaxation of smooth muscle in the walls of the blood vessels, mainly the arterioles. A *vasomotor center* in the medulla of the brain stem regulates these activities, sending its messages through the autonomic nervous system.

The flow of blood into an individual capillary is regulated by a smooth muscle fiber that encircles the entrance to the capillary (see Fig. 14-7). The *precapillary sphincter* widens to allow more blood to enter when tissues need oxygen.

#### RETURN OF BLOOD TO THE HEART

By the time blood arrives in the veins, little force remains from the pumping action of the heart. Also, because the veins tend to expand under pressure, considerable amounts of blood are stored in the venous system. Blood from the extremities is pushed toward the heart by the contraction of skeletal muscles, which compresses the veins and squeezes the blood forward. The valves in the veins ensure that the blood flows toward the heart. Changes in pressures in the abdominal and thoracic cavities during breathing also promote return of blood in the venous system. During inhalation, the diaphragm flattens and puts pressure on the large abdominal veins. At the same time, expansion of the chest causes pressure to drop in the thorax. Together, these actions serve to push and pull blood through these cavities and return it to the heart. As evidence of these effects, if a person stands completely motionless, especially on a hot day when the vessels dilate, enough blood can accumulate in the lower extremities to cause fainting from insufficient oxygen to the brain.

## ▶ Pulse and Blood Pressure

### Meaning of the Pulse

The ventricles pump blood into the arteries regularly about 70 to 80 times a minute. The force of ventricular contraction starts a wave of increased pressure that begins at the heart and travels along the arteries. This wave, called the *pulse,* can be felt in any artery that is relatively close to the surface, particularly if the vessel can be pressed down against a bone. At the wrist the radial artery passes over the bone on the thumb side of the forearm, and the pulse is most commonly obtained here. Other vessels sometimes used for obtaining the pulse are the carotid artery in the neck and the dorsalis pedis on the top of the foot.

Normally, the pulse rate is the same as the heart rate. Only if a heartbeat is abnormally weak, or if the artery is obstructed, may the beat not be detected as a pulse. In checking the pulse of another person, it is important to use your second or third finger. If you use your thumb, you may find that you are getting your own pulse. When taking a pulse, it is important to gauge the strength as well as the regularity and the rate.

Various factors may influence the pulse rate. We will enumerate just a few:

1. The pulse is somewhat faster in small persons than in large persons and usually is slightly faster in women than in men.
2. In a newborn infant the rate may be from 120 to 140 beats/minute. As the child grows, the rate tends to become slower.
3. Muscular activity influences the pulse rate. During sleep the pulse may slow down to 60 beats/minute, while during strenuous exercise

the rate may go up to well over 100 beats/minute. In a person in good condition, the pulse does not remain rapid despite continued exercise.
4. Emotional disturbances may increase the pulse rate.
5. In many infections, the pulse rate increases with the increase in temperature.
6. An excessive amount of secretion from the thyroid gland may cause a rapid pulse.

### Blood Pressure and Its Determination

Blood pressure is the force exerted by the blood against the walls of the vessels. Blood pressure is the product of the output of the heart and the resistance in the vessels. The output of the heart is influenced by:

1. Strength of the contraction of the heart
2. Total blood volume, which controls the volume each beat pushes out.

The resistance in the vessels is affected by:

1. Vasomotor changes. Vasoconstriction increases resistance to flow; vasodilation lowers resistance.
2. Elasticity of blood vessels. Blood vessels lose elasticity and often become hard as a part of aging, thus increasing resistance.
3. Thickness of the blood, called *viscosity.* Increased numbers of red blood cells increase viscosity.

The measurement and careful interpretation of blood pressure may prove a valuable guide in the care and evaluation of a person's health. Because blood pressure decreases as the blood flows from arteries into capillaries and finally into veins, measurements ordinarily are made of arterial pressure only. The instrument used is called a *sphygmomanometer* (sfig-mo-mah-NOM-eh-ter), and two variables are measured:

1. *Systolic pressure,* which occurs during heart muscle contraction, averages around 120 and is expressed in millimeters of mercury (mm Hg).
2. *Diastolic pressure,* which occurs during relaxation of the heart muscle, averages around 80 mm Hg.

The sphygmomanometer is essentially a graduated column of mercury connected to an inflatable cuff. The cuff is wrapped around the subject's upper arm and inflated with air until the brachial artery is compressed and the blood flow cut off. Then, listening with a stethoscope, the investigator slowly lets air out of the cuff until the first pulsations are heard. At this point the pressure in the cuff is equal to the systolic pressure, and this pressure is read off the mercury column. Then, more air is let out until a characteristic muffled sound indicates the point at which the diastolic pressure is to be read. Considerable practice is required to ensure an accurate reading. The blood pressure is reported as a fraction, with the systolic pressure above and the diastolic pressure below, such as 120/80.

## SUMMARY

I. **Blood vessels**
   A. Functional classification
      1. Arteries—carry blood away from heart
         a. Arterioles—small arteries
      2. Veins—carry blood toward heart
         a. Venules—small veins
      3. Capillaries—allow for exchanges between blood and tissues, or blood and air in lungs; connect arterioles and venules
   B. Circuits
      1. Pulmonary circuit—carries blood to and from lungs
      2. Systemic circuit—carries blood to and from rest of body
   C. Structure
      1. Tissue layers
         a. Innermost—single layer of flat epithelial cells (endothelium)
         b. Middle—thicker layer of smooth muscle and elastic connective tissue
         c. Outer—connective tissue
      2. Arteries—all three layers; highly elastic
      3. Arterioles—thinner walls, less elastic tissue, more smooth muscle
      4. Capillaries—only endothelium; single layer of cells
      5. Veins—all three layers; thinner walls than arteries, less elastic tissue

II. **Systemic arteries**
   A. Aorta—largest artery
      1. Divisions
         a. Ascending aorta
            (1) Left and right coronary arteries
         b. Aortic arch
            (1) Brachiocephalic trunk
            (2) Left common carotid artery
            (3) Left subclavian artery
         c. Descending aorta
            (1) Thoracic aorta
            (2) Abdominal aorta

   B. Iliac arteries—final division of aorta; branch to pelvis and legs
   C. Anastomoses—communications between vessels

III. **Systemic veins**
   A. Location
      1. Superficial—near surface
      2. Deep—usually parallel to arteries with same names as corresponding arteries
   B. Superior vena cava—drains upper part of body
   C. Inferior vena cava—drains lower part of body
   D. Venous sinuses—enlarged venous channels
   E. Hepatic portal system—carries blood from abdominal organs to liver, where it is processed before returning to heart

IV. **Physiology of circulation**
   A. Capillary exchange
      1. Primary method—diffusion
      2. Medium—tissue fluid
      3. Blood pressure—drives fluid into tissues
      4. Osmotic pressure—pulls fluid into capillary
   B. Regulation of blood flow
      1. Vasodilation—increase in diameter of blood vessel
      2. Vasoconstriction—decrease in diameter of blood vessel
      3. Vasomotor center—in medulla; controls contraction and relaxation of smooth muscle in vessel wall
      4. Precapillary sphincter—regulates blood flow into capillary
      5. Effects
         a. Control of blood distribution
         b. Regulation of blood pressure

**C.** Return of blood to heart
  **1.** Pumping action of heart
  **2.** Pressure of skeletal muscles on veins
  **3.** Valves in veins
  **4.** Breathing—changes in pressure move blood toward heart
**V. Pulse**—wave of pressure that travels along arteries as ventricles contract

**VI. Blood pressure**
  **A.** Product of cardiac output, vascular resistance
  **B.** Measured in arm with sphygmomanometer
    **1.** Systolic pressure—averages 120 mm Hg
    **2.** Diastolic pressure—averages 80 mm Hg

## QUESTIONS FOR STUDY AND REVIEW

1. Name the three main groups of blood vessels and describe their functions. How has function affected structure?
2. Compare the oxygen content of blood in the pulmonary and systemic vessels.
3. Name the main branches of the aorta.
4. Describe an arterial anastomosis and give several examples.
5. Trace a drop of blood through the shortest possible route from the capillaries of the foot to the capillaries of the head.
6. What are the names and functions of some cranial venous sinuses? Where is the coronary venous sinus and what does it do?
7. What large vessels drain the blood low in oxygen from most of the body into the right atrium? What vessels carry blood high in oxygen into the left atrium?
8. Trace a drop of blood from capillaries in the wall of the small intestine to the right atrium. What is the purpose of going through the liver on this trip?
9. Define *portal system*. Name the largest portal system in the body.
10. What substances diffuse into the tissues from the capillaries? into the capillaries from the tissues?
11. What force pushes fluid out of the capillaries? What force pulls fluid back into the capillaries?
12. What actions help force blood back to the heart?
13. What is meant by *pulse?* Name some places where the pulse is commonly determined.
14. What are some factors that cause an increase in the pulse rate?
15. List five factors that can change blood pressure.
16. What instrument is used for measuring blood pressure? What are the two values usually obtained called, and what is the significance of each?
17. Explain the difference between the terms in the following pairs:
    **a.** *pulmonary circuit* and *systemic circuit*
    **b.** *arteriole* and *venule*
    **c.** *vasodilation* and *vasoconstriction*

# CHAPTER 15

# The Lymphatic System and Immunity

## Behavioral Objectives

After careful study of this chapter, you should be able to:

- List the three major functions of the lymphatic system
- Explain how lymphatic capillaries differ from blood capillaries
- Name the two main lymphatic ducts and describe the area drained by each
- List the major structures of the lymphatic system and give the locations and functions of each
- Describe the composition and function of the reticulo-endothelial system
- Differentiate between nonspecific and specific body defenses and give examples of each
- List several types of inborn immunity
- Define *antigen* and *antibody*
- Compare T cells and B cells with respect to development and type of activity
- Differentiate between natural and artificial acquired immunity
- Differentiate between active and passive immunity
- Differentiate between a vaccine and an immune serum

Memmler, RL., Cohen, BJ, Wood, DL. *STRUCTURE AND FUNCTION OF THE HUMAN BODY, 6/e,*
© 1996 Lippincott-Raven Publishers

# ▶ The Lymphatic System

As noted in the preceding chapter, body cells live in tissue fluid, a liquid derived from the bloodstream. Water and dissolved substances, such as oxygen and nutrients, are constantly filtering through capillary walls into the spaces between cells and constantly adding to the volume of tissue fluid. However, under normal conditions, fluid is also constantly removed so that it does not accumulate in the tissues. Part of this fluid simply returns (by diffusion) to the capillary bloodstream, taking with it some of the end products of cellular metabolism, including carbon dioxide and other substances. A second pathway for the drainage of tissue fluid involves the *lymphatic system.* In addition to the blood-carrying capillaries, there are microscopic vessels called *lymphatic capillaries,* which drain away excess tissue fluid that does not return to the blood capillaries. The relation between the circulatory system and the lymphatic system is shown in Figure 15-1. Another important function of the lymphatic capillaries is to absorb protein from the tissue fluid and return it to the bloodstream. As soon as tissue fluid enters the lymphatic capillary, it is called *lymph.* The lymphatic capillaries join to form the larger lymphatic vessels, or ducts, and these vessels (which we shall have a closer look at in a moment) eventually empty into the veins. However, before the lymph reaches the veins, it flows through a series of filters called *lymph nodes,* where bacteria and other foreign particles are trapped and destroyed. Thus, the lymph nodes may be compared in one way with the oil filter in an automobile.

## Lymphatic Capillaries

The lymphatic capillaries resemble the blood capillaries in that they are made of one layer of flattened (squamous) epithelial cells, also called *endothelium,* which allows for easy passage of soluble materials and water. Gaps between these endothelial cells allow the entrance of proteins and other relatively large suspended particles. Unlike the capillaries of the bloodstream, the lymphatic capillaries begin blindly; that is, they do not serve to bridge two larger vessels. Instead, one end simply lies within a lake of tissue fluid, and the other communicates with a larger lymphatic vessel (see Fig. 15-1).

In the small intestine are some specialized lymphatic capillaries, called *lacteals* (LAK-te-als), which act as a pathway for the transfer of digested fats into the bloodstream. This process is covered in the chapter on the digestive system (see Chap. 17).

## Lymphatic Vessels

The lymphatic vessels are thin walled and delicate and have a beaded appearance because of indentations where valves are located. These valves prevent backflow in the same way as do those found in some veins.

Lymphatic vessels (Fig. 15-2) include *superficial* and *deep* sets. The surface lymphatics are immediately below the skin, often continuing near the superficial veins. The deep vessels are usually larger and accompany the deep veins.

Lymphatic vessels are named according to location. For example, those in the breast are called *mammary* lymphatic vessels, those in the thigh *femoral* lymphatic vessels, and those in the leg *tibial* lymphatic vessels. All the lymphatic vessels form networks, and at certain points they carry lymph into the regional nodes (the nodes that "service" a particular area). For example, nearly all the lymph from the upper extremity and the breast passes through the *axillary lymph nodes,* whereas that from the lower extremity passes through the *inguinal nodes.* Lymphatic vessels carrying lymph away from the regional nodes eventually drain into one of the two terminal vessels, the right lymphatic duct or the thoracic duct, which empty into the bloodstream.

The *right lymphatic duct* is a short vessel about 1.25 cm (1/2 inch) long that receives only the lymph that comes from the upper right quadrant of the body: the right side of the head, neck, and thorax, as well as the right upper extremity. It empties into the right subclavian vein. Its opening into this vein is guarded by two pocket-like semilunar valves to prevent blood from entering the duct. The rest of the body is drained by the thoracic duct.

### THE THORACIC DUCT

The *thoracic duct* is much the larger of the two terminal vessels; it is about 40 cm (16 inches) in length. As shown in Figure 15-2, the thoracic duct receives lymph from all parts of the body except those above the diaphragm on the right side. This duct begins in the posterior part of the abdominal cavity, below the attachment of the diaphragm. The first part of this duct is enlarged to form a cistern, or temporary storage pouch, called the *cisterna chyli*

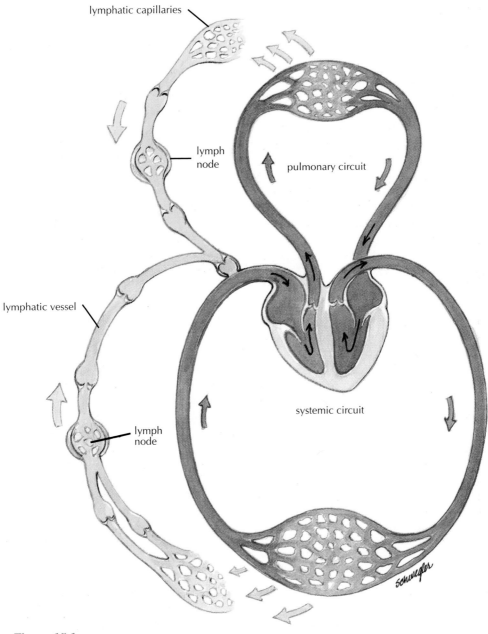

lymphatic capillaries

lymph
node

pulmonary circuit

lymphatic vessel

lymph
node

systemic circuit

Schwegler

**Figure 15-1**

The lymphatic system in relation to the cardiovascular system.

(sis-TER-nah KI-li). *Chyle* (kile) is the milky fluid, formed by the combination of fat globules and lymph, that comes from the intestinal lacteals. Chyle passes through the intestinal lymphatic vessels and the lymph nodes of the mesentery, finally entering the cisterna chyli. In addition to chyle, all the lymph from below the diaphragm empties into the cisterna

chyli by way of the various clusters of lymph nodes and then is carried by the thoracic duct into the bloodstream.

The thoracic duct extends upward through the diaphragm and along the back wall of the thorax up into the root of the neck on the left side. Here it receives the left jugular lymphatic vessels from the

## How Lymphatic Capillaries Work

Lymphatic capillaries are more permeable than blood capillaries because the junctions between the cells are wider and also because there is less basement membrane underlying the cells. Proteins and other large molecules can enter the lymphatic capillaries along with fluid.

Why don't substances that get into the lymphatic capillaries flow right out again? The cells that form the capillary wall overlap slightly to function as one-way valves. Pressure outside the vessel forces materials in. Once fluid enters the capillary, pressure against the cells in the capillary wall blocks its return.

head and neck, the left subclavian vessels from the left upper extremity, and other lymphatic vessels from the thorax and its parts. In addition to the valves along the duct, there are two valves at its opening into the left subclavian vein to prevent the passage of blood into the duct.

### Movement of Lymph

The segments of lymphatic vessels located between the valves contract rhythmically, propelling the lymph along. The rate of these contractions is related to the volume of fluid in the vessel—the more fluid, the more rapid the contractions of the vessel. Lymph is also moved by the same mechanisms that promote venous return of blood to the heart. As skeletal muscles contract during movement they compress the lymphatic vessels and drive lymph forward. Changes in pressures within the abdominal and thoracic cavities due to breathing aid the movement of lymph during passage through these body cavities.

## ▶ Lymphoid Tissue

The above section was just a brief survey of the system of lymph vessels and lymph transport. The lymph nodes were mentioned but they will be described in greater detail now, along with discussion of other organs made of similar tissue. *Lymphoid* (LIM-foyd) *tissue* is distributed throughout the body

and makes up the specialized organs of the lymphatic system. We will look at some of these organs, but first let us consider some properties of lymphoid tissue to see what characteristics these organs have in common.

### Functions

Some of the general functions of lymphoid tissue include the following:

1. Removal of impurities such as carbon particles, cancer cells, disease organisms, and dead blood cells through filtration and phagocytosis.
2. Processing of lymphocytes. Some of these lymphocytes produce antibodies, substances in the blood that aid in combating infection; others attack foreign invaders directly. These cells are discussed later in this chapter.

### Lymph Nodes

The lymph nodes, as we have seen, are designed to filter the lymph once it is drained from the tissues (Fig. 15-3). The lymph nodes are small, rounded masses varying from pinhead size to as long as 2.5 cm (1 inch). Each has a fibrous connective tissue capsule from which partitions (trabeculae) extend into the substance of the node. Inside the node are masses of lymphatic tissue, which include lymphocytes and macrophages, white blood cells active in immunity. At various points in the surface of the node, afferent lymphatic vessels pierce the capsule to carry lymph into the spaces inside the pulp-like nodal tissue. An indented area called the *hilus* (HI-lus) serves as the exit for efferent lymphatic vessels carrying lymph out of the node. At this region other structures, including blood vessels and nerves, connect with the node.

Lymph nodes are seldom isolated. As a rule, they are massed together in groups, the number in each group varying from 2 or 3 to well over 100. Some of these groups are placed deeply, while others are superficial. Those of the most practical importance include the following:

1. *Cervical nodes,* located in the neck, are divided into deep and superficial groups, which drain various parts of the head and neck. They often become enlarged during upper respiratory infections.
2. *Axillary nodes,* located in the axillae (armpits), may become enlarged following infections of the

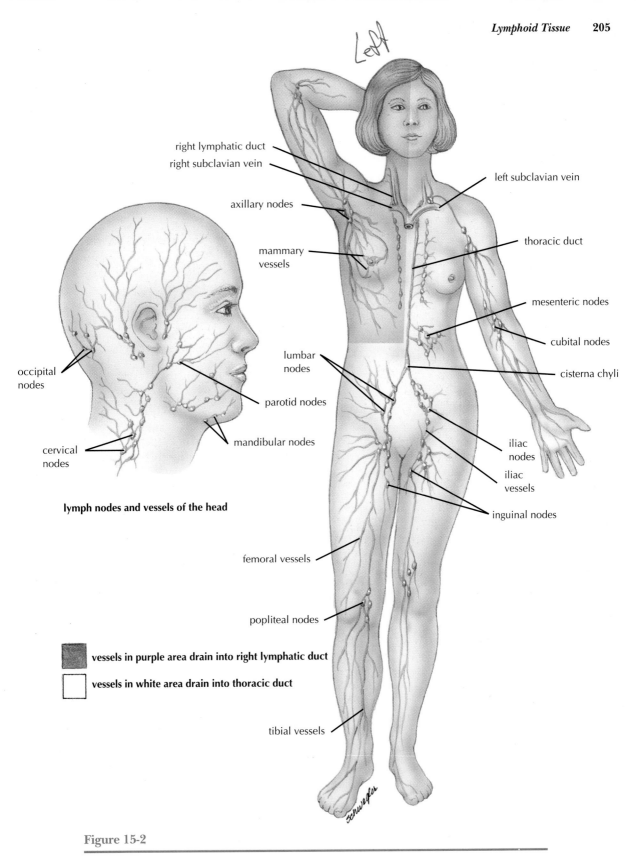

Left

right lymphatic duct

right subclavian vein

axillary nodes

mammary vessels

left subclavian vein

thoracic duct

mesenteric nodes

cubital nodes

cisterna chyli

lumbar nodes

iliac nodes

iliac vessels

inguinal nodes

occipital nodes

parotid nodes

mandibular nodes

cervical nodes

**lymph nodes and vessels of the head**

femoral vessels

popliteal nodes

vessels in purple area drain into right lymphatic duct

vessels in white area drain into thoracic duct

tibial vessels

**Figure 15-2**

Lymphatic system.

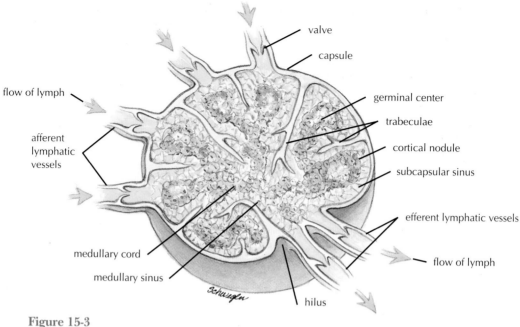

valve

capsule

flow of lymph

germinal center

afferent lymphatic vessels

trabeculae

cortical nodule

subcapsular sinus

efferent lymphatic vessels

flow of lymph

medullary cord

medullary sinus

hilus

**Figure 15-3**

Structure of a lymph node.

upper extremities and the breasts. Cancer cells from the breasts often metastasize (spread) to the axillary nodes.

3. *Tracheobronchial* (tra-ke-o-BRONG-ke-al) *nodes* are found near the trachea and around the larger bronchial tubes. In persons living in highly polluted areas, these nodes become so filled with carbon particles that they are solid black masses resembling pieces of coal.

4. *Mesenteric* (mes-en-TER-ik) *nodes* are found between the two layers of peritoneum that form the mesentery (membrane around the intestine). There are some 100 to 150 of these nodes.

5. *Inguinal nodes,* located in the groin region, receive lymph drainage from the lower extremities and from the external genital organs. When they become enlarged, they are often referred to as *buboes* (BU-bose), from which bubonic plague got its name.

### The Tonsils

There are masses of lymphoid tissue that are designed to filter not lymph, but tissue fluid. Found beneath certain areas of moist epithelium that are exposed to the outside, and hence to contamination, these masses include parts of the digestive, urinary, and respiratory tracts. Associated with the lat-

ter system are those well-known masses of lymphoid tissue called the *tonsils,* which include:

1. The *palatine* (PAL-ah-tine) *tonsils,* oval bodies located at each side of the soft palate. These are generally meant when one refers to "the tonsils."
2. The *pharyngeal* (fah-RIN-je-al) *tonsil,* commonly referred to as *adenoids* (from a general term that means "gland-like"). It is located behind the nose on the back wall of the upper pharynx.
3. The *lingual* (LING-gwal) *tonsils,* little mounds of lymphoid tissue at the back of the tongue.

Any or all of these tonsils may become so loaded with bacteria that the organisms gain the upper hand; removal then is advisable. A slight enlargement of any of them is not an indication for surgery. All lymphoid tissue masses tend to be larger in childhood, so that a physician must consider the patient's age in determining whether these masses are enlarged abnormally. Because the tonsils appear to function in immunity during early childhood, efforts are made *not* to remove them unless absolutely necessary.

### The Thymus

Because of its appearance under the microscope, the *thymus* (THI-mus), located in the upper thorax beneath the sternum, has been considered part of the lymphoid system. However, recent studies suggest that this structure has a much more basic function than was originally thought. It now seems apparent that the thymus plays a key role in the development of the immune system before birth and during the first few months of infancy. Certain lymphocytes must mature in the thymus gland before they can perform their functions in the immune system. These ***T lymphocytes***, or T cells, develop under the effects of the hormone from the thymus gland called ***thymosin*** (THI-mo-sin), which also promotes the growth and activity of lymphocytes in lymphoid tissue throughout the body. Removal causes a decrease in the production of T lymphocytes, as well as a decrease in the size of the spleen and of lymph nodes throughout the body. The thymus is most active during early life. After puberty, the tissue undergoes changes; it shrinks in size and is replaced by connective tissue and fat.

### The Spleen

The spleen is an organ that contains lymphoid tissue designed to filter blood. It is located in the upper left hypochondriac region of the abdomen and normally is protected by the lower part of the rib cage because it is high up under the dome of the diaphragm. The spleen is a soft, purplish, and somewhat flattened organ about 12.5 to 16 cm (5–6 inches) long and 5 to 7.5 cm (2–3 inches) wide. The capsule of the spleen, as well as its framework, is more elastic than that of the lymph nodes. It contains involuntary muscle, which enables the splenic capsule to contract and to withstand some swelling.

The spleen has an unusually large blood supply, considering its size. The organ is filled with a soft pulp, one of the functions of which is to filter out worn-out red blood cells. The spleen also harbors phagocytes, which engulf bacteria and other foreign particles. Round masses of lymphoid tissue are prominent structures inside the spleen, and it is because of these that the spleen is often classified with the other organs made of lymphoid tissue. Some other category might be better, however, because of the other specialized functions of the spleen. Some of these functions are listed below:

1. Cleansing the blood by filtration and phagocytosis.
2. Destroying old, worn-out red blood cells. The iron and other breakdown products of hemoglobin are carried to the liver by the hepatic portal system to be reused or eliminated from the body.
3. Producing red blood cells before birth.
4. Serving as a reservoir for blood, which can be returned to the bloodstream in case of hemorrhage or other emergency.

## ▶ The Reticuloendothelial System

The ***reticuloendothelial*** (reh-tik-u-lo-en-do-THE-le-al) ***system*** consists of related cells concerned with the destruction of worn-out blood cells, bacteria, cancer cells, and other foreign substances that are potentially harmful to the body. They include monocytes, which are relatively large white blood cells (see Fig. 12-1) that are formed in the bone marrow and then circulate in the bloodstream to various parts of the body. On entering the tissues, monocytes develop into ***macrophages*** (MAK-ro-faj-ez), a term that means "big eaters." Some of them are given special names; the ***Kupffer's*** (KOOP-ferz) ***cells,*** for example, are located in the lining of the liver sinusoids (blood channels). Other parts of the reticuloendothelial system are found in the spleen, bone marrow, lymph nodes, and brain. Some macrophages are located in the lungs, where they are called *dust cells* because they ingest solid particles that enter the lungs; others are found in soft connective tissues all over the body.

This widely distributed protective system has been called by several other names, including *tissue macrophage system, mononuclear phagocyte system,* and *monocyte–macrophage system.* These names are descriptive of the type of cells found in this system.

## ▶ Immunity

The reticuloendothelial system is but one of the body's mechanisms for fighting disease and ridding itself of impurities. Because this system is active against a variety of foreign invaders, it is described as a ***nonspecific*** defense. Other nonspecific defenses include the skin and mucous membranes, body secretions such as tears and digestive juices, and the inflammatory response. Our final line of defense against disease, however, is the immune system, which is referred to as ***specific*** because it acts against particular harmful agents. Immunity is a selective process; that is, immunity to one disease does not necessarily cause immunity to another.

There are two main categories of immunity: *inborn* (or inherited) *immunity* and *acquired immunity.* Acquired immunity may be obtained by *natural* or *artificial* means; in addition, acquired immunity may be either *active* or *passive.* Figure 15-4 is a summary of the different types of immunity. Refer to this diagram as we investigate each category in turn.

### Inborn Immunity

Although certain diseases found in animals may be transmitted to humans, many infections, such as chicken cholera, hog cholera, distemper, and other animal diseases, do not affect human beings. However, the constitutional differences that make human beings immune to these disorders also make them susceptible to others that do not affect the lower animals. Such infections as measles, scarlet fever, diphtheria, and influenza do not seem to affect animals in contact with humans experiencing these illnesses. Thus, both humans and animals have what is called a *species immunity* to many of each other's diseases.

Another form of inborn immunity is *racial immunity.* Some racial groups appear to have a greater inborn immunity to certain diseases than other racial groups. For instance, in the United States, blacks are apparently more immune to poliomyelitis, malaria, and yellow fever than are whites. Of course, it is often difficult to tell how much of this variation in resistance to infection is due to environment and how much is due to inborn traits of the various racial groups.

Some members of a given group have a more highly developed *individual immunity* than other members. Newspapers and magazines sometimes feature the advice of an elderly person who is asked to give her secret for living to a ripe old age. One elderly person may say that he practiced temperance and lived a carefully regulated life with the right amount of rest, exercise, and work, whereas the next may boast of her use of alcoholic beverages, constant smoking, lack of exercise, and other kinds of reputedly unhygienic behavior. However, it is possible that the latter person has lived through the onslaughts of toxins and disease organisms, resisted infection, and maintained health in spite of her habits, rather than because of them, thanks to the resistance factors and immunity to disease she inherited.

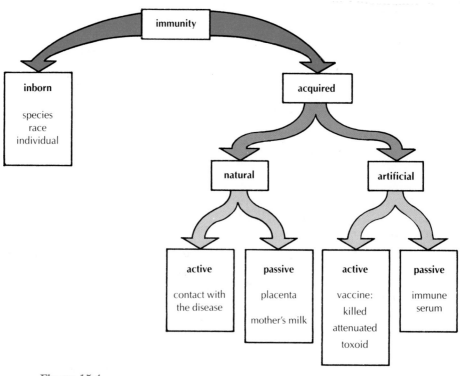

**Figure 15-4**

Types of immunity.

## Acquired Immunity

Unlike inborn immunity, which is due to inherited factors, acquired immunity develops during an individual's lifetime as that person encounters various specific harmful agents.

### ANTIGENS

An *antigen* (AN-te-jen) *(Ag)* is any foreign substance that enters the body and produces an immune response. Most antigens are large protein molecules, but carbohydrates and some lipids may act as antigens. Antigens may be found on the surface of foreign organisms, on the surface of red blood cells and tissue cells, on pollens, in toxins (poisons), and in foods. The critical feature of any substance described as an antigen is that it stimulates the activity of certain lymphocytes classified as T or B cells.

### T CELLS

Both T and B cells come from stem cells in bone marrow, as do all blood cells. They differ, however, in their development and their method of action. Some of the immature stem cells migrate to the thymus gland and become T cells, which constitute about 80% of the lymphocytes in the circulating blood. While in the thymus, these T lymphocytes multiply and become capable of combining with specific foreign antigens, at which time they are described as *sensitized*. These thymus-derived cells produce an immunity that is said to be *cell-mediated immunity*.

There are several types of T cells, each with different functions. Some of these functions are as follows:

1. To destroy foreign cells directly. These are the *killer (cytotoxic) T cells.*
2. To release substances that stimulate other lymphocytes and macrophages and thereby assist in the destruction of foreign cells. It is these *helper T cells* that are infected and destroyed by the AIDS virus.
3. To suppress the immune response in order to regulate it. These *suppressor T cells* may inhibit or destroy active lymphocytes.
4. To remember an antigen and start a rapid response if that antigen is met again. These are *memory T cells.*

The T cell portion of the immune system is generally responsible for defense against cancer cells, certain viruses, and other disease-causing organisms that grow within cells (intracellular parasites), as well as for the rejection of tissue transplanted from another person.

Working with the T cells are *macrophages,* cells derived from blood monocytes. For a T cell to react with an antigen, that antigen must be presented to the T cell on the surface of a macrophage in combination with proteins that the T cell can recognize as belonging to the "self." After combining with T cells, the macrophages release substances called *interleukins* (a name that means "between white blood cells") that stimulate the growth of T cells.

### B CELLS AND ANTIBODIES

An *antibody* (*Ab*), also known as an *immunoglobulin (Ig),* is a substance produced in response to an antigen. Antibodies are manufactured by the second type of lymphocyte active in the immune system. The *B cells*, or B lymphocytes, must mature in the fetal liver or in lymphoid tissue before becoming active in the blood. Exposure to an antigen stimulates B cells to multiply rapidly and produce large numbers (clones) of *plasma cells.* These cells produce specific antibodies that circulate in the blood providing the form of immunity described as *humoral immunity* (the term humoral refers to body fluids). Humoral immunity generally protects against circulating antigens and bacteria that grow outside the cells (extracellular parasites). All antibodies are contained in a portion of the blood plasma called the *gamma globulin* fraction.

Some antibodies remain in the blood to give long-term immunity. In addition, some of the activated B cells do not become plasma cells but, like certain T cells, become memory cells. On repeated contact with an antigen, these cells are ready to produce antibodies immediately. Because of this "immunologic memory," one is usually immune to a childhood disease after having it.

### THE ANTIGEN–ANTIBODY REACTION

The antibody that is produced in response to a specific antigen, such as a bacterial cell or a toxin, has a shape that matches some part of that antigen, much in the same way that the shape of a key matches the shape of its lock. For this reason the antibody can bind specifically to the antigen that caused its production and thereby destroy or inactivate it. In the laboratory this can be seen as a settling (precipitation) or a clumping (agglutination) of the antigen–antibody combination, as is seen in the typing of red blood cells.

The destruction of foreign cells sometimes requires the enzymatic activity of a group of nonspecific proteins in the blood together called *complement.*

If descriptions of the immune system seem complex, bear in mind that from infancy on, your immune system is able to protect you from millions of foreign substances, even those that are synthetic and not found in nature. All the while, the system is kept in check so that it does not usually overreact to produce allergies or mistakenly attack and damage your own body tissues.

### Naturally Acquired Immunity

Immunity may be acquired naturally through contraction of a specific disease. In this case, antibodies manufactured by the infected person's cells act against the infecting agent or its toxins. Each time a person is invaded by the organisms of a disease, his or her cells may manufacture antibodies that provide immunity against the infection. Such immunity may last for years, and in some cases lasts for life. Because the host is actively involved in the production of antibodies, this type of immunity is called *active immunity.*

The infection that calls forth the immunity may be an inapparent infection that is so mild as to cause no symptoms. Nevertheless, it may stimulate the host's cells to produce an active immunity.

Immunity may be acquired naturally by a fetus through the passage of antibodies from the mother through the placenta. Because these antibodies come from an outside source, this type of immunity is called *passive immunity.* The antibodies obtained in this way do not last as long as actively produced antibodies, but they do help protect the infant for about 6 months, at which time the child's own immune system begins to function. Nursing an infant can lengthen this period of protection owing to the presence of specific antibodies in breast milk and colostrum (the first breast secretion). These are the only known examples of naturally acquired passive immunity.

### Artificially Acquired Immunity

A person who is not exposed to repeated small doses of a particular organism has no antibodies against that organism and is defenseless against a heavy infection. Therefore, artificial measures are usually taken to cause persons' tissues to actively manufacture antibodies. One could inject active disease organisms into the tissues, but obviously this would be dangerous. Instead, the harmful agent is treated to reduce its damaging activity and then administered. In this way the tissues are made to produce antibodies without causing a serious illness. This process is known as *vaccination* (vak-sin-A-shun), or *immunization,* and the solution used is called a *vaccine* (vak-SENE). Ordinarily, the administration of a vaccine is a preventive measure designed to provide protection in anticipation of invasion by a certain disease organism.

#### VACCINES

Vaccines can be made with live organisms or with organisms killed by heat or chemicals. If live organisms are used, they must be harmless to humans, such as the cowpox virus used for smallpox immunization, or they must be treated in the laboratory to reduce their ability to cause disease in humans. An organism thus weakened for use in vaccines is described as *attenuated.* A third type of vaccine is made from a form of the toxin produced by a disease organism. The toxin is altered with heat or chemicals to reduce its harmfulness, but it can still function as an antigen to induce immunity. Such an altered toxin is called a *toxoid.*

A final word about the long-term effectiveness of vaccines: in many cases an active immunity acquired by artificial (or even natural) means does not last a lifetime. Circulating antibodies can decline with time. To help maintain a high titer (level) of antibodies in the blood, repeated inoculations, called *booster shots,* are administered at intervals. The number of booster injections recommended varies with the disease and with the environment or range of exposure of the individual.

#### PASSIVE IMMUNIZATION

It takes several weeks to produce a naturally acquired active immunity and even longer to produce an artificial active immunity through the administration of a vaccine. Therefore, a person who receives a large dose of disease organisms and has no established immunity to them stands in great danger. To prevent catastrophe, then, the victim must quickly receive a counteracting dose of borrowed antibodies. This is accomplished through the administration of an ***immune serum,*** or ***antiserum.*** These are prepared in animals (usually horses) or are taken from other humans. The immune serum gives short-lived but effective protection against the invaders in the form of an artificially acquired passive immunity. Some immune sera contain antibodies, known as ***antitoxins,*** that neutralize toxins but have no effect on the toxic organisms themselves.

There are a number of differences between vaccines and antisera. Administration of a vaccine causes antibody formation by the body tissues. In the administration of an antiserum, in contrast, the antibodies are supplied "ready made." However, the passive immunity produced by an antiserum does not last as long as that actively produced by the body tissues. Immune sera are usually used in emergencies, that is, in situations in which there is no time to wait until an active immunity has developed.

# SUMMARY

I. **Lymphatic system**
  A. Functions
    1. Drainage of excess fluid from tissues
    2. Absorption of fats from small intestine
    3. Protection from foreign invaders
  B. Structures
    1. Capillaries—made of endothelium (simple squamous epithelium)
    2. Lymphatic vessels
      a. Superficial
      b. Deep
    3. Right lymphatic duct—drains upper right part of body and empties into right subclavian vein
    4. Thoracic duct—drains remainder of body and empties into left subclavian vein
  C. Movement of lymph
II. **Lymphoid tissue**—distributed throughout body
  A. General functions
    1. Removal of impurities by filtration and phagocytosis
    2. Processing of lymphocytes of immune system
  B. Specialized functions
    1. Nodes—filtration of lymph
    2. Tonsils—filtration of tissue fluid
    3. Thymus
      a. Processing of T lymphocytes (T cells)
      b. Secretion of thymosin—stimulates T lymphocytes in lymphoid tissue
    4. Spleen
      a. Filtration of blood
      b. Destruction of old red cells
      c. Production of red cells before birth
      d. Storage of blood
III. **Reticuloendothelial system**—cells throughout body that get rid of impurities
IV. **Immunity**—specific defenses against disease
  A. Inborn immunity (inherited)—species, racial, individual
  B. Acquired immunity—obtained during life
    1. Immune response
      a. Antigens—stimulate lymphocyte activity
      b. Antibodies—inactivate disease organisms or neutralize toxins
      c. T cells (T lymphocytes)
        (1) Processed in thymus
        (2) Produce cell-mediated immunity
        (3) Macrophages—present antigen to T cells
      d. B cells (B lymphocytes)
        (1) Mature in lymphoid tissue
        (2) Develop into plasma cells that produce circulating antibodies
        (3) Produce humoral immunity
    2. Naturally acquired immunity
      a. Active—acquired through contact with the disease
      b. Passive—acquired from antibodies obtained through placenta and mother's milk
    3. Artificially acquired immunity
      a. Active—immunization with vaccines
      b. Passive—administration of immune serum (antiserum)

## QUESTIONS FOR STUDY AND REVIEW

1. What is lymph? How is it formed?
2. Briefly describe the system of lymph circulation.
3. Describe the lymphatic vessels with respect to design, appearance, and location.
4. Name the two main lymphatic ducts. What part of the body does each drain, and into what blood vessel does each empty?
5. What is the cisterna chyli and what are its purposes?
6. Name two functions of lymphoid tissue.
7. Describe the structure of a typical lymph node.
8. What are the neck nodes called and what are some of the causes of enlargement of these lymph nodes?
9. What parts of the body are drained by vessels entering the axillary nodes and what conditions cause enlargement of these nodes?
10. What parts of the body are drained by lymphatics that pass through the inguinal lymph nodes?
11. What are the different tonsils called and where are they located? What is the purpose of these and related structures?
12. What is the function of the thymus?
13. Give the location of the spleen and name several of its functions.
14. Describe the reticuloendothelial system.
15. Define *immunity* and describe the basic immune process.
16. Give three examples of inborn immunity.
17. What is the basic difference between inborn and acquired immunity?
18. Define *antigen* and *antibody*. Why is the reaction between and antigen and antibody described as specific?
19. How do T and B cells differ? In what ways are they the same?
20. Name four types of T cells.
21. What is the role of macrophages in immunity?
22. Outline the various categories of acquired immunity and give an example of each.
23. What is a vaccine? a toxoid? What is a booster shot?
24. What is an immune serum? an antitoxin?

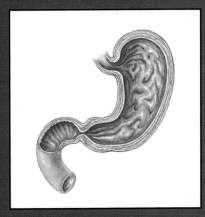

# UNIT V

# Energy—
# Supply and Use

**I**t is the purpose of the four chapters of this unit to show how oxygen and nutrients are processed, taken up by the body fluids and used by the cells to yield energy. This unit also describes how the stability of body functions (homeostasis) is maintained and how waste products are excreted.

# CHAPTER 16

# Respiration

## Behavioral Objectives

After careful study of this chapter, you should be able to:

- Define *respiration* and describe the three phases of respiration

- Name all the structures of the respiratory system

- Explain the mechanism for pulmonary ventilation

- List the ways in which oxygen and carbon dioxide are transported in the blood

- Describe the ways in which respiration is regulated

Memmler, RL, Cohen, BJ, Wood, DL. *STRUCTURE AND FUNCTION OF THE HUMAN BODY*, 6/e,
© 1996 Lippincott-Raven Publishers

**M**ost people think of respiration simply as the process by which air moves into and out of the lungs, that is, *breathing.* By scientific definition, respiration is the process by which oxygen is obtained from the environment and delivered to the cells. Carbon dioxide is transported to the outside in a reverse pathway.

Respiration includes three phases:

1. *Pulmonary ventilation,* which is the exchange of air between the atmosphere and the air sacs of the lungs. This is normally accomplished by the inspiration and expiration of breathing.
2. The *diffusion* of gases, which includes the passage of oxygen from the air sacs into the blood and the passage of carbon dioxide out of the blood
3. The *transport* of oxygen to the cells and the transport of carbon dioxide from the cells to the lungs. This is accomplished by the circulating blood.

In the process of *cellular respiration,* oxygen is taken into the cell and used to break down nutrients with the release of energy. Carbon dioxide is the main waste product of cellular respiration.

The respiratory system is an intricate arrangement of spaces and passageways that conduct air into the lungs (Fig. 16-1). These spaces include the *nasal cavities;* the *pharynx* (FAR-inks), which is common to the digestive and respiratory systems; the voice box, or *larynx* (LAR-inks); the windpipe, or *trachea* (TRA-ke-ah); and the *lungs* themselves, with their conducting tubes and air sacs. The entire system might be thought of as a pathway for air between the atmosphere and the blood.

## ▶ The Respiratory System

### The Nasal Cavities

Air makes its initial entrance into the body through the openings in the nose called the *nostrils.* Immediately inside the nostrils, located between the roof of the mouth and the cranium, are the two spaces known as the *nasal cavities.* These two spaces are separated from each other by a partition, the *nasal septum.* The septum and the walls of the nasal cavities are constructed of bone covered with mucous membrane. On the lateral (side) walls of each nasal cavity are three projections called the *conchae* (KONG-ke). The conchae greatly increase the sur-

face over which air must travel on its way through the nasal cavities.

The lining of the nasal cavities is a mucous membrane, which contains many blood vessels that bring heat and moisture to it. The cells of this membrane secrete a large amount of fluid—up to 1 quart each day. The following changes are produced in the air as it comes in contact with the lining of the nose:

1. Foreign bodies, such as dust particles and microorganisms, are filtered out by the hairs of the nostrils or caught in the surface mucus.
2. Air is warmed by the blood in the vascular membrane.
3. Air is moistened by the liquid secretion.

For these protective changes to occur, it is preferable to breathe through the nose than through the mouth.

The *sinuses* are small cavities lined with mucous membrane in the bones of the skull. The sinuses communicate with the nasal cavities, and they are highly susceptible to infection.

### The Pharynx

The muscular *pharynx,* or throat, carries air into the respiratory tract and carries foods and liquids into the digestive system. The upper portion, located immediately behind the nasal cavity, is called the *nasopharynx* (na-zo-FAR-inks); the middle section, located behind the mouth, is called the *oropharynx* (o-ro-FAR-inks); and the lowest portion is called the *laryngeal* (lah-RIN-je-al) *pharynx.* This last section opens into the larynx toward the front and into the esophagus toward the back.

### The Larynx

The *larynx,* or voice box, (Fig. 16-2) is located between the pharynx and the trachea. It has a framework of cartilage, one of which is the thyroid cartilage that protrudes in the front of the neck. The projection formed by the thyroid cartilage is popularly called the *Adam's apple* because it is considerably larger in the male than in the female. On both sides at the upper end of the larynx are folds of mucous membrane used in producing speech. These are the vocal folds, or *vocal cords* (Fig. 16-3). They are set into vibration by the flow of air from the lungs. A difference in the size of the larynx is what accounts for the difference between male and female voices; because a man's larynx is larger than a

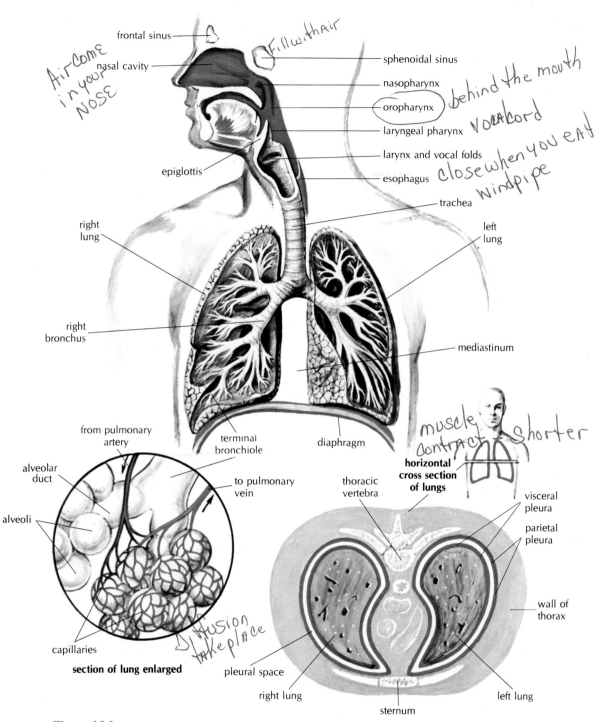

**Figure 16-1**

Respiratory system.

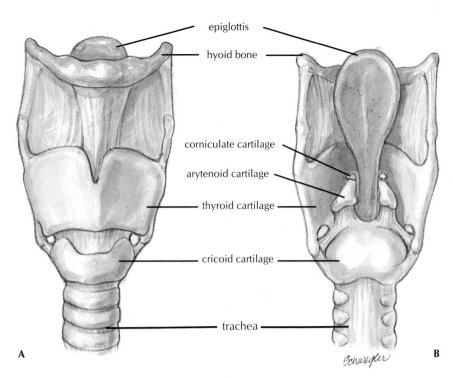

epiglottis

hyoid bone

corniculate cartilage

arytenoid cartilage

thyroid cartilage

cricoid cartilage

trachea

A                                    B

**Figure 16-2**

The larynx. **(A)** Anterior view. **(B)** Posterior view.

woman's, his voice is lower in pitch. The nasal cavities, the sinuses, and the pharynx all serve as resonating chambers for speech, just as the cabinet does for a stereo speaker.

The space between the vocal cords is called the *glottis* (GLOT-is), and the little leaf-shaped cartilage that covers the larynx during swallowing is called the *epiglottis* (ep-ih-GLOT-is). The glottis and epiglottis help keep food out of the remainder of the respiratory tract. As the larynx moves upward and forward

during swallowing, the epiglottis moves downward, covering the opening into the larynx. The glottis assists by closing during swallowing. You can feel the larynx move upward toward the epiglottis during this process by placing the flat ends of your fingers on your larynx as you swallow.

### The Trachea, or Windpipe

The *trachea* is a tube that extends from the lower edge of the larynx to the upper part of the chest

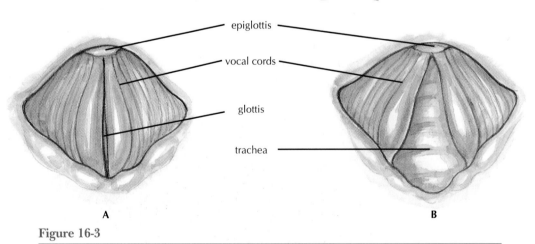

epiglottis

vocal cords

glottis

trachea

A                                    B

**Figure 16-3**

The vocal cords viewed from above. **(A)** The glottis in closed position. **(B)** The glottis in open position.

above the heart. It has a framework of cartilages to keep it open. These cartilages, shaped somewhat like a tiny horseshoe or the letter C, are found along the entire length of the trachea. The open sections in the cartilages are lined up in the back so that the esophagus can bulge into this region during swallowing. The purpose of the trachea is to conduct air between the larynx and the lungs.

## The Bronchi

At its inferior end, the trachea divides into two primary, or main stem, *bronchi* (BRONG-ki) which enter the lungs. The right bronchus is considerably larger in diameter than the left and extends downward in a more vertical direction. Therefore, if a foreign body is inhaled, it is likely to enter the right lung. Each bronchus enters the lung at a notch or depression called the *hilus* (HI-lus) or *hilum* (HI-lum). Blood vessels and nerves also connect with the lung in this region.

### THE LINING OF THE AIR PASSAGEWAYS

The bronchi and other conducting passageways of the respiratory tract are lined with a special type of epithelium (Fig. 16-4). Basically, it is simple columnar epithelium, but the cells are arranged in such a way that they appear stratified. The tissue is thus described as *pseudostratified* meaning "falsely stratified." These epithelial cells have cilia to filter out impurities and to create movement of fluids within the conducting tubes. The cilia beat to drive impurities toward the throat where they can be eliminated by coughing, sneezing, or blowing the nose.

## The Lungs

The *lungs* are the organs in which the diffusion of gases takes place through the extremely thin and delicate lung tissues. The two lungs, set side by side in the thoracic (chest) cavity, are constructed in the following manner:

Each primary bronchus enters the lung at the hilus and immediately subdivides. The right bronchus divides into three secondary bronchi, each of which enters one of the three lobes of the right lung. The left bronchus gives rise to two secondary bronchi, which enter the two lobes of the left lung. Because the subdivisions of the bronchi resemble the branches of a tree, they have been given the common name *bronchial tree*. The bronchi subdivide again and again becoming progressively smaller as they branch through lung tissue. The smallest of these conducting tubes are called *bronchioles* (BRONG-ke-oles). The bronchi contain small bits of cartilage, which give firmness to the walls and serve to hold the passageways open so that air can pass in and out easily. However, as the bronchi become smaller, the cartilage decreases in amount. In the bronchioles there is no cartilage at all; what remains is mostly smooth muscle, which is under the control of the autonomic (involuntary) nervous system.

At the end of the *terminal bronchioles,* the smallest subdivisions of the bronchial tree, are the clusters of tiny air sacs in which most gas exchange takes place. These sacs are known as *alveoli* (al-VE-o-li). The wall of each alveolus is made of a single-cell layer of squamous (flat) epithelium. This very thin wall provides easy passage for the gases entering and leaving the blood as it circulates through the millions of tiny capillaries covering the alveoli. Certain cells in the alveolar wall produce *surfactant* (sur-FAK-tant), a substance that prevents the alveoli from collapsing by reducing the surface tension ("pull") of the fluids that line them. There are millions of alveoli in the human lung. The resulting surface area in contact with gases approximates 60 square meters,

### Figure 16-4

Microscopic view of the ciliated epithelium that lines the respiratory passageways (Cormack DH: Essential Histology, plate 4-1D. Philadelphia, JB Lippincott, 1993)

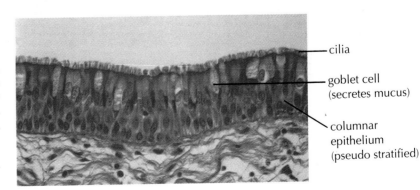

cilia

goblet cell (secretes mucus)

columnar epithelium (pseudo stratified)

about three times as much lung tissue as is necessary for life. Because of the many air spaces, the lung is light in weight; normally a piece of lung tissue dropped into a glass of water will float. Figure 16-5 shows a microscopic view of lung tissue.

The pulmonary circuit brings blood to and from the lungs. In the lungs the blood passes through the capillaries around the alveoli, where the gas exchange takes place (see Fig. 16-7).

## ▶ The Lung Cavities

The lungs occupy a considerable portion of the thoracic cavity, which is separated from the abdominal cavity by the muscular partition known as the *diaphragm.* Each lung is covered by a continuous doubled sac, known as the *pleura.* Each layer of the pleura is named according to its location. The portion of the pleura that is attached to the chest wall is the *parietal pleura,* and the portion that is attached to the surface of the lung is called the *visceral pleura.* Each closed sac completely surrounds the lung, except in the place where the bronchus and blood vessels enter the lung, known as the *root* of the lung. Between the two layers of the pleura is the *pleural space,* in which there is a thin film of fluid that lubricates the membranes. The effect is the same as two pieces of glass joined by a film of water: that is, they slide easily on each other but strongly resist separation. Thus, the lungs are able to move and enlarge effortlessly in response to changes in the thoracic volume that occur during breathing.

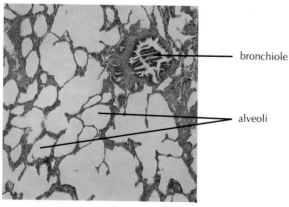

**Figure 16-5**

Lung tissue as seen through a microscope. (Courtesy of Dana Morse Bittus and B. J. Cohen)

The region between the lungs, the *mediastinum* (me-de-as-TI-num), contains the heart, great blood vessels, esophagus, trachea, and lymph nodes.

## ▶ Physiology of Respiration

### Pulmonary Ventilation

Ventilation is the movement of air into and out of the lungs, as in breathing. There are two phases of ventilation (Fig. 16-6):

1. *Inhalation* is the drawing of air into the lungs.
2. *Exhalation* is the expulsion of air from the lungs.

In *inhalation,* the active phase of breathing, the respiratory muscles contract to enlarge the thoracic cavity. The diaphragm is a strong, dome-shaped muscle attached to the body wall around the base of the rib cage. The contraction and flattening of the diaphragm cause a piston-like downward motion that results in an increase in the vertical dimension of the chest. During quiet breathing, the movement of the diaphragm accounts for most of the increase in thoracic volume. The muscles between the ribs (intercostal muscles) also participate. During exertion, the rib cage is moved further up and out by contraction of muscles in the neck and chest wall.

As the thoracic cavity increases in size, gas pressure within the cavity decreases. When the pressure drops to slightly below atmospheric pressure, air is drawn into the lungs, as by suction.

In *exhalation,* the passive phase of breathing, the muscles of respiration relax, allowing the ribs and diaphragm to return to their original positions. The tissues of the lung are elastic and recoil during exhalation. During forced exhalation, the internal intercostal muscles and the muscles of the abdominal wall contract, pulling the bottom of the rib cage in and down, pushing the abdominal viscera upwards against the relaxed diaphragm.

### Air Movement

Air enters the respiratory passages and flows through the ever-dividing tubes of the bronchial tree. As the air traverses this passage, it moves more and more slowly through the great number of bronchial tubes until there is virtually no forward flow as it reaches the alveoli. Here the air moves by diffusion, which soon equalizes any differences in the amounts of gases present. Each breath causes relatively little change in the gas composition of the

bronchiole

alveoli

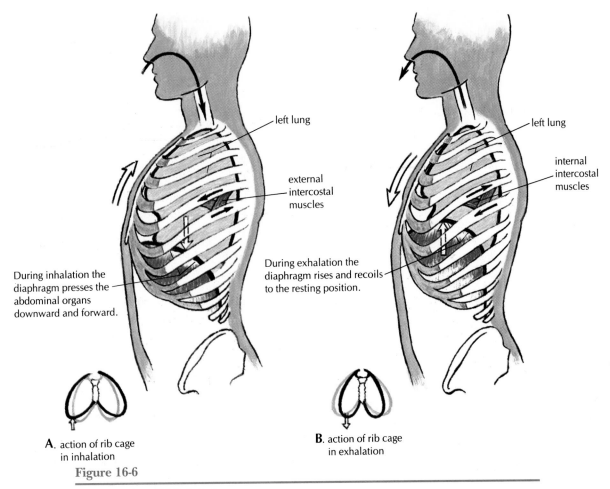

**A.** action of rib cage in inhalation

left lung

external intercostal muscles

During inhalation the diaphragm presses the abdominal organs downward and forward.

**B.** action of rib cage in exhalation

left lung

internal intercostal muscles

During exhalation the diaphragm rises and recoils to the resting position.

**Figure 16-6**

**(A)** Inhalation. **(B)** Exhalation.

alveoli, but normal continuous breathing ensures the presence of adequate oxygen and the removal of carbon dioxide.

Table 16-1 gives the definitions of and average values for some of the breathing volumes and capacities that are important in any evaluation of respiratory function. A lung *capacity* is a sum of volumes.

### Gas Exchanges

The barrier that separates the air in the alveolus from the blood in the capillary is very thin and moist, ideally suited for the exchange of gases by diffusion. Normally, inspired air contains about 21% oxygen and 0.04% carbon dioxide; expired air has only 16% oxygen and 3.5% carbon dioxide. A two-way diffusion takes place through the walls of the alveoli.

Recall that *diffusion* refers to the movement of molecules from an area in which they are in higher concentration to an area where they are in lower concentration. Blood entering the lung capillaries after returning from the tissues is relatively low in oxygen. Therefore, oxygen diffuses from the alveolus, where its concentration is higher, into the blood. Again, based on relative concentration, carbon dioxide diffuses out of the blood into the air of the alveolus (Fig. 16-7).

### Gas Transport

Almost all the oxygen that diffuses into the capillary blood in the lungs is bound to the *hemoglobin* of the red blood cells. A very small percentage is carried in solution in the plasma. The hemoglobin molecule is a large protein with four small iron-

**TABLE 16-1**
**Lung Volumes and Capacities**

| Volume | Definition | Average Value |
|--------|-----------|---------------|
| Tidal volume | The amount of air moved into or out of the lungs in quiet, relaxed breathing | 500 mL |
| Residual volume | The volume of air that remains in the lungs after maximum ~~inhalation~~ exhalation | 1200 mL |
| Vital capacity | The volume of air that can be expelled from the lungs by maximum exhalation following maximum inhalation | 4800 mL |
| Total lung capacity | The total volume of air that can be contained in the lungs after maximum inhalation | 6000 mL |
| Functional residual capacity | The amount of air remaining in the lungs after normal exhalation | 2400 mL |

containing "heme" regions. The oxygen is bound to these heme portions. Arterial blood (in systemic arteries and pulmonary veins) is 97% saturated with oxygen, whereas venous blood (in systemic veins and pulmonary arteries) is about 70% saturated with oxygen. This 27% difference represents the oxygen that has been taken up by the cells. Note that in respira-

## The Air We Breathe

The air we breathe comes under close inspection because of our increasing awareness of air pollution, both indoors and out. The air in many populated areas is polluted by automobile exhaust and industrial smoke. Air in agricultural areas may include smoke from burning fields and drift of aerial pesticide applications far from the intended target. Lumber operations may burn scrap. Wind carries dust and pollen everywhere. Much work is being carried out to decrease these sources of airborne pollution.

Air inside homes or offices often carries a less visible burden of pollution. Many products give off invisible gases that tend to accumulate in modern buildings, which usually are tightly sealed. Unless adequate amounts of clean air are supplied, the occupants often complain of vague illnesses. Attention needs to be given to processing the air not only for temperature, but also for humidity. Most importantly, the air must be filtered so that undesirable pollutants are removed.

tory studies, gas concentrations are expressed as pressure in millimeters mercury (mmHg), just as blood pressure is reported. Because air is a mixture of gases, each gas exerts a partial pressure that is abbreviated as $P_{O_2}$ and $P_{CO_2}$ for oxygen and carbon dioxide respectively.

To enter the cells, oxygen must separate from hemoglobin. Normally the bond between oxygen and hemoglobin is easily broken, with oxygen being released as blood travels into areas in which the oxygen concentration is relatively low.

The carbon dioxide produced in the tissues is transported to the lungs in three ways:

1. About 10% is dissolved in the plasma and the fluid in red blood cells.
2. About 20% is combined with the protein portion of hemoglobin and plasma proteins.
3. About 70% is transported as an ion, known as a **bicarbonate ion,** which is formed when carbon dioxide dissolves in blood fluids.

The bicarbonate ion is formed slowly in the plasma but much more rapidly inside the red blood cells, where an enzyme called **carbonic anhydrase** increases the speed of the reaction. The bicarbonate formed in the red blood cells moves to the plasma and then is carried to the lungs. Here, the process is reversed as bicarbonate reenters the red blood cells and releases carbon dioxide for diffusion into the alveoli and exhalation.

Carbon dioxide is important in regulating the pH (acid–base balance) of the blood. As a bicarbonate ion is formed from carbon dioxide in the

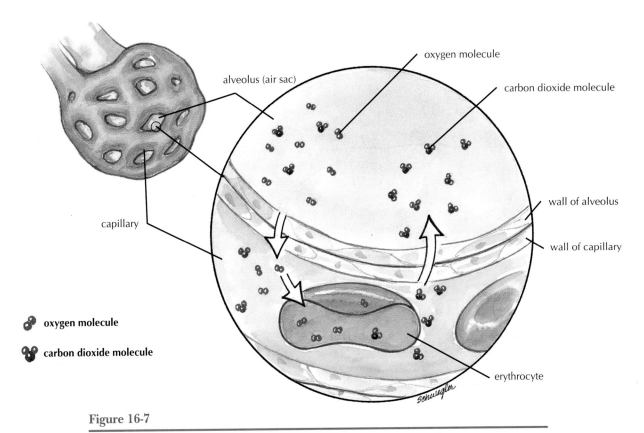

oxygen molecule

carbon dioxide molecule

alveolus (air sac)

capillary

wall of alveolus

wall of capillary

erythrocyte

**oxygen molecule**

**carbon dioxide molecule**

**Figure 16-7**

Diagram showing the diffusion of gas molecules in the lungs.

plasma, a hydrogen ion (H+) is also produced. Therefore, the blood becomes more acidic as the amount of carbon dioxide in the blood increases. The exhalation of carbon dioxide shifts the pH of the blood more toward the alkaline (basic) range. The bicarbonate ion is also an important buffer in the blood, acting chemically to help keep the pH of body fluids at a steady pH of 7.4.

### Regulation of Respiration

Regulation of respiration is a complex process that must keep pace with moment-to-moment changes in cellular oxygen requirements and carbon dioxide production. Regulation depends primarily on the respiratory control centers located in the medulla and pons of the brain stem. The main control center in the medulla sets the basic pattern of respiration. This can be modified by centers in the pons. Breathing is regulated so that levels of oxygen, carbon dioxide, and acid are kept within normal limits.

From the respiratory center in the medulla, motor nerve fibers extend into the spinal cord. From the cervical (neck) part of the cord, these nerve fibers continue through the *phrenic* (FREN-ik) *nerve* to the diaphragm. The diaphragm and the other muscles of respiration are voluntary in the sense that they can be regulated by messages from the higher brain centers, notably the cortex. It is possible for a person to deliberately breathe more rapidly or more slowly or to hold his breath and not breathe at all for a time. Usually we breathe without thinking about it, while the respiratory centers in the medulla and pons do the controlling.

Of vital importance in the control of respiration are *chemoreceptors* (ke-mo-re-SEP-tors), which, like the receptors for taste and smell, are sensitive to chemicals dissolved in body fluids. The receptors that regulate respiration are found in structures called the *carotid* and *aortic bodies,* as well as outside the medulla of the brain stem. The carotid bodies are located near the bifurcation (forking) of the common carotid arteries in the neck, whereas the aortic bodies are located in the aortic arch. These bodies contain many small blood vessels and sensory neurons, which are sensitive to increases in carbon dioxide and acidity (H+) and to decreases in oxygen supply.

Changes in the composition of the blood generate nerve impulses that are then sent to the brain. Because there is usually an ample reserve of oxygen in the blood, carbon dioxide has the most immediate effect in regulating respiration. When carbon dioxide levels increase, breathing must be increased to blow off the excess. It is only when oxygen levels fall considerably that this gas becomes a controlling factor. The receptor cells outside the medulla respond to the concentration of hydrogen ion in cerebrospinal fluid (CSF) as governed by the amount of carbon dioxide that is dissolved in the blood.

### Respiratory Rates

Normal rates of breathing vary from 12 to 20 times per minute for adults. In children, rates may vary from 20 to 40 times per minute, depending on age and size. In infants, the respiratory rate may be more than 40 times per minute. To determine the respiratory rate, one counts the subject's breathing for at least 30 seconds, observing in such a way that the person is unaware that a count is being made.

## SUMMARY

I. **Respiration**—supply of oxygen to the cells and removal of carbon dioxide
  A. Pulmonary ventilation—the movement of air into and out of the lungs
  B. Diffusion of gases
  C. Transport of oxygen and carbon dioxide
II. **Respiratory system**
  A. Nasal cavities—filter, warm, and moisten air
  B. Pharynx (throat)—carries air into respiratory tract and food into digestive tract
  C. Larynx (voice box)—contains vocal cords
    1. Epiglottis—covers larynx on swallowing to help prevent food from entering
  D. Trachea (windpipe)
  E. Bronchi—branches of trachea that enter lungs and then subdivide
    1. Bronchioles—smallest subdivisions
  F. Lungs
    1. Pleura—membrane that encloses lung and lines chest wall
    2. Mediastinum—space and organs between lungs
  G. Alveoli—tiny air sacs in lungs
III. **Physiology of respiration**
  A. Pulmonary ventilation
    1. Inhalation—drawing of air into lungs
    2. Exhalation—expulsion of air from lungs
    3. Lung volumes—used to evaluate respiratory function (*e.g.*, vital capacity, total lung capacity, etc.)
  B. Gas exchange
    1. Diffusion of gases from area of higher concentration to area of lower concentration
    2. In lungs—oxygen enters blood and carbon dioxide leaves
    3. In tissues—oxygen leaves blood and carbon dioxide enters
  C. Gas transport
    1. Oxygen—almost totally bound to heme portion of hemoglobin in red blood cells; separates from hemoglobin when oxygen concentration is low (in tissues)
    2. Carbon dioxide—most carried as bicarbonate ion; regulates pH of blood
  D. Regulation of respiration
    1. Controlled by centers in medulla and pons
    2. Chemoreceptors — detect changes in blood composition

## QUESTIONS FOR STUDY AND REVIEW

1. What is the definition of *respiration* and what are its three phases?
2. Trace the pathway of air from the outside into the blood.
3. What is the purpose of the cilia on the cells that line the respiratory passageways?
4. What are the advantages of breathing through the nose?
5. Describe the lung cavities and pleura.
6. What muscles are used for inhalation? forceful exhalation?

7. How is oxygen transported in the blood? What causes its release to the tissues?
8. How is carbon dioxide transported in the blood?
9. How does the pH of the blood change as the level of carbon dioxide increases? decreases?
10. What are chemoreceptors and how do they function to regulate breathing?
11. Define five volumes or capacities used to measure breathing.

# CHAPTER 17

# Digestion

## Selected Key Terms

The following terms are defined in the Glossary:

absorption

bile

chyle

chyme

colon

defecation

deglutition

digestion

duodenum

emulsify

enzyme

esophagus

hydrolysis

lacteal

mastication

peristalsis

peritoneum

saliva

sphincter

villi

## Behavioral Objectives

After careful study of this chapter, you should be able to:

- Name the two main functions of the digestive system
- Describe the four layers of the digestive tract wall
- Describe the layers of the peritoneum
- Name and describe the organs of the digestive tract
- Name and describe the accessory organs of digestion
- List the functions of each organ involved in digestion
- Explain the role of enzymes in digestion and give examples of enzymes
- Name the digestion products of fats, proteins, and carbohydrates
- Define *absorption*
- Define *villi* and state how villi function in absorption
- Describe how bile functions in digestion
- Name the ducts that carry bile into the digestive tract
- List the main functions of the liver
- Explain the use of feedback in regulating digestion and give several examples

Memmler, RL, Cohen, BJ, Wood, DL. *STRUCTURE AND FUNCTION OF THE HUMAN BODY*, 6/e,
© 1996 Lippincott-Raven Publishers

# ▶ What the Digestive System Does

Every body cell needs a constant supply of nutrients to provide energy and building blocks for the manufacture of body substances. Food as we take it in, however, is too large to enter the cells. It must first be broken down into particles small enough to pass through the cell membrane. This process is known as *digestion.* After digestion, food must be carried to the cells in every part of the body by the circulation. The transfer of food into the circulation is called *absorption.* Digestion and absorption are the two chief functions of the digestive system.

For our purposes the digestive system may be divided into two groups of organs:

1. The *digestive tract,* a continuous passageway beginning at the mouth, where food is taken in, and terminating at the anus, where the solid waste products of digestion are expelled from the body.
2. The *accessory organs,* which are necessary for the digestive process but are not a direct part of the digestive tract. They release substances into the digestive tract through ducts.

## The Wall of the Digestive Tract

Although modified for specific tasks in different organs, the wall of the digestive tract, from the esophagus to the anus, is similar in structure throughout. Follow the diagram of the small intestine in Figure 17-1 as we describe the layers of this wall from the innermost to the outermost surface. First is the *mucous membrane,* so called because its epithelial layer contains many mucus-secreting cells. The type of epithelium is simple columnar. In Figure 17-2, a microscopic view of the small intestine, the mucus-secreting cells appear as clear areas mixed in with the epithelial cells.

The layer of connective tissue beneath this, the *submucosa,* contains blood vessels and some of the nerves that help regulate digestive activity. Next are two layers of *smooth muscle.* The inner layer has circular fibers, and the outer layer has longitudinal fibers. The alternate contractions of these muscles create the wavelike movement that propels food through the digestive tract and mixes it with digestive juices. This movement is called *peristalsis* (per-ih-STAL-sis).

The outermost layer of the wall consists of fibrous connective tissue. Most of the abdominal organs have an additional layer of *serous membrane* that is part of the peritoneum (per-i-to-NE-um).

## The Peritoneum

The abdominal cavity is lined with a thin, shiny serous membrane that also covers most of the abdominal organs (Fig. 17-3). The portion of this membrane that lines the abdomen is called the *parietal peritoneum;* that covering the organs is called the *visceral peritoneum.* In addition to these single-layered portions of the peritoneum there are a number of double-layered structures that carry blood vessels, lymph vessels, and nerves, and sometimes act as ligaments supporting the organs.

The *mesentery* (MES-en-ter-e) is a double-layered portion of the peritoneum shaped somewhat like a fan. The handle portion is attached to the back wall, and the expanded long edge is attached to the small intestine. Between the two layers of membrane that form the mesentery are the blood vessels, lymphatic vessels, and nerves that supply the intestine. The section of the peritoneum that extends from the colon to the back wall is the *mesocolon* (mes-o-KO-lon).

A large double layer of the peritoneum containing much fat hangs like an apron over the front of the intestine. This *greater omentum* (o-MEN-tum) extends from the lower border of the stomach into the pelvic part of the abdomen and then loops back up to the transverse colon. There is also a smaller membrane, called the *lesser omentum,* that extends between the stomach and the liver.

## The Digestive Tract

As we study the organs of the digestive system, locate each in Figure 17-4.

The digestive tract is a muscular tube extending through the body. It is composed of several parts: the *mouth, pharynx, esophagus, stomach, small intestine,* and *large intestine.* The digestive tract is sometimes called the *alimentary tract,* derived from a Latin word that means "food." It is more commonly referred to as the *gastrointestinal (GI) tract* because of the major importance of the stomach and intestine in the process of digestion.

## The Mouth

The mouth, also called the *oral cavity,* is where a substance begins its travels through the digestive tract (Fig. 17-5). The mouth has three digestive functions:

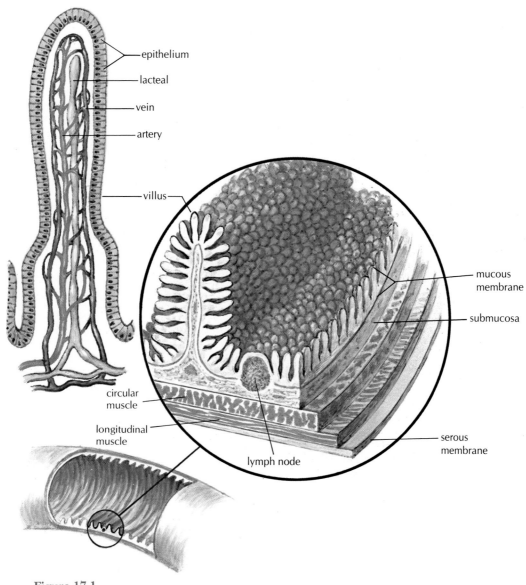

**Figure 17-1**

Diagram of the wall of the small intestine showing numerous villi. At the left is an enlarged drawing of a single villus.

1. To receive food, a process called *ingestion*
2. To prepare food for digestion
3. To begin the digestion of starch

Into this space projects a muscular organ, the tongue, which is used for chewing and swallowing, and is one of the principal organs of speech. The tongue has on its surface a number of special organs, called *taste buds,* by means of which taste sensations (bitter, sweet, sour, or salty) can be differentiated.

The oral cavity also contains the teeth. A child between 2 and 6 years of age has 20 teeth; an adult with a complete set of teeth has 32. Among these, the cutting teeth, or *incisors,* occupy the front part of the oral cavity, whereas the larger grinding teeth, the *molars,* are in the back.

### DECIDUOUS, OR BABY, TEETH
The first eight *deciduous* (de-SID-u-us) teeth to appear through the gums are the incisors. Later the

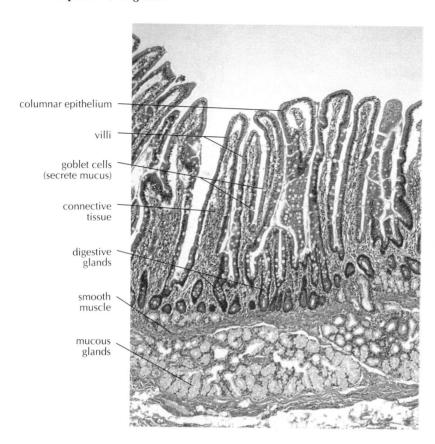

columnar epithelium

villi

goblet cells
(secrete mucus)

connective
tissue

digestive
glands

smooth
muscle

mucous
glands

**Figure 17-2**

Microscopic view of the duodenum (Cormack DH: Essential
Histology, plate 13-48. Philadelphia, JB Lippincott, 1993)

*canines* (eyeteeth) and molars appear. Usually, the 20
baby teeth have all appeared by the time a child has
reached the age of 2 or 2½ years. During the first 2
years the permanent teeth develop within the jawbones from buds that are present at birth. The first
permanent tooth to appear is the important 6-year
molar. This permanent tooth comes in before the
baby incisors are lost. Because decay and infection
of adjacent deciduous molars may spread to and involve new, permanent teeth, deciduous teeth need
proper care.

### PERMANENT TEETH

As a child grows, the jawbones grow, making
space for additional teeth. After the 6-year molars
have appeared, the baby incisors loosen and are replaced by permanent incisors. Next, the baby canines (cuspids) are replaced by permanent canines,
and finally, the baby molars are replaced by the bi-
cuspids (premolars) of the permanent teeth.

Now the larger jawbones are ready for the appearance of the 12-year, or second, permanent
molar teeth. During or after the late teens, the third
molars, or so-called *wisdom teeth,* may appear. In

some cases the jaw is not large enough for these
teeth or there are other abnormalities, so that the
third molars may have to be removed. Figure 17-6
shows the parts of a molar.

### THE SALIVARY GLANDS

While food is in the mouth, it is mixed with
*saliva,* one purpose of which is to moisten the food
and facilitate the processes of chewing, or *mastication* (mas-tih-KA-shun), and swallowing, or *deglutition* (deg-lu-TISH-un). Saliva also helps keep the
teeth and mouth clean and reduce bacterial growth.

This watery mixture contains mucus and an
enzyme called *salivary amylase* (AM-ih-laze), which
begins the digestive process by converting starch
to sugar. It is manufactured mainly by three pairs
of glands that function as accessory organs:

1. The *parotid* (pah-ROT-id) *glands,* the largest of
   the group, are located below and in front of the
   ear.
2. The *submandibular* (sub-man-DIB-u-lar), or *sub-
   maxillary* (sub-MAK-sih-ler-e), *glands* are located
   near the body of the lower jaw.

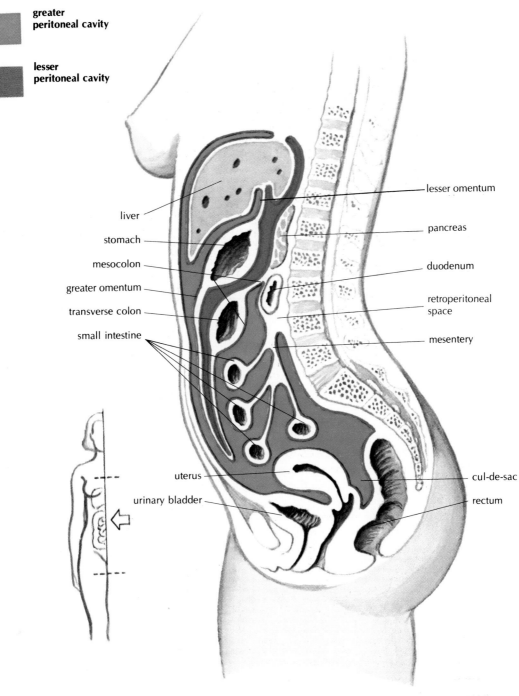

greater
peritoneal cavity

lesser
peritoneal cavity

lesser omentum

pancreas

duodenum

retroperitoneal
space

mesentery

cul-de-sac

rectum

liver

stomach

mesocolon

greater omentum

transverse colon

small intestine

uterus

urinary bladder

**Figure 17-3**

Diagram of the abdominal cavity showing the peritoneum.

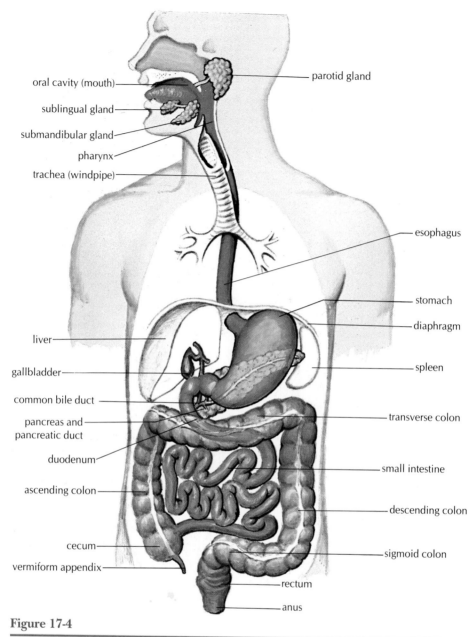

**Figure 17-4**

Digestive system.

3. The *sublingual* (sub-LING-gwal) *glands* are under the tongue.

All these glands empty by means of ducts into the oral cavity.

### The Pharynx and Esophagus

The *pharynx* (FAR-inks) is commonly referred to as the *throat*. The oral part of the pharynx is visible when you look into an open mouth and depress the tongue. The palatine tonsils may be seen at either side. The pharynx also extends upward to the nasal cavity and downward to the level of the larynx. The *soft palate* is tissue that forms the back of the roof of the oral cavity. From it hangs a soft, fleshy, V-shaped mass called the *uvula* (U-vu-lah).

In swallowing, a small portion of chewed food mixed with saliva, called a *bolus,* is pushed by the tongue into the pharynx. When the food reaches the pharynx, swallowing occurs rapidly by an involun-

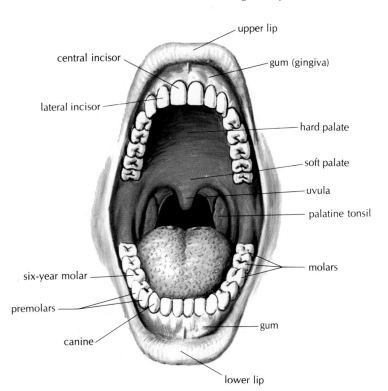

**Figure 17-5**

The mouth, showing the teeth and tonsils.

tary reflex action. At the same time, the soft palate and uvula are raised to prevent food and liquid from entering the nasal cavity, and the tongue is raised to seal the back of the oral cavity. The entrance of the trachea is guarded during swallowing by a leaf-shaped cartilage, the *epiglottis,* which covers the opening of the larynx. The swallowed food is then moved by peristalsis into the *esophagus* (eh-SOF-ah-gus), a muscular tube about 25 cm (10 inches) long that carries food into the stomach. No additional digestion occurs in the esophagus.

### The Stomach

The stomach is an expanded J-shaped organ in the upper left region of the abdominal cavity (Fig. 17-7). In addition to the two muscle layers already described, it has a third, inner oblique (angled) layer that aids in grinding food and mixing it with digestive juices. The left-facing arch of the stomach is the *greater curvature,* whereas the right surface forms the *lesser curvature.* Each end of the stomach is guarded by a muscular ring, or *sphincter* (SFINK-ter), that permits the passage of substances in only one direction.

Between the esophagus and the stomach is the *lower esophageal sphincter (LES).* This valve has also

been called the *cardiac sphincter* because it separates the esophagus from the region of the stomach that is close to the heart. We are sometimes aware of the existence of this sphincter; sometimes it does not relax as it should, producing a feeling of being unable to swallow past that point. Between the distal, or far, end of the stomach and the small intestine is the *pyloric* (pi-LOR-ik) *sphincter.* The region of the stomach leading into this sphincter, the *pylorus* (pi-LOR-us), is important in regulating how rapidly food moves into the small intestine.

The stomach serves as a storage pouch, digestive organ, and churn. When the stomach is empty, the lining forms many folds called *rugae* (RU-je). These folds disappear as the stomach expands. (It may be stretched to hold one half of a gallon of food and liquid.) Special cells in the lining of the stomach secrete substances that mix together to form *gastric juice,* the two main components of which are:

1. Hydrochloric acid (HCl), a strong acid that softens the connective tissue in meat and destroys foreign organisms.
2. Pepsin, a protein-digesting enzyme. This enzyme is produced in an inactive form and is activated

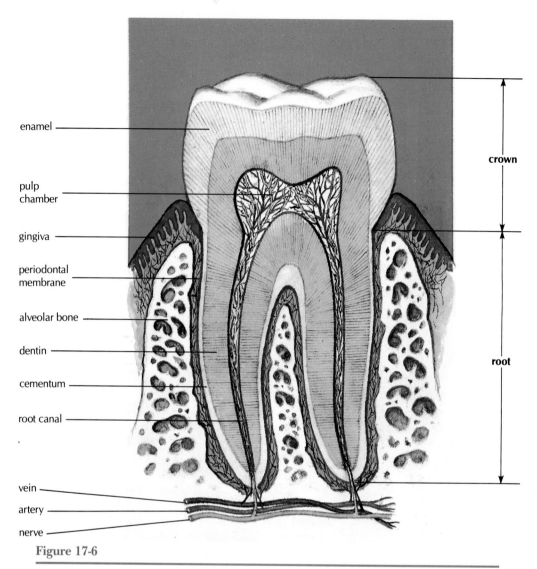

enamel

pulp chamber

gingiva

periodontal membrane

alveolar bone

dentin

cementum

root canal

vein

artery

nerve

**crown**

**root**

**Figure 17-6**

A molar tooth. (Cohen B: Medical Terminology: An Illustrated Guide, 2nd ed, p. 208. Philadelphia, JB Lippincott, 1994)

only when food enters the stomach and HCl is produced.

The semi-liquid mixture of gastric juice and food that leaves the stomach to enter the small intestine is called *chyme* (kime).

### The Small Intestine

The small intestine is the longest part of the digestive tract. It is known as the small intestine because, although it is longer than the large intestine, it is smaller in diameter, with an average width of about 2.5 cm (1 inch). When relaxed to its full length, the small intestine is about 6 m (20 feet) long. The first 25 cm (10 inches) or so of the small intestine make up the *duodenum* (du-o-DE-num). Beyond the duodenum are two more divisions: the *jejunum* (je-JU-num), which forms the next two fifths of the small intestine, and the *ileum* (IL-e-um), which constitutes the remaining portion.

The wall of the duodenum contains glands that secrete large amounts of mucus to protect the small intestine from the strongly acid chyme entering from the stomach. Cells of the small intestine also

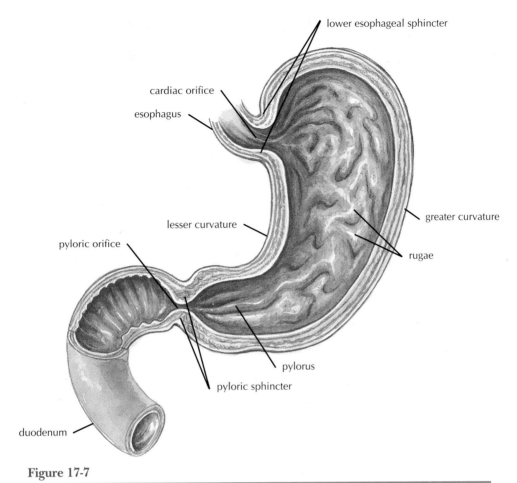

cardiac orifice

esophagus

lower esophageal sphincter

greater curvature

lesser curvature

rugae

pyloric orifice

pylorus

pyloric sphincter

duodenum

**Figure 17-7**

Longitudinal section of the stomach and a portion of the duodenum showing the interior.

secrete enzymes that digest proteins and carbohydrates. In addition, digestive juices from the liver and pancreas enter the small intestine through a small opening in the duodenum. Most of the digestive process takes place in the small intestine under the effects of these juices.

Most absorption of digested food also occurs through the walls of the small intestine. To increase the surface area of the organ for this purpose, the mucosa is formed into millions of tiny, finger-like projections, called *villi* (VIL-li) (see Fig. 17-1), which give the inner surface a velvety appearance. In addition, each epithelial cell has small projecting folds of the cell membrane known as *microvilli*. These create a remarkable increase in the total surface area available in the small intestine for the absorption of nutrients.

## The Large Intestine

Any material that cannot be digested as it passes through the digestive tract must be eliminated from the body. In addition, most of the water secreted into the digestive tract for proper digestion must be reabsorbed into the body to prevent dehydration. The storage and elimination of undigested waste and the reabsorption of water are the functions of the large intestine.

The large intestine is about 6.5 cm (2.5 inches) in diameter and about 1.5 m (5 feet) long. The outer longitudinal muscle fibers form three separate bands on the surface. These bands draw up the wall of the organ to give it its distinctive puckered appearance.

The large intestine begins in the lower right region of the abdomen. The first part is a small pouch

## Surface Area and Transport

Whenever materials pass from one system to another, they must travel through a cellular membrane. A major factor in how much transport can occur per unit of time is the total surface area of the membrane.

The problem of packing a large amount of surface into a small space is solved in the body by folding the membranes. We do the same thing. Imagine trying to store a bed sheet in the closet without folding it!

In the small intestine, where digested food must move into the bloodstream, there is folding of membranes down to the level of the single cells. The organ as a whole is coiled to fit its length into the abdominal cavity. The inner wall of the organ forms circular folds called *plicae circulares*. The projecting villi provide more surface area than a flat membrane would. Even further, the individual cells that line the small intestine have microvilli, tiny fingerlike folds of the cell membrane, that increase surface area tremendously.

Can you name other portions of the digestive tract or other systems that show this folding pattern and explain why it is there?

called the **cecum** (SE-kum). Between the ileum of the small intestine and the cecum is a sphincter, the **ileocecal** (il-e-o-SE-kal) **valve,** that prevents food from traveling backward into the small intestine. Attached to the cecum is a small, blind tube containing lymphoid tissue; it is called the **vermiform** (VER-mihform) **appendix** (*vermiform* means "wormlike"). Inflammation of this tissue as a result of infection or obstruction is **appendicitis.**

The second portion, the **ascending colon,** extends upward along the right side of the abdomen toward the liver. The large intestine then bends and extends across the abdomen, forming the **transverse colon.** At this point it bends sharply and extends downward on the left side of the abdomen into the pelvis, forming the **descending colon.** The lower part of the colon bends posteriorly in an S shape and continues downward as the **sigmoid colon.** The sigmoid colon empties into the **rectum,** which serves as a temporary storage area for indigestible or unabsorbable food residue (see Fig. 17-4). A narrow portion of the dis-

tal large intestine is called the **anal canal.** This leads to the outside of the body through an opening called the **anus** (A-nus).

Large quantities of mucus, but no enzymes, are secreted by the large intestine. At intervals, usually after meals, the involuntary muscles within the walls of the large intestine propel solid waste material, called *feces* or stool, toward the rectum. This material is then eliminated from the body by both voluntary and involuntary muscle actions, a process called **defecation** (def-e-KA-shun).

While the food residue is stored in the large intestine, bacteria that normally live in the colon act on it to produce vitamin K and some of the B-complex vitamins. Systemic antibiotic therapy may destroy these bacteria and others living in the large intestine, causing undesirable side effects.

## ▶ The Accessory Structures

### The Liver

The liver, often referred to by the word root *hepat,* is the largest glandular organ of the body (Fig. 17-8). It is located in the upper right portion of the abdominal cavity under the dome of the diaphragm. The lower edge of a normal-sized liver is level with the lower margin of the ribs. The human liver is the same reddish brown color as the animal liver seen in the supermarket. It has a large right lobe and a smaller left lobe; the right lobe includes two inferior smaller lobes. The liver is supplied with blood through two vessels: the portal vein and the hepatic artery. These vessels deliver about 1½ quarts of blood to the liver every minute. The hepatic artery carries oxygenated blood, whereas the portal system of veins carries blood that is rich in the end products of digestion. This most remarkable organ has so many functions that only some of its major activities can be listed here:

1. The storage of glucose (simple sugar) in the form of **glycogen,** an animal starch. When the blood sugar level falls below normal, liver cells convert glycogen to glucose and release it into the bloodstream; this serves to restore the normal concentration of blood sugar.
2. The formation of blood plasma proteins, such as albumin, globulins, and clotting factors
3. The synthesis of **urea** (u-RE-ah), a waste product of protein metabolism. Urea is released into the

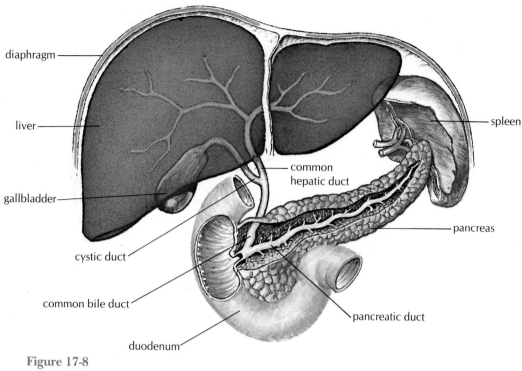

diaphragm

liver

gallbladder

cystic duct

common bile duct

duodenum

spleen

common
hepatic duct

pancreas

pancreatic duct

**Figure 17-8**

blood and transported to the kidneys for elimination.

4. The modification of fats so they can be used more efficiently by cells all over the body
5. The manufacture of bile
6. The destruction of old red blood cells. The pigment released from these cells in both the liver and the spleen is eliminated in the bile. This pigment (bilirubin) gives the stool its characteristic dark color.
7. The ***detoxification*** (de-tok-sih-fih-KA-shun) (removal of the poisonous properties) of harmful substances such as alcohol and certain drugs
8. The storage of some vitamins and iron.

The main digestive function of the liver is the production of ***bile.*** The salts contained in bile act like a detergent to ***emulsify*** fat, that is, to break up fat into small droplets that can be acted on more effectively by digestive enzymes. Bile also aids in the absorption of fat from the small intestine. Bile leaves the lobes of the liver by two ducts that merge to form the ***common hepatic duct.*** After collecting bile from the gallbladder, this duct, now called the ***common bile duct,*** delivers bile into the duodenum. These and the other accessory ducts are shown on Figure 17-8.

### The Gallbladder

The gallbladder is a muscular sac on the inferior surface of the liver that serves as a storage pouch for bile. Although the liver may manufacture bile continuously, the body is likely to need it only a few times a day. Consequently, bile from the liver flows into the hepatic ducts and then up through the ***cystic*** (SIS-tik) ***duct*** connected with the gallbladder. When chyme enters the duodenum, the gallbladder contracts, squeezing bile through the cystic duct and into the common bile duct leading to the duodenum.

### The Pancreas

The pancreas is a long gland that extends from the duodenum to the spleen. The pancreas produces enzymes that digest fats, proteins, carbohydrates, and nucleic acids. The protein-digesting enzymes are produced in inactive forms, which must be converted to active forms in the small intestine by other enzymes. The pancreas also produces large amounts of alkaline fluid, which neutralizes the chyme in the small intestine, thus protecting the lining of the digestive tract. These juices collect in a main duct that joins the common bile duct or empties into the duodenum near the common bile duct.

t persons also have an additional smaller duct that opens into the duodenum.

The pancreas also functions as an endocrine gland, producing the hormones insulin and glucagon that regulate sugar metabolism. These secretions of the islet cells are released directly into the blood.

## ▶ The Process of Digestion

Although the different organs of the digestive tract are specialized for digesting different types of food, the basic chemical process of digestion is the same for fats, proteins, and carbohydrates. In every case this process requires enzymes. Enzymes are proteins that speed the rate of chemical reactions but are not themselves changed or used up in these reactions. An enzyme is highly specific; that is, it acts only in a certain type of reaction involving a certain type of food molecule. For example, the carbohydrate-digesting enzyme amylase only splits starch into the disaccharide (double sugar) maltose. Another enzyme is required to split maltose into two molecules of the monosaccharide (simple sugar) glucose. Other enzymes split fats into their building blocks, glycerol and fatty acids, and still others split proteins into their building blocks, amino acids.

Because water is added to these molecules as they are split by enzymes, the process is called *hydrolysis* (hi-DROL-ih-sis), which means "splitting by means of water." You can now understand why a large amount of water is needed in digestion. Water is needed not only to produce digestive juices and to dilute food so that it can move more easily through the digestive tract, but also is used in the chemical process of digestion itself.

Let us see what happens to a mass of food from the time it is taken into the mouth to the moment that it is ready to be absorbed.

In the mouth the food is chewed and mixed with saliva, softening it so that it can be swallowed easily. Salivary amylase initiates the process of digestion by changing some of the starches into sugars.

When the food reaches the stomach, it is acted upon by gastric juice, which contains hydrochloric acid and enzymes. The hydrochloric acid has the important function of liquefying the food. In addition, it activates the enzyme pepsin, which is secreted by the cells of the gastric lining in an inactive form. Once activated by hydrochloric acid, pepsin works to digest protein; nearly every type of protein in the diet is first digested by this enzyme. Two other enzymes are also secreted by the stomach, but they are of no importance in adults. The food, gastric juice, and mucus, the latter of which is also secreted by cells of the gastric lining, are mixed to form the semi-liquid substance chyme. Chyme is moved from the stomach to the small intestine for further digestion.

In the duodenum the chyme is mixed with the greenish yellow bile delivered from the liver and the gallbladder through the common bile duct. Rather than containing enzymes, the bile contains salts that split (emulsify) fats into smaller particles to allow the powerful secretion from the pancreas to act on them most efficiently. The pancreatic juice contains a number of enzymes, including the following:

1. *Lipase.* Following the physical division of fats into tiny particles by the action of bile, the powerful pancreatic lipase does almost all the digesting of fats. In this process fats are usually broken down into two simpler compounds, glycerol (glycerine) and fatty acids, which are more readily absorbable. If pancreatic lipase is absent, fats are expelled with the feces in undigested form.
2. *Amylase.* This enzyme changes starch to sugar.
3. *Trypsin* (TRIP-sin). This enzyme splits proteins into amino acids, which are small enough to enter the bloodstream.

The intestinal juice contains a number of enzymes, including three that act on complex sugars to transform them into the simpler form in which they are absorbed. These are *maltase, sucrase,* and *lactase.* It must be emphasized that most of the chemical changes in foods occur in the intestinal tract because of the pancreatic juice, which could probably adequately digest all foods even if no other digestive juice were produced. When pancreatic juice is absent, serious digestive disturbances always occur.

Table 17-1 summarizes the main substances used in digestion. Note that, except for HCl, sodium bicarbonate, and bile salts, all the substances listed are enzymes.

### Absorption
The means by which the digested food reaches the blood is known as *absorption.* Most absorption takes place through the mucosa of the small intestine by means of the villi (see Fig. 17-1). Within each villus are a small artery and a small vein bridged with

**TABLE 17-1**
**Digestive Juices Produced by the Organs of the Digestive Tract and the Accessory Organs**

| Organ | Main Digestive Juices Secreted | Action |
|---|---|---|
| Salivary glands | Salivary amylase* | Begins starch digestion |
| Stomach | Hydrochloric acid (HCl) | Breaks down proteins |
| | Pepsin* | Begins protein digestion |
| Small intestine | Peptidases* | Digest proteins to amino acids |
| | Lactase, maltase, sucrase* | Digest disaccharides to monosaccharides |
| Pancreas | Sodium bicarbonate | Neutralizes HCl |
| | Amylase* | Digests starch |
| | Trypsin* | Digests protein to amino acids |
| | Lipases* | Digest fats to fatty acids and glycerol |
| | Nucleases* | Digest nucleic acids |
| Liver | Bile salts | Emulsify fats |

*Enzymes

capillaries. Simple sugars, amino acids, some simple fatty acids, and water are absorbed into the blood through the capillary walls in the villi. From here they pass by way of the portal system to the liver, to be stored or released and used as needed.

Most fats have an alternative method of reaching the blood. Instead of entering the blood capillaries, they are absorbed by lymphatic capillaries in the villi that are called *lacteals*. The fat droplets give the lymph a milky appearance (the word *lacteal* means "like milk"). The mixture of lymph and fat globules that is drained from the small intestine after a quantity of fat has been digested is called *chyle* (kile). It circulates in the lymph and eventually enters the bloodstream near the heart.

Minerals and vitamins taken in with food are also absorbed from the small intestine. The minerals and some of the vitamins dissolve in water and are absorbed directly into the blood. Other vitamins are incorporated in fats and are absorbed along with the fats. Some B vitamins and vitamin K are produced by the action of bacteria in the colon and are absorbed from the large intestine.

### Control of Digestion

As food moves through the digestive tract, its rate of movement and the activity of each organ it passes through must be carefully regulated. If food moves too slowly or digestive secretions are inadequate, the body will not get enough nourishment. If food moves too rapidly or excess secretions are produced, digestion may be incomplete or the lining of the digestive tract may be damaged.

There are two types of control over digestion: nervous and hormonal. Both illustrate the principles of feedback control. The nerves that control digestive activity are located in the submucosa and between the muscle layers of the organ walls. Instructions for action come from the autonomic (visceral) nervous system. In general, parasympathetic stimulation increases activity and sympathetic stimulation decreases activity. Excess sympathetic stimulation, as in stress, can block the movement of food through the digestive tract and inhibit secretion of the mucus that is so important in protecting the lining of the digestive tract.

The hormones involved in regulation of digestion are produced by the digestive organs themselves. Let us look at some examples of these controls.

The sight, smell, thought, taste, or feel of food in the mouth stimulates, through the nervous system, the secretion of saliva and the release of gastric juice. Once in the stomach, food stimulates the re-

## TABLE 17-2
## Hormones Active in Digestion

| Hormone | Source | Action |
|---------|--------|--------|
| Gastrin | Stomach | Stimulates release of gastric juice |
| Gastric-inhibitory peptide (GIP) | Small intestine | Inhibits release of gastric juice |
| Secretin | Duodenum | Stimulates release of water and bicarbonate from pancreas; stimulates release of bile from liver; inhibits the stomach |
| Cholecystokinin (CCK) | Duodenum | Stimulates release of digestive enzymes from pancreas; stimulates release of bile from gallbladder; inhibits the stomach |

lease into the blood of the hormone *gastrin,* which promotes stomach secretions and movement. When chyme enters the duodenum, nerve impulses inhibit movement of the stomach so that food will not move too rapidly into the small intestine. This action is a good example of negative feedback. At the same time, hormones released from the duodenum feed back to the stomach to reduce its activity. Two of these hormones, *secretin* (se-KRE-tin) and *cholecystokinin* (ko-le-sis-to-KI-nin), have the additional function of stimulating the liver, pancreas, or gallbladder to release their juices. Table 17-2 lists some of the hormones that help regulate the digestive process.

### Hunger and Appetite

Hunger is the desire for food and can be satisfied by the ingestion of a filling meal. Hunger is regulated by centers in the hypothalamus that can be modified by input from higher brain centers. Therefore, cultural factors and memories of past food in-

take can influence hunger. Strong, mildly painful contractions of the empty stomach may stimulate a feeling of hunger. Messages received by the hypothalamus reduce hunger as food is chewed and swallowed and begins to fill the stomach. The short-term regulation of food intake works to keep the amount of food taken in within the limits of what can be processed by the intestine. The long-term regulation of food intake maintains appropriate blood levels of certain nutrients.

Appetite differs from hunger in that, although it is basically a desire for food, it often has no relationship to the need for food. Even after an adequate meal that has relieved hunger, a person may still have an appetite for additional food. A chronic loss of appetite, called *anorexia* (an-o-REK-se-ah), may be due to a great variety of physical and mental disorders. Because the hypothalamus and the higher brain centers are involved in the regulation of hunger, it is possible that emotional and social factors contribute to the development of anorexia.

## SUMMARY

I. **Digestive system**
  A. Functions—digestion and absorption
  B. Structure of wall—mucosa, submucosa, smooth muscle, serosa
  C. Peritoneum—serous membrane that lines abdominal cavity and folds over organs
   1. Peritonitis—peritoneal inflammation
II. **Digestive tract**
  A. Mouth
   1. Functions
    a. Ingestion of food

  b. Mastication of food
  c. Lubrication of food
  d. Digestion of starch with salivary amylase
   2. Structures
    a. Tongue
    b. Deciduous (baby) teeth—20 (incisors, canines, molars)
    c. Permanent teeth—32 (incisors, canines, premolars, molars)

**B.** Pharynx (throat)—moves portion of food (bolus) into esophagus by reflex swallowing (deglutition)

**C.** Esophagus—long muscular tube that carries food to stomach by peristalsis

**D.** Stomach functions
  1. Storage of food
  2. Breakdown of food by churning
  3. Liquefaction of food with hydrochloric acid (HCl) to form chyme
  4. Digestion of protein with enzyme pepsin

**E.** Small intestine
  1. Functions
    **a.** Digestion of food
    **b.** Absorption of food through villi (small projections of intestinal lining)
  2. Divisions—duodenum, jejunum, ileum

**F.** Large intestine
  1. Functions
    **a.** Storage and elimination of waste (defecation)
    **b.** Reabsorption of water
  2. Divisions—cecum; ascending, transverse, descending, and sigmoid colons; rectum; anus

**III. Accessory organs**
  **A.** Liver functions
    1. Storage of glucose
    2. Formation of blood plasma proteins
    3. Synthesis of urea
    4. Modification of fats
    5. Manufacture of bile
    6. Destruction of old red blood cells
    7. Detoxification of harmful substances
    8. Storage of vitamins and iron
  **B.** Gallbladder—stores bile until needed for digestion
  **C.** Pancreas
    1. Secretes powerful digestive juice
    2. Secretes neutralizing fluid

**IV. Process of digestion**
  **A.** Products
    1. Simple sugars (monosaccharides) from carbohydrates
    2. Amino acids from proteins
    3. Glycerol and fatty acids from fats
  **B.** Absorption—movement of digested food into the circulation
  **C.** Control
    1. Nervous control
      **a.** Parasympathetic system—generally increases activity
      **b.** Sympathetic system—generally decreases activity
    2. Hormonal control
      **a.** Stimulation of digestive activity
      **b.** Feedback to inhibit responses

## QUESTIONS FOR STUDY AND REVIEW

1. Trace the path of a mouthful of food through the digestive tract.
2. What is the peritoneum? Name the two layers and describe their locations. Name four double-layered peritoneal structures.
3. Differentiate between deciduous and permanent teeth with respect to kinds and numbers.
4. What is the pharynx? What is above and below it?
5. What is peristalsis? Name some structures in which it occurs.
6. Name two purposes of the acid in gastric juice.
7. Name the principal digestive enzymes. Where is each formed? What does each do?
8. Where does absorption occur, and what structures are needed for absorption?
9. What types of digested materials are absorbed into the blood?
10. What types of digested materials are absorbed into the lymph?
11. Name the accessory organs of digestion and the functions of each.
12. Name five nondigestive functions of the liver.
13. Give examples of negative feedback in the control of digestion.
14. Name several hormones that regulate digestion.
15. How is hunger regulated?

# CHAPTER 18

# Metabolism, Nutrition, and Body Temperature

## Selected Key Terms

The following terms are defined in the Glossary:

**anabolism**

**catabolism**

**glucose**

**glycogen**

**malnutrition**

**metabolic rate**

**mineral**

**oxidation**

**vitamin**

## Behavioral Objectives

After careful study of this chapter, you should be able to:

- Differentiate between catabolism and anabolism
- Differentiate between the anaerobic and aerobic phases of catabolism and give the end products and the relative amount of energy released from each
- Explain the role of glucose in metabolism
- Compare the energy contents of fats, proteins, and carbohydrates
- Define *essential amino acid*
- Explain the roles of minerals and vitamins in nutrition and give examples of each
- Define *metabolic rate*
- Name several factors that affect the metabolic rate
- Explain how heat is produced in the body
- List the ways in which heat is lost from the body
- Describe the role of the hypothalamus in regulating body temperature

Memmler, RL, Cohen, BJ, Wood, DL. *STRUCTURE AND FUNCTION OF THE HUMAN BODY*, 6/e,
© 1996 Lippincott-Raven Publishers

# ❯ Metabolism

T he end products of digestion are used for all the cellular activities of the body which, all together, make up *metabolism.* These activities fall into two categories:

1. *Catabolism,* which is the breakdown of complex compounds into simpler compounds. It includes the digestion of food into usable small molecules and the release of energy from these molecules within the cell.
2. *Anabolism,* which is the building of simple compounds into substances needed for cellular activities and the growth and repair of tissues.

Through the steps of catabolism and anabolism there is a constant turnover of body materials as energy is consumed, cells function and grow, and waste products are generated.

## How Cells Obtain Energy From Food

Energy is released from nutrients in a series of reactions called *cellular respiration,* each step of which requires a specific enzyme as a catalyst. Early studies on cellular respiration were done with *glucose* as the starting compound. Glucose is a simple sugar that is the main energy source for the body. The first steps in the breakdown of glucose do not require oxygen, that is, they are *anaerobic.* This phase of catabolism occurs in the cytoplasm of the cell. It yields a very small amount of energy which is used to make ATP (adenosine triphosphate), the energy compound of the cells. Because the breakdown is incomplete, some organic end product results. In muscle cells operating briefly under anaerobic conditions, lactic acid accumulates while the cells build up an oxygen debt, as described in Chapter 7. However, our cells must use nutrients more completely to generate enough energy for survival. Further oxygen-requiring, or *aerobic,* reactions occur within the mitochondria of the cell and transfer most of the energy remaining in the nutrients to ATP.

In the course of these aerobic reactions, carbon dioxide is released and oxygen is combined with hydrogen from the food molecules to form water. The carbon dioxide is eliminated by the lungs in breathing. Because of the type of chemical reactions involved, and because oxygen is used in the final steps, this process of cellular respiration is called *oxidation.* Although the oxidation of food is often compared to the burning of fuel, this comparison is inaccurate. Burning results in a sudden and often wasteful release of energy in the form of heat and light. In contrast, metabolic oxidation occurs in steps, and much of the energy released is stored as ATP for later use by the cells; some of the energy is released as heat, which is used to maintain body temperature, as discussed later in this chapter.

### METABOLIC RATE

*Metabolic rate* refers to the rate at which energy is released from food in the cells. It is affected by a person's size, body fat, sex, age, activity, and hormones, especially thyroid hormone (thyroxine). *Basal metabolism* is the amount of energy needed to maintain life functions while the body is at rest. Metabolic rate is high in children and adolescents and decreases with age.

### The Use of Nutrients

As noted, the main source of energy in the body is glucose. Other carbohydrates in the diet are converted to glucose for metabolism. Glucose circulates in the blood to provide energy for all the cells. Reserves of glucose are stored in liver and muscle cells as *glycogen* (GLI-ko-jen), a compound built from glucose molecules. When it is needed for energy, glycogen is broken down to glucose. Glycerol and fatty acids (from fat digestion) and amino acids (from protein digestion) can also be used for energy, but they enter the breakdown process at different points. Fat in the diet yields more than twice as much energy as protein or carbohydrate (it is more "fattening"); fat yields about 9 kilocalories (kcal) of energy per gram, whereas protein and carbohydrate each yield about 4 kcal per gram. Food that is ingested in excess of need is converted to fat and stored in adipose tissue.

Before oxidation, the amino acids must have their nitrogen (amine) groups removed. This occurs in the liver, where the nitrogen groups are then formed into urea by combination with carbon dioxide. The bloodstream transports the urea to the kidneys to be eliminated.

### Anabolism

Food molecules are built into body materials by other metabolic steps, all of which are catalyzed by enzymes. Ten of the 20 amino acids needed to build proteins can be manufactured internally by metabolic reactions. The remaining 10 amino acids cannot be made by the body and therefore must be taken in with the diet; these are the ***essential amino acids.*** Most animal proteins supply all of the essential amino acids and are described as ***complete proteins.*** Vegetables may be lacking in one or more of the essential amino acids. Persons on strict vegetarian diets must learn to combine foods, such as beans with rice or wheat, to obtain all the essential amino acids. The few essential fatty acids that are needed are easily obtained through a healthful, balanced diet.

### Minerals and Vitamins

In addition to needing fats, proteins, and carbohydrates, the human body requires minerals and vitamins. ***Minerals*** are elements needed for body structure, fluid balance, and such activities as muscle contraction, nerve impulse conduction, and blood clotting. A list of the main minerals needed in a proper diet is given in Table 18-1. Some additional minerals not listed are also required for good health. Those needed in extremely small amounts are referred to as ***trace elements.***

***Vitamins*** are organic substances needed in very small quantities. Vitamins are parts of enzymes or other essential substances used in cell metabolism, and vitamin deficiencies lead to a variety of nutritional diseases. A list of vitamins is given in Table 18-2.

## ▶ Some Practical Aspects of Nutrition

Good nutrition is absolutely essential for the maintenance of health. If any vital food material is missing from the diet, the body will suffer from ***malnu-***

---

**TABLE 18-1**
**Minerals**

| Mineral | Functions | Sources | Deficiencies |
|---------|-----------|---------|--------------|
| Potassium (K) | Nerve and muscle activity | Fruits, meats, seafood, milk | Muscular and neurologic disorders |
| Sodium (Na) | Body fluid balance; nerve impulse conduction | Most foods and table salt | Weakness, cramps, diarrhea, dehydration |
| Calcium (Ca) | Formation of bones and teeth, blood clotting, nerve conduction, muscle contraction | Dairy products, eggs, green vegetables, legumes (peas and beans) | Rickets, tetany, bone demineralization |
| Phosphorus (P) | Formation of bones and teeth; found in ATP, nucleic acids, cell membrane | Beef, egg yolk, dairy products | Bone demineralization, abnormal metabolism |
| Iron (Fe) | Oxygen carrier (hemoglobin) | Meat, eggs, spinach, prunes, cereals, legumes | Anemia, dry skin, indigestion |
| Iodine (I) | Thyroid hormones | Seafood, iodized salt | Hypothyroidism, goiter |
| Magnesium (Mg) | Catalyst for enzyme reactions, carbohydrate metabolism | Green vegetables, grains | Spasticity, arrhythmia, vasodilation |
| Manganese (Mn) | Catalyst in actions of calcium and phosphorus | Legumes, nuts, cereals, green leafy vegetables | Possible reproductive disorders |
| Copper (Cu) | Necessary for absorption and oxidation of vitamin C and iron and in formation of hemoglobin | Liver, fish, oysters, legumes, nuts | Anemia |
| Cobalt (Co) | Part of vitamin $B_{12}$ involved in blood cell production and in synthesis of insulin | Animal products | Pernicious anemia |
| Zinc (Zn) | Promotes carbon dioxide transport and energy metabolism; found in enzymes | Many foods | Alopecia (baldness) possibly related to diabetes |
| Fluorine (F) | Prevents tooth decay | Water and many foods | Dental caries |

**TABLE 18-2**
**Vitamins**

| Vitamins | Functions | Sources | Deficiencies |
|---|---|---|---|
| Retinol (A) | Required for healthy epithelial tissues and for eye pigments | Yellow vegetables, fish, liver oils, dairy products | Night blindness; dry, scaly skin |
| Thiamin ($B_1$) | Required for enzymes involved in oxidation of food | Pork, cereal, grains, meats, legumes | Beriberi, a disease of nerves (neuritis) |
| Riboflavin ($B_2$) | Needed for enzymes required for oxidation of food | Milk, eggs, kidney, liver, green leafy vegetables | Skin and tongue disorders |
| Niacin (nicotinic acid) | Involved in oxidation of food | Yeast, lean meat, liver, grains, legumes, | Pellagra with dermatitis, diarrhea, mental disorders |
| Pyridoxine ($B_6$) | Involved in metabolism of various food substances and transport of amino acids | Liver, milk, cereals, eggs, yeast, vegetables | Irritability, convulsions, muscle twitching, skin disorders |
| Pantothenic acid | Essential for normal growth | Yeast, liver, eggs | Sleep disturbances, lack of coordination, skin lesions |
| Cyanocobalamin ($B_{12}$) | Production of blood cells (hemopoiesis) | Meat, liver, milk, eggs | Pernicious anemia |
| Biotin | Involved in fat and glycogen formation, amino acid metabolism | Peanuts, liver, tomatoes, eggs | Lack of coordination, dermatitis, fatigue |
| Folic acid and folates | Required for amino acid metabolism, maturation of red blood cells | Vegetables, liver | Anemia, digestive disorders, neural tube defects in the embryo |
| Ascorbic acid (C) | Maintains healthy skin and mucous membranes; involved in synthesis of collagen | Citrus fruits, green vegetables | Scurvy, poor bone and wound healing |
| Calciferol (D) | Aids in absorption of calcium from intestinal tract | Fish liver oils, dairy products | Rickets, bone deformities |
| Tocopherol (E) | Protects cell membranes | Seeds, green vegetables | In question |

*trition.* One commonly thinks of a malnourished person as one who does not have enough to eat, but malnutrition can also occur from eating too much of the wrong foods. The simplest and perhaps best advice for maintaining a healthful diet is to eat a wide variety of fresh, wholesome foods daily. The U.S. Department of Agriculture (USDA) presently recommends a distribution of food in the diet as shown in Figure 18-1. The "Food Guide Pyramid" emphasizes grains, fruits, and vegetables with limited intake of fats, oils, and sugars.

Because proteins, unlike carbohydrates and fats, are not stored, protein foods should be taken in on a regular basis. The average U.S. diet, however, contains more than enough protein and too much fat, especially animal fat. It is currently recommended that calories from the three types of food be distributed as follows:

| | |
|---|---|
| Fat | 30% |
| Carbohydrate | 58% |
| Protein | 12% |

The carbohydrates should be mainly complex, naturally occurring carbohydrates with a minimum of refined sugars. Less than half of the fats in the diet should be *saturated fats.* These are fats, mostly from animal sources, that are solid at room temperature, such as butter and lard. Also included in this group are fats that are artificially hardened (hydrogenated), such as vegetable shortening and margarine, which it is best to avoid. Evidence shows that these manufactured fat products may be just as harmful, if not more so, than natural saturated fats.

## Food as Good Medicine

Does an apple a day keep the doctor away? This adage reflects what people have long known—that the foods we eat have a profound effect on health. That apple has fiber, which is known to reduce the risk of colon cancer by diluting carcinogens in the bowel and reducing their contact with the bowel wall. It is also lacking in fats, which should be controlled in the diet, especially saturated fats. People with diets rich in fruits and vegetables have lower incidences of diabetes, cardiovascular disease, and several types of cancer. Vegetables in the mustard family, such as broccoli, cabbage, and kale, appear to contain compounds that inhibit the growth of cancer cells. Studies are now underway to identify which components of the various vegetables may provide this protection.

*Unsaturated fats* are mainly from plants. They are liquid at room temperature and are generally referred to as *oils*. Diets high in saturated fats are associated with a higher than normal incidence of cancer, heart disease, and cardiovascular problems, although the relation between these factors is not clear.

A weight loss diet should follow the same guidelines as given above with a reduction in portion sizes. Exercise is also recommended.

The need for mineral and vitamin supplements to the diet is a subject of controversy. Some researchers maintain that adequate amounts of these substances can be obtained from a varied, healthful diet; others hold that pollution, depletion of the soils, and the storage, refining, and processing of foods make supplementation beneficial. Most agree, however, that children, elderly persons, pregnant and lactating women, and teenagers, who often do not get enough of the proper foods, would profit from additional minerals and vitamins.

When required, supplements should be selected by a physician or nutritionist to fit the particular needs of the individual. Megavitamin dosages may cause unpleasant reactions and in some cases are hazardous. Vitamins A and D have both been found to cause serious toxic effects; a relatively small excess of vitamin D may result in the appearance of dangerous symptoms.

The subject of allergies has received much attention recently. Some persons develop clear allergic (hypersensitive) symptoms if they eat certain foods. The most common food allergens are wheat, yeast, milk, and eggs, but almost any food might cause an allergic reaction in a given individual. Although food allergies may be responsible for symptoms with no other known cause, most persons can probably eat all types of foods.

## ▶ Body Temperature

Heat is an important byproduct of the many chemical activities constantly going on in tissues all over the body. Heat is always being lost through a variety of outlets. However, because of a number of regulatory devices, under normal conditions body temperature remains constant within quite narrow limits. Maintenance of a constant temperature despite both internal and external influences is one phase of homeostasis, the tendency of all body processes to maintain a normal state despite forces that tend to alter them.

### Heat Production

Heat is produced when oxygen combines with nutrients in the cells. Thus, heat is a byproduct of the cellular reactions that generate energy. The amount of heat produced by a given organ varies with the kind of tissue and its activity. While at rest, muscles may produce as little as 25% of total body heat, but when a number of muscles contract, heat production may be multiplied hundreds of times owing to the increase in metabolic rate. Under basal conditions (at rest), the abdominal organs, particularly the liver, produce about 50% of total body heat. While the body is at rest, the brain produces 15% of body heat, and an increase in nervous tissue activity produces very little increase in heat production. The largest amount of heat, therefore, is produced in the muscles and the glands. Although it would seem from this description that some parts of the body would tend to become much warmer than others, the circulating blood distributes the heat fairly evenly.

The rate at which heat is produced is affected by a number of factors, which may include activity level, hormone production, food intake, and age. Hormones, such as thyroxine from the thyroid gland and epinephrine (adrenaline) from the medulla of

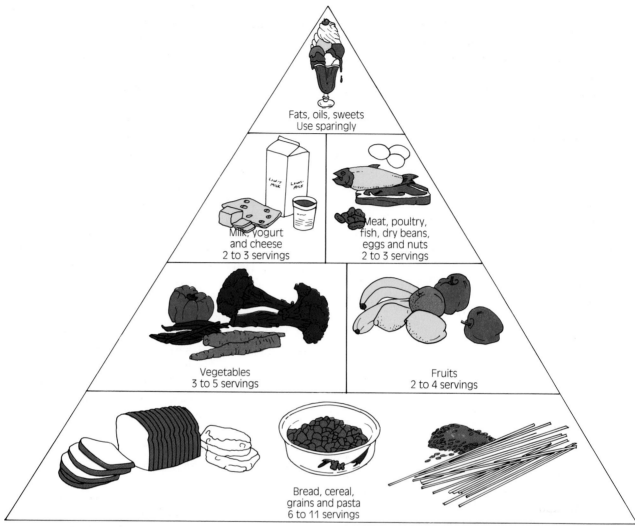

**Figure 18-1**

Guide to daily food choices. (Taylor C, Lillis C, LeMone P: Fundamentals of Nursing: The Art and Science of Nursing Care, 2nd ed, p. 805. Philadelphia, JB Lippincott, 1993)

the adrenal gland, may increase the rate of heat production. The intake of food is also accompanied by increased heat production. The reasons for this are not entirely clear, but a few causes are known. More fuel is poured into the blood with food intake and is therefore more readily available for cellular metabolism. The glandular structures and the muscles of the digestive system generate additional heat as they set to work. This does not account for all the increase, however, nor does it account for the much greater increase in metabolism following a meal containing a large amount of protein. Whatever the reasons, the intake of food definitely increases the chemical activities that go on in the body and thus adds to heat production.

## Heat Loss

More than 80% of heat loss occurs through the skin. The remaining 15% to 20% is dissipated by the respiratory system and with the urine and feces. Networks of blood vessels in the dermis (deeper part) of the skin can bring considerable quantities of blood near the surface so that heat can be dissipated to the outside. This can occur in several ways. Heat can be transferred to the surrounding air in the process of **conduction**. Heat also travels from its source in the form of heat waves or rays, a process termed **radiation.** If the air is moving so that the layer of heated air next to the body is constantly being carried away and replaced with cooler air (as by an electric fan), the process is known as **convection**. Finally, heat loss may be produced by *evaporation,* the process by which liquid changes to the vapor state. Rub some alcohol on your arm; it evaporates rapidly, and in so doing uses so much heat from the skin that your arm feels cold. Perspiration does the same thing, although not as quickly. The rate of heat loss through evaporation depends on the humidity of the surrounding air. When this exceeds 60% or so, perspiration does not evaporate so readily, making one feel generally miserable unless one can resort to some other means of heat loss such as convection caused by a fan.

If the temperature of the surrounding air is lower than that of the body, excessive heat loss is prevented by both natural and artificial means. Clothing checks heat loss by trapping "dead air" in both its material and its layers. This noncirculating air is a good insulator. An effective natural insulation against cold is the layer of fat under the skin. Even when skin temperature is low, this fatty tissue prevents the deeper tissues from losing much heat. On the average, this layer is slightly thicker in females than in males. Naturally there are individual variations, but as a rule the degree of insulation depends on the thickness of this layer of subcutaneous fat.

Other factors that play a part in heat loss include the volume of tissue compared with the amount of skin surface. A child loses heat more rapidly than an adult. Such parts as fingers and toes are affected most by exposure to cold because they have a great amount of skin compared with total tissue volume.

## Temperature Regulation

Given that body temperature remains almost constant in spite of wide variations in the rate of heat production or loss, there must be mechanisms for regulating temperature. Many areas of the body take part in this process, but the most important heat-regulating center is the area inside the brain, located just above the pituitary gland, the *hypothalamus.* Some of the cells in the hypothalamus control the production of heat in the body tissues, whereas another group of cells controls heat loss. This control comes about in response to the heat brought to the brain by the blood as well as to nerve impulses from temperature receptors in the skin. If these two factors indicate that too much heat is being lost, impulses are sent quickly from the hypothalamus to the autonomic (involuntary) nervous system, which in turn causes constric-

tion of the skin blood vessels to reduce heat loss. Other impulses are sent to the muscles to cause shivering, a rhythmic contraction of many body muscles, which results in increased heat production. Furthermore, the output of epinephrine may be increased if conditions call for it. Epinephrine increases cell metabolism for a short period, and this in turn increases heat production. Also, the smooth muscle around the hair roots contracts, forming "goose-flesh," a reaction that conserves heat in many animals but only little heat in humans.

If there is danger of overheating, the hypothalamus will transmit impulses that (1) stimulate the sweat glands to increase their activity and (2) dilate the blood vessels in the skin so that there is increased blood flow with a correspondingly greater loss of heat. The hypothalamus may also encourage the relaxation of muscles and thus minimize the production of heat in these organs.

Muscles are especially important in temperature regulation because variations in the amount of activity in these large masses of tissue can readily increase or decrease the total amount of heat produced. Because muscles form roughly one third of the bulk of the body, either an involuntary or a purposeful increase in the activity of this big group of organs can form enough heat to offset a considerable decrease in the temperature of the environment.

Very young and very old persons are limited in their ability to regulate body temperature when exposed to extremes in environment. The body temperature of a newborn infant will decrease if the infant is exposed to a cool environment for a long period. Elderly persons also are not able to produce enough heat to maintain body temperature in a cool environment. Heat loss mechanisms are not fully developed in the newborn; the elderly do not lose as much heat from their skin. Both groups should be protected from extreme temperatures.

## Normal Body Temperature

The normal temperature range obtained by either a mercury or an electronic thermometer may extend from 36.2°C to 37.6°C (97°F to 100°F). Body temperature varies with the time of day. Usually it is lowest in the early morning, because the muscles have been relaxed and no food has been taken in for several hours. Temperature tends to be higher in the late afternoon and evening because of physical activity and consumption of food.

Normal temperature also varies in different parts of the body. Skin temperature obtained in the axilla (armpit) is lower than mouth temperature, and mouth temperature is a degree or so lower than rectal temperature. It is believed that, if it were possible to place a thermometer inside the liver, it would register a degree or more higher than rectal temperature. The temperature within a muscle might be even higher during its activity.

Although the Fahrenheit scale is used in the United States, in most parts of the world temperature is measured with the *Celsius* (SEL-se-us) thermometer. On this scale the ice point is at 0° and the normal boiling point of water is at 100°, the interval between these two points being divided into 100 equal units. The Celsius scale is also called the *centigrade scale* (think of 100 cents). See Appendix 2 for a comparison of the Celsius and Fahrenheit scales and formulas for converting from one to the other.

## SUMMARY

I. **Metabolism**—life-sustaining reactions that occur in the living cell
   A. Catabolism—breakdown of complex compounds into simpler compounds
      1. Cellular respiration—a series of reactions in which food is oxidized for energy
         a. Anaerobic phase—does not require oxygen
            (1) Location—cytoplasm
            (2) Yield—small amount of energy
            (3) End product—organic (*i.e.,* lactic acid)
         b. Aerobic phase—requires oxygen
            (1) Location—mitochondria
            (2) Yield—almost all remaining energy in food
            (3) End products—carbon dioxide and water
      2. Metabolic rate—rate at which energy is released from food in the cells
         a. Basal metabolism—amount of energy needed to maintain life functions while at rest
      3. Use of nutrients
         a. Glucose—main energy source
         b. Fats—highest energy yield
         c. Proteins—used for energy after removal of nitrogen
   B. Anabolism—building of simple compounds into materials needed by the body
   C. Minerals—elements needed for body structure and cell activities
      1. Trace elements—elements needed in extremely small amounts

   D. Vitamins—organic substances needed in small amounts
II. **Practical aspects of nutrition**
   A. Malnutrition—too little food or inadequate amounts of specific foods
   B. Components of healthy diet
      1. USDA Food Guide Pyramid
      2. Less than 15% of total daily calories in form of saturated fats
III. **Body temperature**
   A. Heat production—most heat produced in muscles and glands; distributed by the circulation
   B. Heat loss
      1. Skin
      2. Urine
      3. Feces
      4. Breathing
   C. Temperature regulation
      1. Hypothalamus—main temperature regulating center; responds to temperature of blood and temperature receptors in skin
      2. Results of decrease in body temperature
         a. Constriction of blood vessels in skin
         b. Shivering
         c. Increased release of epinephrine
      3. Results of increase in body temperature
         a. Dilation of skin vessels
         b. Sweating
         c. Relaxation of muscles
   D. Normal body temperature—ranges from 36.2°C to 37.6°C; varies with time of day and location measured

## QUESTIONS FOR STUDY AND REVIEW

1. Define *cellular respiration*.
2. In what part of the cell does anaerobic respiration occur? What are its end products?
3. In what part of the cell does aerobic respiration occur? What are its end products?
4. Define *metabolic rate*. What factors affect the metabolic rate?
5. Approximately how many kilocalories are released from a gram of butter as compared with a gram of egg white? with a gram of sugar?
6. Gelatin is not a complete protein. What are the dangers of eating flavored gelatin as a sole source of protein?
7. Name the foods that are emphasized in the USDA Food Guide Pyramid; which are to be eaten sparingly?
8. How is heat produced in the body? What structures produce the most heat during increased activity?

9. Name four factors that affect heat production.
10. How is heat lost from the body?
11. Name four ways in which heat escapes to the environment.
12. In what ways is heat kept in the body?
13. Name the main temperature regulator and describe what it does when the body is too hot and when it is too cold. What parts do muscles play?
14. What is the normal body temperature range? How does it vary with respect to the time of day and the part of the body?
15. Differentiate between the terms in each of the following pairs:
    a. *catabolism* and *anabolism*
    b. *aerobic* and *anaerobic*
    c. *mineral* and *vitamin*
    d. *saturated fats* and *unsaturated fats*

# CHAPTER 19

# The Urinary System and Body Fluids

## Behavioral Objectives

After careful study of this chapter, you should be able to:

- List the systems that eliminate waste and name the substances eliminated by each
- Describe the parts of the urinary system and give the functions of each
- Trace the path of a drop of blood as it flows through the kidney
- Describe a nephron
- Describe the functions of renin and angiotensin
- Name the four processes involved in urine formation and describe the functions of each
- Identify the role of ADH in urine formation
- Name three normal constituents of urine
- Compare intracellular and extracellular fluids
- List four types of extracellular fluids
- Name the systems that are involved in water balance
- Define *electrolytes* and describe their importance
- Describe three methods for regulating the pH of body fluids

Memmler, RL, Cohen, BJ, Wood, DL. *STRUCTURE AND FUNCTION OF THE HUMAN BODY*, 6/e,
© 1996 Lippincott-Raven Publishers

The urinary system is also called the *excretory system* because one of its main functions is to remove waste products from the blood and eliminate them from the body. Other functions of the urinary system include regulation of the volume of body fluids and balance of the pH and electrolyte composition of these fluids.

Although the focus of this chapter is the urinary system, certain aspects of other systems will also be discussed because body systems work interdependently to maintain a balance, or homeostasis. The chief excretory mechanisms of the body, and some of the substances that they eliminate, are listed below:

1. The *urinary system* excretes water, waste products containing nitrogen, and salts. These are all constituents of the urine.
2. The *digestive system* eliminates water, some salts, bile, and the residue of digestion, all of which are contained in the feces. The liver is important in eliminating the products of red blood cell destruction and in breaking down certain drugs and toxins.
3. The *respiratory system* eliminates carbon dioxide and water. The latter appears as vapor, as can be demonstrated by breathing on a windowpane.
4. The skin, or *integumentary system,* excretes water, salts, and very small quantities of nitrogenous wastes. These all appear in perspiration, though evaporation of water from the skin may go on most of the time without our being conscious of it.

## ▶ Organs of the Urinary System

The main parts of the urinary system, shown in Figure 19-1, are as follows:

1. Two *kidneys.* These organs extract wastes from the blood, balance body fluids, and form urine.
2. Two *ureters* (u-RE-ters). These tubes conduct urine from the kidneys to the urinary bladder.
3. A single *urinary bladder.* This reservoir receives and stores the urine brought to it by the two ureters.
4. A single *urethra* (u-RE-thrah). This tube conducts urine from the bladder to the outside of the body for elimination.

## ▶ The Kidneys

### Location of the Kidneys

The two kidneys lie against the muscles of the back in the upper abdomen. They are up under the dome of the diaphragm and are protected by the lower ribs and the rib (costal) cartilages. Each kidney is enclosed in a membranous capsule that is made of fibrous connective tissue; this is loosely adherent to the kidney itself. In addition, there is a crescent of fat called the *adipose capsule* around the perimeter of the organ. This capsule is one of the chief supporting structures of the kidney. The kidneys, as well as the ureters, lie behind the peritoneum. Thus, they are not in the peritoneal cavity but rather in an area known as the *retroperitoneal* (ret-ro-per-ih-to-NE-al) *space.*

The blood supply to the kidney is illustrated in Figure 19-2. Blood is brought to the kidney by a short branch of the abdominal aorta called the *renal artery.* After entering the kidney, the renal artery subdivides into smaller and smaller branches, which eventually make contact with the functional units of the kidney, the *nephrons* (NEF-ronz) (Figs.19-3 and 19-4). Blood leaves the kidney by vessels that finally merge to form the *renal vein*, which carries blood into the inferior vena cava for return to the heart.

### Structure of the Kidneys

The kidney is a somewhat flattened organ about 10 cm (4 inches) long, 5 cm (2 inches) wide, and 2.5 cm (1 inch) thick. On the inner, or medial, border there is a notch called the *hilus,* at which region the renal artery, the renal vein, and the ureter connect with the kidney. The outer, or lateral, border is convex (curved outward), giving the entire organ a bean-shaped appearance.

The kidney is divided into three regions: the renal cortex, the renal medulla, and the renal pelvis (see Fig. 19-3). The renal *cortex* is the outer portion of the kidney. The renal *medulla* contains the tubes that collect urine. These tubes form a number of cone-shaped structures called *pyramids.* The tips of the pyramids point toward the *renal pelvis*, a funnel-shaped basin that forms the upper end of the ureter. Cuplike extensions of the renal pelvis surround the tips of the pyramids and collect urine; these extensions are called *calyces* (KA-lih-seze) (sing., *calyx* [KA-liks]). The urine that collects in the pelvis then passes down the ureters to the bladder.

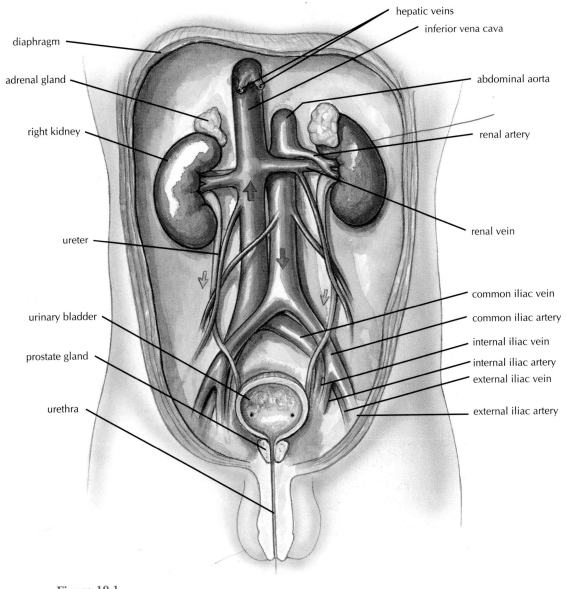

**Figure 19-1**

Urinary system, with blood vessels.

The kidney is a glandular organ; that is, most of the tissue is epithelium with just enough connective tissue to serve as a framework. As is the case with most organs, the most fascinating aspect of the kidney is too small to be seen with the naked eye. This basic unit of the kidney, where the kidney's work is actually done, is the nephron (see Fig. 19-4). The nephron is essentially a tiny coiled tube with a bulb at one end. This bulb, called *Bowman's capsule,* surrounds a cluster of capillaries called the *glomerulus* (glo-MER-u-lus) (pl., *glomeruli* [glo-MER-u-li]). Each kidney contains about 1 million nephrons; if all these coiled tubes were separated, straightened out, and laid end to end, they would span some 120 kilometers (75 miles)! Figure 19-5 is a microscopic view of kidney tissue showing several glomeruli, each surrounded by a Bowman's capsule. It also shows sections through the tubular portions of the nephrons.

A small blood vessel, called the *afferent arteriole,* supplies the glomerulus with blood; another small

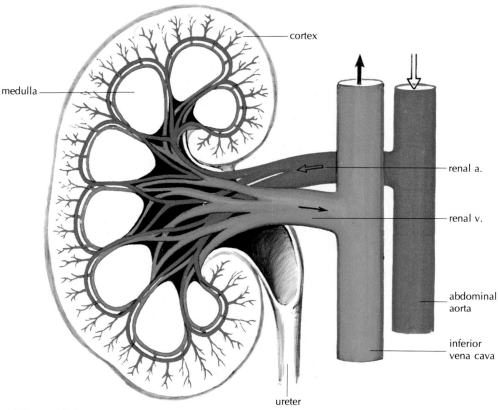

cortex

medulla

renal a.

renal v.

abdominal
aorta

inferior
vena cava

ureter

**Figure 19-2**

Blood supply and circulation of the kidney.

vessel, called the *efferent arteriole,* carries blood from the glomerulus to capillaries that surround the tubular portion of the nephron. These small vessels, *peritubular capillaries,* are named for their location around the nephron.

The tubular part of the nephron consists of several portions. The coiled portion leading from Bowman's capsule is called the *proximal convoluted* (KON-vo-lu-ted) *tubule (PCT).* The tubule then uncoils to form a hairpin-shaped segment called the *loop of Henle.* Continuing from the loop, the tubule coils once again into the *distal convoluted tubule (DCT),* so called because it is farther along the tubule from Bowman's capsule than is the PCT. The distal end of each tubule empties into a collecting tubule, which then continues through the medulla toward the renal pelvis. The glomerulus, Bowman's capsule, and the proximal and distal convoluted tubules of most nephrons are within the renal cortex. The loops of Henle extend varying distances into the medulla.

The distal convoluted tubule curls back toward the glomerulus between the afferent and efferent arterioles. At the point at which the distal convoluted tubule contacts the afferent arteriole, there are specialized glandular cells that form the *juxtaglomerular* (juks-tah-glo-MER-u-lar) *apparatus* (Fig. 19-6), abbreviated as *JG apparatus.* The name means "near the glomerulus," which describes the location of the cells in this structure. The function of the JG apparatus is described below.

### Functions of the Kidneys

The kidneys are involved in the following processes:

1. Excretion of unwanted substances such as waste products from cell metabolism, excess salts, and toxins. One product of amino acid metabolism is nitrogen-containing waste material, a chief form of which is *urea* (u-RE-ah). The kidneys provide a specialized mechanism for the elimination of these nitrogenous (ni-TROJ-en-us) wastes.

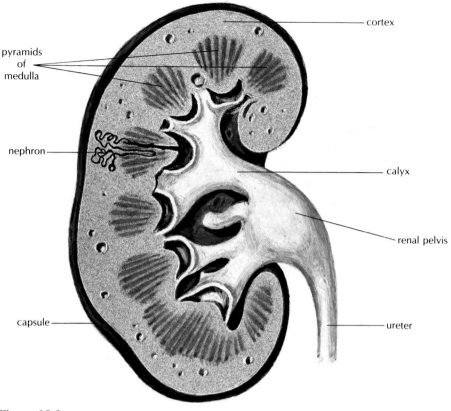

cortex

pyramids
of
medulla

nephron

calyx

renal pelvis

capsule

ureter

**Figure 19-3**

Longitudinal section through the kidney showing its internal structure and a much enlarged diagram of a nephron. There are more than 1 million nephrons in each kidney.

2. Maintenance of water balance. Although the amount of water consumed in a day can vary tremendously, the kidneys can adapt to these variations so that the volume of body water remains remarkably stable from day to day. Water is constantly lost in many ways: from the skin, from the respiratory system during exhalation, and from the intestinal tract. Normally, the amount of water taken in or produced (intake) is approximately equal to the amount lost (output).

3. Regulation of the acid–base balance of body fluids. Acids are constantly being produced by cell metabolism, and certain foods can cause acids or bases to be formed in the body. Bases in the form of antacids, such as baking soda, may also be ingested. However, if the body is to function normally, a certain critical proportion of acids and bases must be maintained at all times.

4. Production of *renin* (RE-nin), an enzyme which is important in the regulation of blood pressure. If blood pressure falls too low for effective filtration through the glomerulus, the cells of the juxtaglomerular (JG) apparatus release renin into the blood. Here, the enzyme activates a protein called *angiotensin* (an-je-o-TEN-sin) that causes blood vessels to constrict, thus raising blood pressure.

5. Production of the hormone *erythropoietin* (eh-rith-ro-POY-eh-tin) *(EPO),* which stimulates the production of red blood cells in the red bone marrow. The hormone is produced when the kidneys do not get enough oxygen.

### Renal Physiology
### GLOMERULAR FILTRATION

The process of urine formation begins in the glomerulus and Bowman's capsule. The membranes that form the walls of the glomerular capillaries are sieve-like and permit the free flow of water and soluble materials through them. Like other capillary walls, these are impermeable (im-PER-me-abl) to

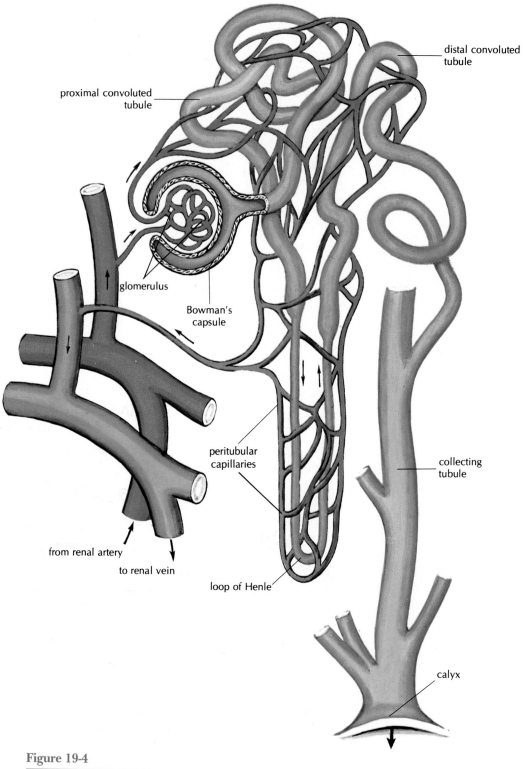

proximal convoluted
tubule

distal convoluted
tubule

glomerulus

Bowman's
capsule

peritubular
capillaries

from renal artery

to renal vein

loop of Henle

collecting
tubule

calyx

**Figure 19-4**

Simplified diagram of a nephron.

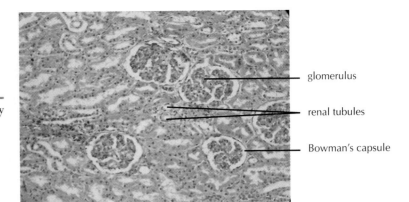

**Figure 19-5**

Microscopic view of the kidney. (Courtesy of Dana Morse Bittus and B. J. Cohen)

blood cells and large protein molecules, so these substances remain in the blood (Fig. 19-7).

Because the diameter of the afferent arteriole is slightly larger than the diameter of the efferent arteriole, blood can enter the glomerulus more easily than it can leave. Thus, the pressure of the blood in the glomerulus is about three to four times as high as it is in other body capillaries. To understand this effect, think of placing your thumb over the end of a garden hose as water comes through. Because the diameter of the opening is made smaller, water is forced out under higher pressure. As a result of

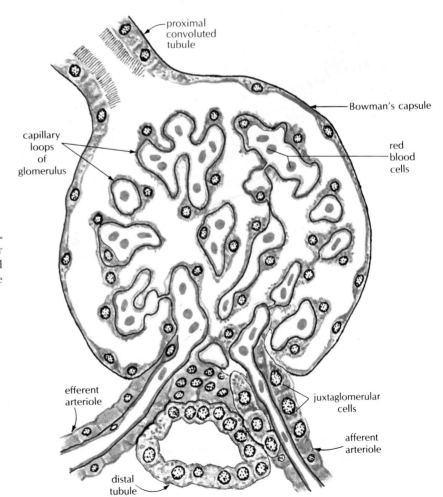

**Figure 19-6**

Structure of the juxtaglomerular apparatus. Note how the distal convoluted tubule contacts the arterioles.

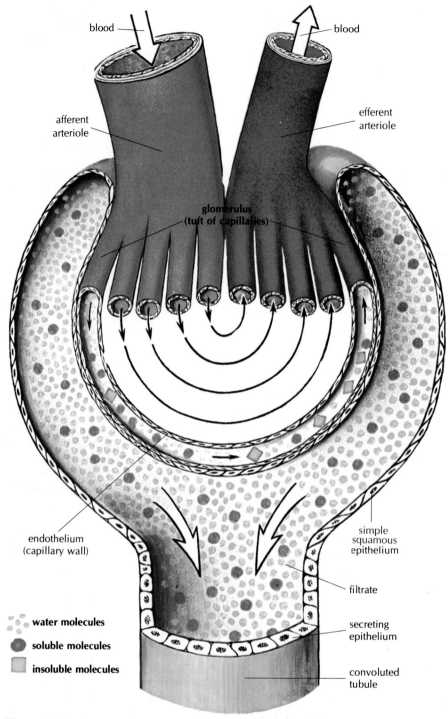

blood

blood

afferent
arteriole

efferent
arteriole

**glomerulus
(tuft of capillaries)**

endothelium
(capillary wall)

simple
squamous
epithelium

filtrate

water molecules

soluble molecules

insoluble molecules

secreting
epithelium

convoluted
tubule

**Figure 19-7**

Diagram showing the process of filtration in the formation of urine. The high pressure inside the capillaries of the glomerulus forces dissolved substances (but not plasma proteins) and much water into the space inside Bowman's capsule. The smaller caliber of the efferent vessel as compared with that of the larger afferent vessel causes this pressure.

this increased pressure in the glomerulus, materials are constantly being "squeezed" out of the blood into Bowman's capsule. This process is known as *glomerular filtration.* The fluid that enters Bowman's capsule, called the *glomerular filtrate,* begins its journey along the tubular system of the nephron. In addition to water and the normal soluble substances in the blood, other substances, such as drugs, may also be filtered and become part of the glomerular filtrate.

### TUBULAR REABSORPTION

About 160 to 180 liters of filtrate are formed each day in the kidneys. However, only 1 to 1.5 liters of urine are eliminated daily. Clearly, most of the water that enters the nephron is not excreted with the urine but rather is returned to the circulation. In addition to water, many substances that are needed by the body, such as nutrients and ions, also pass into the nephron as part of the filtrate. These must be returned to the body as well. So the process of filtration that occurs in Bowman's capsule is followed by a process of *reabsorption.* As the filtrate travels through the tubular system of the nephron, water and other needed substances leave the tubule by diffusion and active transport and enter the surrounding tissue fluids. They then enter the blood in the peritubular capillaries and return to the circulation. Most of the urea and other nitrogenous waste material is kept within the tubule to be eliminated with the urine.

### TUBULAR SECRETION

Before the filtrate leaves the body as urine, a final adjustment in composition is made by the process of *tubular secretion.* In this process, some substances are actively moved from the blood into the nephron. Of particular importance is the secretion of potassium ions and hydrogen ions. The long term regulation of acid-base balance is carried out by the kidney through tubular secretion of hydrogen ions.

### CONCENTRATION OF THE URINE

The amount of water that is eliminated with the urine is regulated by a complex mechanism within the nephron that is influenced by *antidiuretic hormone* (ADH) from the posterior pituitary gland. The process is called the *counter-current mechanism* because it involves fluid traveling in opposite directions within the loop of Henle. The counter-current mechanism is illustrated in Figure 19-8. Its essentials are described below.

## The Transport Maximum

The kidney works very efficiently to return valuable substances to the blood following glomerular filtration. However, the carriers that are needed for active transport of these substances can become overloaded. There is a limit to the amount of each substance that can be reabsorbed in a given time period. The limit of this rate of reabsorption is called the *transport maximum (Tm)* for that substance, and it is measured in milligrams (mg) per minute.

If a substance is present in excess in the blood, it may exceed its transport maximum and then, because it cannot be totally reabsorbed, some will be excreted in the urine. For example, if the concentration of glucose in the blood exceeds 180 mg/mL, glucose will begin to appear in the urine, a condition called *glucosuria* (glu-ko-SU-re-ah). The most common cause of glucosuria is uncontrolled diabetes mellitus.

As the filtrate passes through the loop of Henle, salts, especially sodium, are actively pumped out by the cells of the nephron, with the result that the interstitial fluid of the medulla becomes increasingly concentrated. Because the nephron is not very permeable to water at this point, the fluid within the nephron becomes increasingly dilute. As the fluid passes through the more permeable DCT and through the collecting tubule, water is drawn out by the concentrated fluids around the nephron and returned to the blood. (Remember, according to the laws of diffusion, water follows salt.) In this manner the urine becomes more concentrated and its volume is reduced.

The role of ADH is to make the walls of the DCT and collecting tubule more permeable to water so that more water will be reabsorbed and less will be excreted with the urine. The release of ADH is regulated by a feedback system. As the blood becomes more concentrated, the hypothalamus causes more ADH to be released from the posterior pituitary; as the blood becomes more dilute, less ADH is released. In the disease diabetes insipidus there is inadequate secretion of ADH from the hypothalamus. This results in the elimination of large amounts of very dilute urine accompanied by excessive thirst.

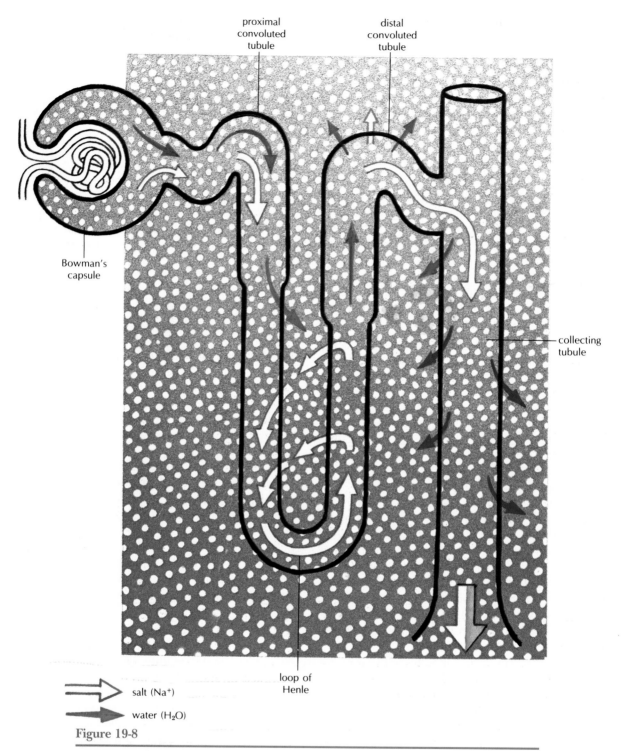

proximal
convoluted
tubule

distal
convoluted
tubule

Bowman's
capsule

collecting
tubule

loop of
Henle

salt (Na⁺)

water (H₂O)

**Figure 19-8**

Loop of Henle, where the proportions of waste and water in urine are regulated according to the body's constantly changing needs. The concentration of urine is determined by means of intricate exchanges of water and salt, constituting the counter-current mechanism.

To summarize the above processes involved in urine formation:

1. Glomerular filtration allows all diffusible materials to pass from the blood into the nephron.
2. Tubular reabsorption moves useful substances back into the blood while keeping waste products in the nephron to be eliminated in the urine.
3. Tubular secretion moves additional substances from the blood into the nephron for elimination. Movement of hydrogen ions is one means by which the pH of body fluids is balanced.
4. The counter-current mechanism concentrates the urine and reduces the volume excreted. The pituitary hormone ADH allows more water to be reabsorbed from the nephron.

Although this story seems quite complex, and there appears to be a great deal of back-and-forth exchanges, these processes together allow the kidney to "fine tune" body fluids. As the filtrate makes its slow journey through the twists and turns of the nephron, there is ample time for exchanges to take place between the kidney tubules and the circulating blood.

## ▶ Organs That Store and Transport Urine

### The Ureters

The two ureters are long, slender, muscular tubes that extend from the kidney basin down to and through the lower part of the urinary bladder. Their length naturally varies with the size of the individual and so may be anywhere from 25 cm to 32 cm (10–13 inches) long. Nearly 2.5 cm (1 inch) of its lower part enters the bladder by passing obliquely (at an angle) through the bladder wall. The ureters are entirely extraperitoneal, being located behind and, at the lower part, below the peritoneum.

The wall of the ureter includes a lining of epithelial cells, a relatively thick layer of involuntary muscle, and finally an outer coat of fibrous connective tissue. The lining is continuous with that of the renal pelvis and the bladder. The muscles of the ureters are capable of the same rhythmic contraction (peristalsis) found in the digestive system. Urine is moved along the ureter from the kidneys to the bladder by gravity and by peristalsis at frequent intervals. Because of the oblique direction of the last part of each ureter through the lower bladder wall,

compression of the ureters by the full bladder prevents backflow of urine.

### The Urinary Bladder
#### CHARACTERISTICS OF THE BLADDER

When it is empty, the urinary bladder (Fig. 19-9) is located below the parietal peritoneum and behind the pubic joint. When it is filled, it pushes the peritoneum upward and may extend well into the abdominal cavity proper. The urinary bladder is a temporary reservoir for urine, just as the gallbladder is a storage sac for bile.

The bladder wall has many layers. It is lined with mucous membrane; the lining of the bladder, like that of the stomach, is thrown into folds called *rugae* when the organ is empty. Beneath the mucosa is a layer of connective tissue. Then follows a three-layered coat of involuntary muscle tissue that is capable of great stretching. Finally, there is an incomplete coat of peritoneum that covers only the upper portion of the bladder. When the bladder is empty, the muscular wall becomes thick and the entire organ feels firm. As the bladder fills, the muscular wall becomes thinner, and the organ may increase from a length of 5 cm (2 inches) up to as much as 12.5 cm (5 inches) or even more. A moderately full bladder holds about 470 mL (1 pint) of urine.

In the floor of the bladder is the *trigone* (TRI-gone), a triangle formed by the openings of the two ureters and the urethra (see Fig. 19-9). As the bladder fills with urine, it expands upward leaving the trigone at the base stationary. This prevents the stretching of the ureteral openings and the possible back flow of urine into the ureters.

#### URINATION

The process of expelling (voiding) urine from the bladder is called *urination* or *micturition* (mik-tu-RISH-un). Near the outlet of the bladder, a circle of smooth muscle forms the *internal sphincter,* which contracts involuntarily to prevent emptying of the bladder. As the bladder fills, stretch receptors send impulses to a center in the lower part of the spinal cord. From this center, motor impulses are sent out to the bladder musculature and the organ is emptied. In the infant this emptying is automatic (a reflex action). As a child matures, higher brain centers gain control over the reflex action and over a voluntary external sphincter that is located below the internal sphincter. The time of urination can

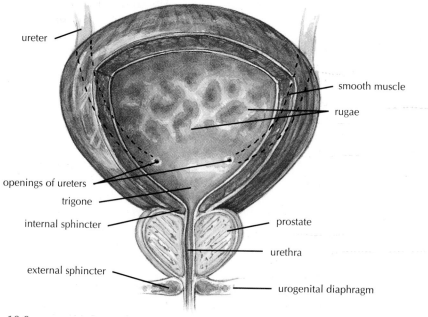

ureter

smooth muscle

rugae

openings of ureters

trigone

internal sphincter

prostate

urethra

external sphincter

urogenital diaphragm

**Figure 19-9**

Interior of the urinary bladder, shown in the male. The trigone is a triangle in the floor of the bladder marked by the openings of the ureters and the urethra.

then be voluntarily controlled unless the bladder becomes too full.

### The Urethra

The *urethra,* the tube that extends from the bladder to the outside, is the means by which the bladder is emptied. The urethra differs in men and women; in the male it is part of the reproductive system and the urinary system, and it is much longer than is the female urethra.

The urethra in the female is a thin-walled tube about 4 cm (1.5 inches) long. It is located behind the pubic joint and is embedded in the muscle of the front wall of the vagina. The external opening, called the *urethral meatus,* is located just in front of the vaginal opening between the labia minora.

The male urethra is about 20 cm (8 inches) long. Early in its course, it passes through the prostate gland, where it is joined by the two ducts carrying the male sex cells. From here it leads through the *penis* (PE-nis), the male organ of copulation, to the outside. The male urethra, then, serves the dual purpose of conveying the sex cells and draining the bladder, whereas the female urethra performs only the latter function.

### The Effects of Aging on the Urinary System

Aging causes the kidneys to lose some of their ability to concentrate urine. With aging, progressively more water is needed to excrete the same amount of waste. Older persons find it necessary to drink more water than young persons, and they eliminate larger amounts of urine (polyuria) even at night (nocturia). Beginning at about age 40 there is a decrease in the number and size of the nephrons. Often more than 50% of them are lost before age 80. There may be an increase in blood urea nitrogen (BUN) without serious symptoms. Elderly persons are more susceptible than young persons to infections of the urinary system. Childbearing may cause damage to the musculature of the pelvic floor, resulting in urinary tract problems in later years. Enlargement of the prostate, common in older men, may cause obstruction and a back pressure in the ureters and kidneys. If this condition is untreated, it will cause permanent damage to the kidneys. Age changes may predispose to but do not cause incontinence (inability to control urination). Most elderly persons (60% in nursing homes and up to 90% living independently) have no incontinence.

## The Urine

### Normal Constituents

Urine is a yellowish liquid that is about 95% water and 5% dissolved solids and gases. The amount of these dissolved substances is indicated by its *specific gravity.* The specific gravity of pure water, used as a standard, is 1.000. Because of the dissolved materials it contains, urine has a specific gravity that normally varies from 1.002 (very dilute urine) to 1.040 (very concentrated urine). When the kidneys are diseased, they lose the ability to concentrate urine, and the specific gravity no longer varies as it does when the kidneys function normally.

Some of the dissolved substances normally found in the urine are the following:

1. *Nitrogenous waste products,* including urea, uric acid, and *creatinine* (kre-AT-ih-nin)
2. *Electrolytes,* including sodium chloride (as in common table salt) and different kinds of sulfates and phosphates. Electrolytes are excreted in appropriate amounts to keep their blood concentration constant.
3. *Yellow pigment,* which is derived from certain bile compounds. Pigments from foods and drugs also may appear in the urine.

Some abnormal substances found in urine include glucose, ketones (signs of diabetes mellitus), albumin, blood (signs of kidney inflammation), and pus (a sign of kidney infection).

## Body Fluids

The normal proportion of body water varies from 50% to 70% of a person's weight. It is highest in the young and in thin, muscular individuals. As the amount of fat increases, and as a person ages, the percentage of water in the body decreases. Water is important to living cells as a solvent, as a transport medium, and as a participant in metabolic reactions.

Various electrolytes (salts), nutrients, gases, waste, and special substances such as enzymes and hormones are dissolved or suspended in body water. The composition of body fluids is an important factor in homeostasis. Whenever the volume or chemical makeup of these fluids deviates even slightly from normal, disease results. The constancy of body fluids is maintained in ways that include the following:

1. The thirst mechanism, which maintains the volume of water at a constant level
2. Kidney activity, which regulates the volume and composition of body fluids
3. Hormones, which serve to regulate fluid volume and electrolytes
4. Regulators of pH (acidity), including buffers, respiration, and kidney function

### Fluid Compartments

Although body fluids have much in common no matter where they are located, there are some important differences between fluid inside and fluid outside cells. Accordingly, fluids are grouped into two main compartments:

1. *Intracellular fluid* is contained within the cells. About two thirds to three fourths of all body fluids are in this category.
2. *Extracellular fluid* includes all body fluids outside of cells. In this group are included the following:
   a. *Blood plasma,* which constitutes about 4% of an individual's body weight
   b. *Interstitial* (in-ter-STISH-al) *fluid,* or more simply, tissue fluid. This fluid is located in the spaces between the cells of tissues all over the body. It is estimated that tissue fluid constitutes about 15% of body weight.
   c. *Lymph*
   d. *Fluid in special compartments,* such as cerebrospinal fluid, the aqueous and vitreous humors of the eye, serous fluid, and synovial fluid. Together these make up about 1% to 3% of total body fluids. Fluids are not locked into one compartment. There is a constant interchange between compartments as fluids are transferred across semipermeable cell membranes by diffusion and osmosis (Fig. 19-10).

### Intake and Output of Water

In a person whose health is normal, the quantity of water taken in (intake) is approximately equal to the quantity lost (output). The quantity of water consumed in a day varies considerably. The average adult in a comfortable environment takes in about 2500 mL of water daily. About half of this quantity comes from drinking water and other beverages, and about half comes from foods—fruits, vegetables, and soups.

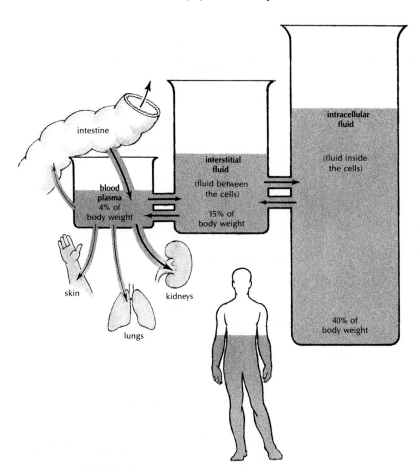

**Figure 19-10**

Main fluid compartments showing the relative percentage of body fluid in each.

Water is constantly being lost from the body by the following routes:

1. The ***skin.*** Although sebum and keratin help prevent dehydration, water is constantly evaporating from the surface of the skin. Larger amounts of water are lost from the skin in the form of sweat when it is necessary to cool the body.
2. The ***lungs*** expel water along with carbon dioxide.
3. The ***intestinal tract*** eliminates water along with the feces.
4. The ***kidneys*** excrete the largest quantity of water lost each day. About 1 to 1.5 liters of water are eliminated daily in the urine.

### SENSE OF THIRST

The control center for the sense of thirst is located in the hypothalamus. This center plays a major role in the regulation of total fluid volume. A decrease in fluid volume or an increase in the concentration of body fluids stimulates the thirst center and thus causes an individual to drink water or other fluids containing large amounts of water. Dryness of the mouth also causes a sensation of thirst.

### ELECTROLYTES AND THEIR FUNCTIONS

Electrolytes are important constituents of body fluids. These are compounds that separate into positively and negatively charged ions and carry an electric current in solution. A few of the most important ions are reviewed below. The first three are positive ions (cations), and the last two are negative ions (anions).

1. ***Sodium*** is chiefly responsible for maintaining osmotic balance and body fluid volume. It is the main positive ion in extracellular fluids. Sodium is required for nerve impulse conduction and is important in maintaining acid–base balance.
2. ***Potassium*** is also important in the transmission of nerve impulses and is a major positive ion in in-

tracellular fluids. Potassium is involved in cellular enzyme activities, and it helps regulate the chemical reactions by which carbohydrate is converted to energy and amino acids are converted to protein.

3. *Calcium* is required for bone formation, muscle contraction, nerve impulse transmission, and blood clotting.
4. *Phosphate* is essential in the metabolism of carbohydrates, bone formation, and acid–base balance. Phosphates are found in the cell membrane and in the nucleic acids (DNA and RNA).
5. *Chloride* is essential for formation of the hydrochloric acid of the gastric juice.

Electrolytes must be kept in the proper concentration in both intracellular and extracellular fluids. Although some electrolytes are lost in the feces and through the skin as sweat, the job of balancing electrolytes is left mainly to the kidneys, as described earlier in this chapter.

Several hormones are involved in this process. Aldosterone produced by the adrenal cortex promotes the reabsorption of sodium (and water) and the elimination of potassium. Calcium and phosphate levels are regulated by hormones from the parathyroid and thyroid glands. Parathyroid hormone increases blood calcium levels by causing the bones to release calcium and by causing the kidneys to reabsorb calcium. The thyroid hormone calcitonin lowers blood calcium by causing calcium to be deposited in the bones.

## The Acid–Base Balance

The pH scale is a measure of how acidic or basic (alkaline) a solution is. Body fluids are slightly alkaline at approximately pH 7.4. They must be kept within a narrow range of pH, or damage, even death, will result. Several systems act together to maintain acid–base balance. These are:

1. *Buffer systems.* Buffers are substances that prevent sharp changes in hydrogen ion (H+) concentration and thus maintain a relatively constant pH. They do so by accepting or releasing these ions as needed. The main buffer systems in the body are bicarbonate buffers, phosphate buffers, and proteins, such as hemoglobin and plasma proteins.
2. *Respiration.* The role of respiration in controlling pH was described in Chapter 16. Recall that the release of carbon dioxide from the lungs acts to make the blood more alkaline (increase the pH) by reducing the amount of carbonic acid formed. In contrast, a reduction in the release of carbon dioxide will act to make the blood more acidic (decrease the pH).
3. *Kidney function.* The kidneys serve to regulate pH by reabsorbing or eliminating hydrogen ions as needed. Much of the hydrogen that is eliminated by the kidneys is transported into the nephron by tubular secretion. Uncontrolled shifts in pH may result in *acidosis,* a condition caused by a drop in the pH of body fluids, or *alkalosis,* caused by an increase in pH. Either condition is dangerous and may be fatal.

## SUMMARY

I. **Urinary system**
   A. Main function is excretion—removal and elimination of waste materials from blood
      1. Other systems that eliminate waste
         a. Digestive system—eliminates undigested food, water, salts, bile
         b. Respiratory system—eliminates carbon dioxide, water
         c. Skin—eliminates water, salts, nitrogen waste
II. **Kidneys**—located in upper abdomen, against the back
   A. Structure
      1. Cortex—outer portion
      2. Medulla—inner portion
      3. Pelvis
         a. Upper end of ureter
         b. Calyces—cuplike extensions that receive urine
      4. Nephron
         a. Functional unit of kidney
         b. Parts—Bowman's capsule, proximal convoluted tubule, Henle's loop, distal convoluted tubule
      5. Blood supply
         a. Renal artery
            (1) Afferent arteriole—enters Bowman's capsule

**(2)** Efferent arteriole—leaves Bowman's capsule

**(3)** Glomerulus—coil of capillaries in Bowman's capsule

**(4)** Peritubular capillaries—surround nephron

    **b.** Renal vein

**B.** Functions

    **1.** Excretion of waste, excess salts, toxins

    **2.** Water balance

    **3.** Regulation of pH

    **4.** Production of renin—enzyme that regulates blood pressure via angiotensin

    **5.** Production of erythropoietin—hormone that stimulates red blood cell production

**C.** Physiology

    **1.** Filtration—pressure in glomerulus forces water and soluble substances out of blood and into Bowman's capsule

      **a.** Glomerular filtrate—material that leaves blood and enters the nephron

    **2.** Reabsorption—most of filtrate leaves nephron by diffusion and active transport and returns to blood through peritubular capillaries

    **3.** Tubular secretion—materials moved from blood into nephron for excretion

    **4.** Counter-current mechanism—method for concentrating urine based on movement of ions out of nephron

      **a.** ADH

        **(1)** Hormone from posterior pituitary

        **(2)** Promotes reabsorption of water

**III. Organs that store and transport urine**

**A.** Ureters—carry urine from the kidneys to the bladder

**B.** Urinary bladder—stores urine until it is eliminated

    **1.** Micturition—urination

      **a.** Reflex control—impulses from central nervous system in response to stretching of bladder wall

      **b.** Voluntary control—external sphincter around urethra

**C.** Urethra—carries urine out of body

    **1.** Female urethra—4.0 cm long; opens in front of vagina

    **2.** Male urethra—20 cm long; carries both urine and semen

**IV. Urine**

**A.** Normal constituents—water, nitrogenous waste, electrolytes, pigments

**B.** Abnormal constituents—glucose, ketones, albumin, blood, pus

**V. Body fluids**

**A.** Fluid compartments

    **1.** Intracellular fluid—contained within the cells

    **2.** Extracellular fluid—outside the cells

      **a.** Blood plasma

      **b.** Interstitial (tissue) fluid

      **c.** Lymph

      **d.** Fluid in special compartments

**B.** Water balance

    **1.** Output—through skin, lungs, intestinal tract, kidneys

    **2.** Intake—through food and beverages

      **a.** Thirst center—in hypothalamus

**C.** Electrolytes—release ions in solution

    **1.** Regulation

      **a.** Kidneys—main regulators

      **b.** Hormones

        **(1)** Aldosterone—reabsorption of sodium, excretion of potassium

        **(2)** Parathyroid hormone—increase in blood calcium level

        **(3)** Calcitonin—decrease in blood calcium level

**D.** Acid–base balance—normal pH is 7.4

    **1.** Mechanisms to regulate pH

      **a.** Buffers—maintain constant pH

      **b.** Kidney—regulates amount of hydrogen ion excreted

      **c.** Respiration—release of carbon dioxide increases pH (less acid); retention of carbon dioxide decreases pH (more acid)

# QUESTIONS FOR STUDY AND REVIEW

1. Name the body systems that have excretory functions.
2. Name the organs of the urinary system.
3. Where are the kidneys located?
4. Describe the external appearance of the kidneys and tell what tissues form most of the kidney structure.
5. Describe the structure of a nephron.
6. Name the blood vessels associated with the nephron.
7. Compare the afferent arteriole and the efferent arteriole.
8. What happens to the glomerular filtrate as it passes through the nephron?
9. How does ADH affect kidney function?
10. What is the function of the juxtaglomerular apparatus?
11. What structures empty into the kidney pelvis and what drains the pelvis?
12. What is micturition? How is it controlled?
13. Describe the female urethra and tell how it differs from the male urethra in structure and function.
14. What are some of the effects of aging on the urinary system?
15. What is urea? Where and how is it formed?
16. Name three types of substances dissolved in urine.
17. List four ways in which body fluids are regulated.
18. In a healthy person, what is the ratio of fluid intake to output?
19. Name five common ions in the body. How is each used?
20. Name three hormones involved in electrolyte balance and explain what each does.
21. Name the three main buffer systems in the body.
22. How does the respiratory system help regulate pH?

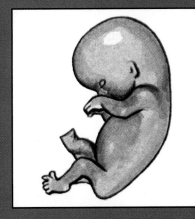

# UNIT **VI**

# Perpetuating
# Life

**T**he last unit includes two
chapters on the struc-
tures and functions related to
reproduction and inheritance.
The reproductive system is
not necessary for the continuation of the life of the individual, but rather is needed for
the continuation of the human
species. The germ cells and
their genes have been studied
intensively during recent
years and are part of the
rapidly developing science of
heredity.

# CHAPTER 20

# The Male and Female Reproductive Systems

## Behavioral Objectives

After careful study of this chapter, you should be able to:

- Name the male and female gonads and describe the function of each
- State the purpose of meiosis
- List the accessory organs of the male and female reproductive tracts and cite the function of each
- Describe the composition and function of semen
- Draw and label a spermatozoon
- List in the correct order the hormones produced during the menstrual cycle and cite the source of each
- Describe the functions of the main male and female sex hormones
- Explain how negative feedback regulates reproductive function in both males and females
- Describe the changes that occur during and after menopause
- Define *contraception* and cite the main methods of contraception in use

Memmler, RL, Cohen, BJ, Wood, DL. *STRUCTURE AND FUNCTION OF THE HUMAN BODY*, 6/e,
© 1996 Lippincott-Raven Publishers

The chapters in this unit deal with what is cer tainly one of the most interesting and mysterious attributes of life: the ability to reproduce. The simplest forms of life, one-celled organisms, usually need no partner to reproduce; they simply divide by themselves. This form of reproduction is known as *asexual* (nonsexual) reproduction.

In most animals, however, reproduction is *sexual,* meaning that there are two kinds of individuals, males and females, each of which has specialized cells designed specifically for the perpetuation of the species. These specialized sex cells are known as *germ cells,* or *gametes* (GAM-etes). In the male they are called *spermatozoa* (sper-mah-to-ZO-ah), and in the female they are called *ova* (O-vah). Germ cells are characterized by having half as many chromosomes as are found in any other cell in the body. During their formation they go through a special process of cell division, called *meiosis* (mi-O-sis), that halves the number of chromosomes. In humans, meiosis reduces the chromosome number from 46 to 23.

Although the reproductive apparatuses of males and females are different, the organs of both sexes may be divided into two groups: primary and accessory.

1. The primary organs are the *gonads* (GO-nads), or sex glands; they produce the germ cells and manufacture hormones. The male gonads are the *testes* (TES-teze), and the female gonads are the *ovaries* (O-vah-reze).
2. The *accessory organs* include a series of ducts that provide for the transport of germ cells, as well as various exocrine glands.

## ▶ The Male Reproductive System

### The Testes

The male gonads (testes) are normally located outside the body proper, suspended between the thighs in a sac called the *scrotum* (SKRO-tum) (Fig. 20-1). The testes are egg-shaped organs measuring about 3.7 to 5 cm (1.5–2 inches) in length and approximately 2.5 cm (1 inch) in each of the other two dimensions. During embryonic life the testes develop from tissue near the kidney. A month or two before birth, the testes normally descend (move downward) through the *inguinal* (ING-gwih-nal) *canal* in the abdominal wall into the scrotum. Each testis then remains suspended by a *spermatic cord* (Fig. 20-2) that extends through the inguinal canal. This cord contains blood vessels, lymphatic vessels, nerves, and the tube (ductus deferens) that transports spermatozoa away from the testis. Each gland must descend completely if it is to function normally; to produce spermatozoa, the testes must be kept at the temperature of the scrotum, which is lower than that of the abdominal cavity.

The bulk of the specialized tissue of the testes consists of tiny coiled *seminiferous* (seh-mih-NIF-er-us) *tubules.* Cells in the walls of these tubules produce spermatozoa. Between the tubules are the specialized *interstitial* (in-ter-STISH-al) *cells* that secrete the male sex hormone *testosterone* (tes-TOS-teh-rone). Figure 20-3 is a microscopic view of the testis in cross section showing the seminiferous tubules, interstitial cells, and developing spermatozoa.

After being secreted by the testes, testosterone is absorbed directly into the bloodstream. This hormone has two functions. The first is maintenance of the reproductive structures, including development of the spermatozoa. A second involves the development of *secondary sex characteristics,* traits that characterize males and females but are not directly concerned with reproduction. In males they include a deeper voice, broader shoulders, narrower hips, a greater percentage of muscle tissue, and more body hair than are found in females.

### The Duct System

The tubes that carry the spermatozoa (also referred to simply as *sperm cells* or *sperm*) begin with the tubules inside the testis itself. From these tubes the cells are collected by a greatly coiled tube 6 meters (20 feet) long, called the *epididymis* (ep-ih-DID-ih-mis), which is located inside the scrotal sac. While they are temporarily stored in the epididymis, the sperm cells mature and become motile, that is, able to move, or "swim," by themselves. The epididymis finally extends upward as the *ductus deferens* (DEF-er-enz), also called the *vas deferens.* The ductus deferens, contained in the spermatic cord, continues through the inguinal canal into the abdominal cavity. Here, it separates from the remainder of the spermatic cord and curves behind the urinary bladder. The ductus deferens then joins with the duct of the *seminal vesicle* (VES-ih-kl) on the same side to form the *ejaculatory* (e-JAK-u-lah-to-re) *duct.* The two

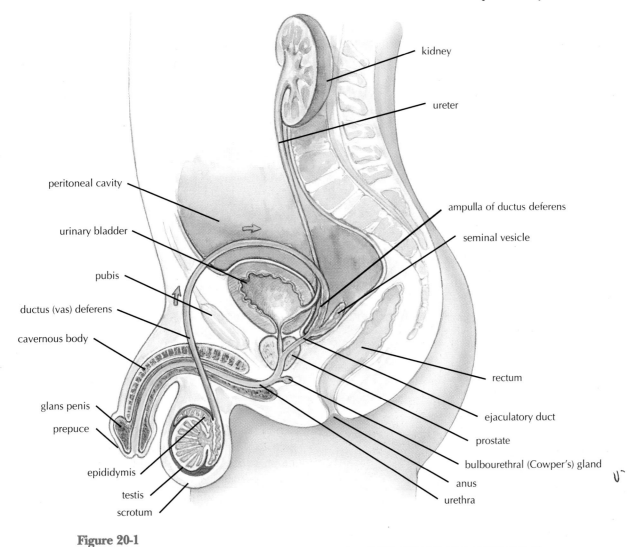

**Figure 20-1**

Male genitourinary system. The arrows indicate the course of sperm cells through the duct system.

ejaculatory ducts enter the body of the prostate gland where they empty into the urethra.

### Formation of Semen

***Semen*** (SE-men) is the mixture of spermatozoa and various secretions that is expelled from the body. The secretions serve to nourish and transport the spermatozoa, neutralize the acidity of the vaginal tract, and lubricate the reproductive tract during sexual intercourse. The glands discussed below contribute secretions to the semen.

#### THE SEMINAL VESICLES

The seminal vesicles are tortuous muscular tubes with small outpouchings. They are about 7.5 cm

(3 inches) long and are attached to the connective tissue at the back of the urinary bladder. The glandular lining produces a thick, yellow, alkaline secretion containing large quantities of simple sugar and other substances that provide nourishment for the sperm. The seminal fluid forms a large part of the volume of the semen.

#### THE PROSTATE GLAND

The prostate gland lies immediately below the urinary bladder, where it surrounds the first part of the urethra. Ducts from the prostate carry its secretions into the urethra. The thin, alkaline prostatic secretion helps neutralize the acidity of the vaginal tract and enhance the motility of the spermatozoa.

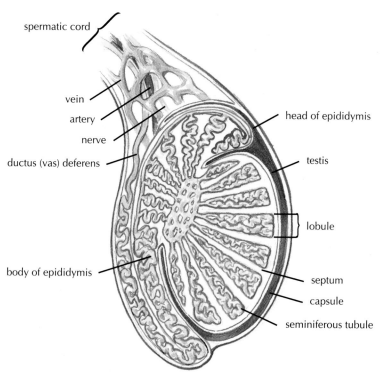

**Figure 20-2**

Structure of the testis, also showing the epididymis and spermatic cord.

The prostate gland is also supplied with muscular tissue, which upon signal from the nervous system contracts to aid in the expulsion of the semen from the body.

### MUCUS-PRODUCING GLANDS

The largest of the mucus-producing glands in the male reproductive system are *Cowper's,* or *bulbourethral* (bul-bo-u-RE-thral), *glands,* a pair of pea-sized organs located in the pelvic floor just below the prostate gland. The ducts of these glands extend about 2.5 cm (1 inch) from each side and empty into the urethra before it extends within the penis. Other

very small glands secrete mucus into the urethra as it passes through the penis. The mucus from all of these glands serves mainly as a lubricant.

### The Urethra and the Penis

The male urethra, as we saw earlier, serves the dual purpose of conveying urine from the bladder and carrying the reproductive cells and their accompanying secretions to the outside. The ejection of semen into the receiving canal (vagina) of the female is made possible by the *erection,* or stiffening and enlargement, of the penis, through which the longest part of the urethra extends. The penis is

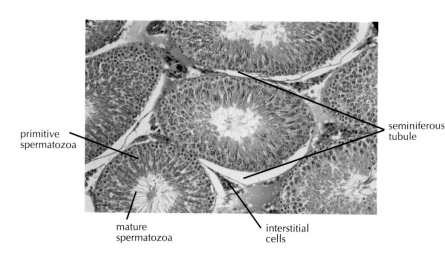

**Figure 20-3**

Microscopic view of the testis. (Courtesy of Dana Morse Bittus and B. J. Cohen)

made of a sponge-like tissue containing many blood spaces that are relatively empty when the organ is flaccid but fill with blood and distend when the penis is erect. The spongy tissue is subdivided into three segments, each called a ***corpus*** (meaning "body") (Fig. 20-4). A single, ventrally located ***corpus spongiosum*** contains the urethra. On either side is a larger ***corpus cavernosum*** (pl., *corpora cavernosa*). The penis and scrotum are referred to as the *external genitalia* of the male.

Reflex centers in the spinal cord order the contraction of the smooth muscle tissue in the prostate gland, followed by the contraction of skeletal muscle in the pelvic floor. This provides the force needed for ***ejaculation*** (e-jak-u-LA-shun), the expulsion of semen through the urethra to the outside.

### The Spermatozoa

Spermatozoa are tiny individual cells (Fig. 20-5). They are so small that at least 200 million are contained in the average ejaculation. Spermatozoa are continuously manufactured in the testes. They develop within the seminiferous tubules with the aid of special cells called ***Sertoli,*** or "nurse," ***cells.*** These cells help nourish and protect the developing sperm cells.

The individual sperm cell has an oval head that is largely a nucleus containing chromosomes. Covering the head like a cap is the ***acrosome*** (AK-rosome), which contains enzymes that help the spermatozoon penetrate the ovum. The whiplike tail (flagellum) enables the sperm cell to make its way through the various passages until it reaches the ovum of the female. A middle region contains many mitochondria to provide energy for movement. Out of the millions of spermatozoa in an ejaculation, only one, if any, fertilizes the ovum. The remainder

of the cells live only a few hours, up to a maximum of 3 days.

## ◗ Hormonal Control of Male Reproduction

The activities of the testes are under the control of two hormones produced by the anterior pituitary gland (hypophysis). One of these, ***follicle stimulating hormone (FSH),*** stimulates the Sertoli cells and promotes the formation of spermatozoa. Testosterone is also needed in this process. The other hormone is ***luteinizing hormone (LH),*** called ***interstitial cell–stimulating hormone*** (***ICSH***) in males. It stimulates the interstitial cells to produce testosterone. These pituitary hormones are named for their activity in female reproduction, although they are chemically the same in both males and females.

The pituitary gland is regulated by a region of the brain just above it called the *hypothalamus.* Starting at puberty, the hypothalamus begins to secrete hormones that trigger the release of FSH and LH. These hormones are secreted continuously in the male. The activity of the hypothalamus is in turn regulated by a negative feedback mechanism involving testosterone. As the level of testosterone in the blood increases, the hypothalamus secretes less releasing hormone; as the level of testosterone decreases, the hypothalamus secretes more releasing hormone (see Fig. 11-2).

### The Effects of Aging on Male Reproduction

A gradual decrease in the production of testosterone and spermatozoa begins as early as age 20

**Figure 20-4**

Cross section of the penis.

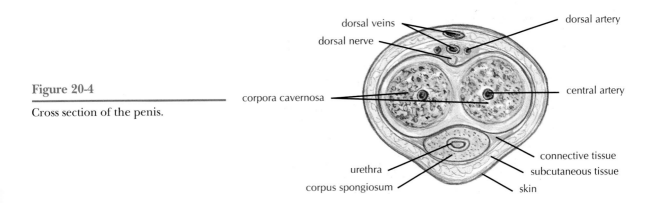

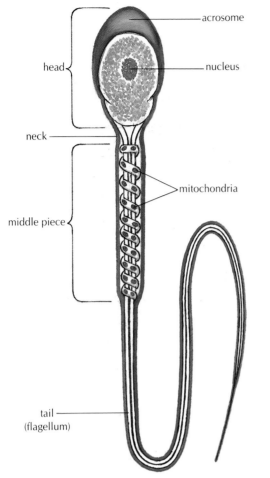

**Figure 20-5**

Diagram of a human spermatozoon showing major structural features. (Chaffee EE, Lytle IM: Basic Physiology and Anatomy, 4th ed, p. 549. Philadelphia, JB Lippincott, 1980)

and continues throughout life. Secretions from the prostate and seminal vesicles decrease in amount and become less viscous. In a few men (less than 10%) sperm cells remain late in life, even to age 80.

## ▶ The Female Reproductive System

### The Ovaries

In the female the counterparts of the testes are the two *ovaries*, where the female sex cells, or *ova*, are formed (Fig. 20-6). The ovaries are small, somewhat flattened oval bodies measuring about 4 cm (1.6 inches) in length, 2 cm (0.8 inch) in width, and 1 cm (0.4 inch) in depth. Like the testes, the ovaries

descend, but only as far as the pelvic portion of the abdomen. Here, they are held in place by ligaments, including the broad ligament, the ovarian ligament, and others, that attach them to the uterus and the body wall.

The outer layer of each ovary is made of a single layer of epithelium. Beneath this layer the ova are produced. The ova begin a complicated process of maturation, or "ripening," which takes place in small fluid-filled clusters of cells called *ovarian follicles* (o-VA-re-an FOL-ih-kls) or *Graafian* (GRAF-e-an) *follicles* (Fig. 20-7). The cells of the ovarian follicle walls secrete estrogen. When an ovum has ripened, the ovarian follicle ruptures, and the ovum is discharged from the surface of the ovary and makes its way to the nearest *oviduct* (O-vih-dukt). The oviducts are two tubes, one of which is in the vicinity of each ovary. The rupture of a follicle allowing the escape of the egg cell is called *ovulation* (ov-u-LA-shun).

The ovaries of a newborn female contain a large number of potential ova. Each month during the reproductive years, several ripen, but usually only one is released. Once this ovum has been expelled, the follicle is transformed into a solid glandular mass called the *corpus luteum* (LU-te-um), which means "yellow body." This structure secretes both estrogen and progesterone. Commonly, the corpus luteum shrinks and is replaced by scar tissue. When a pregnancy occurs, however, this structure remains active for about two months.

## In Vitro Fertilization

The term *in vitro* literally means "in glass." It is used to describe tests or procedures done in the laboratory. In vitro fertilization refers to fertilization of an egg cell outside the mother's body in a laboratory dish. The first such procedure was done successfully in 1978.

The mother must be given hormones to cause ovulation of several eggs. These are then withdrawn with a needle and fertilized with the father's sperm. After a few divisions the fertilized egg is placed in the uterus for development. Additional fertilized eggs can be frozen to repeat the procedure in case of failure or for later pregnancies. Other possibilities for in vitro fertilization include using donor eggs, donor sperm cells, or both.

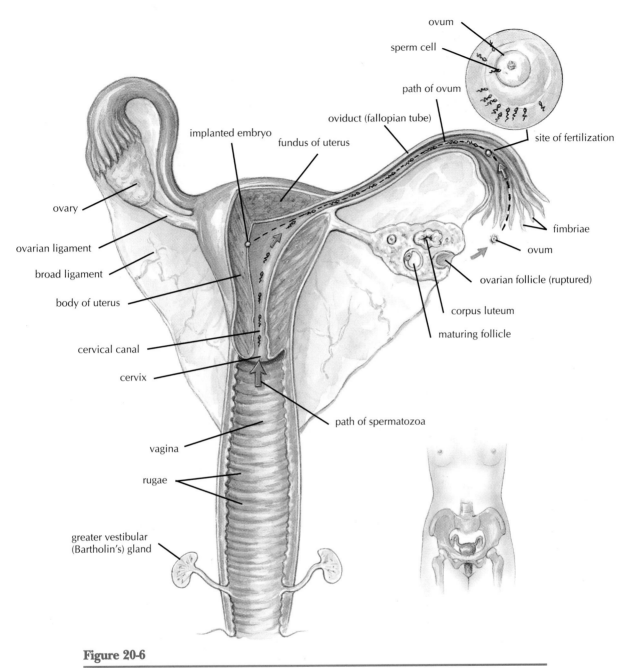

**Figure 20-6**

Female reproductive system.

## The Oviducts

The tubes that transport the ova in the female reproductive system, the oviducts, are also known as *uterine* (U-ter-in) *tubes,* or *fallopian* (fah-LO-pe-an) *tubes.* They are small, muscular structures, nearly 12.5 cm (5 inches) long, extending from a point near the ovaries to the uterus (womb). There is no direct connection between the ovaries and these tubes. The ova are swept into the oviducts by a current in the peritoneal fluid produced by the small, fringe-like extensions called *fimbriae* (FIM-bre-e) that are located at the edges of the abdominal openings of the tubes.

Unlike the spermatozoon, the ovum cannot move by itself. Its progress through the oviduct toward the uterus depends on the sweeping action of

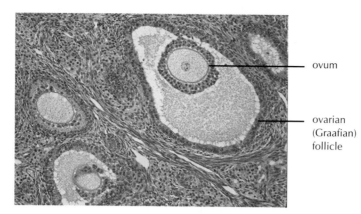

ovum

ovarian
(Graafian)
follicle

**Figure 20-7**

Microscopic view of the ovary showing egg cells developing within ovarian follicles. (Courtesy of Dana Morse Bittus and B. J. Cohen)

cilia in the lining of the tubes and on peristalsis of the tubes. It takes about 5 days for an ovum to reach the uterus from the ovary.

### The Uterus

The organ to which the oviducts lead is the *uterus* (U-ter-us), and it is within this structure that the fetus grows to maturity.

The uterus is a pear-shaped, muscular organ about 7.5 cm (3 inches) long, 5 cm (2 inches) wide, and 2.5 cm (1 inch) deep. The upper portion rests on the upper surface of the urinary bladder; the lower portion is embedded in the pelvic floor between the bladder and the rectum. The wider upper portion is called the body, or *corpus;* the lower, narrower part is the *cervix* (SER-viks), or neck. The small, rounded part above the level of the tubal entrances is known as the *fundus* (FUN-dus) (see Fig. 20-6). The cavity inside the uterus is shaped somewhat like a capital T, but it is capable of changing shape and dilating as a fetus develops. The cervix leads to the *vagina* (vah-JI-nah), the lower part of the birth canal, which opens to the outside of the body.

The lining of the uterus is a specialized epithelium known as *endometrium* (en-do-ME-tre-um), and it is this layer that is involved in menstruation.

The *broad ligaments* support the uterus, extending from each side of the organ to the lateral body wall. Along with the uterus, these two portions of peritoneum form a partition dividing the female pelvis into anterior and posterior areas. The ovaries are suspended from the broad ligaments, and the oviducts lie within the upper borders. Blood vessels that supply these organs are found between the layers of the broad ligament.

### The Vagina

The vagina is a muscular tube about 7.5 cm (3 inches) long connecting the uterine cavity with the outside. It receives the cervix, which dips into the upper vagina in such a way that a circular recess is formed, giving rise to areas known as *fornices* (FOR-nih-seze). The deepest of these spaces, behind the cervix, is the *posterior fornix* (FOR-niks) (Fig. 20-8). This recess in the posterior vagina lies adjacent to the lowest part of the peritoneal cavity, a narrow passage between the uterus and the rectum named the *cul-de-sac* (from a French term meaning "bottom of the sack"). A rather thin layer of tissue separates the posterior fornix from this region.

The lining of the vagina is a wrinkled mucous membrane something like that found in the stomach. The folds (rugae) permit enlargement so that childbirth usually does not tear the lining. In addition to being a part of the birth canal, the vagina is the organ that receives the penis during sexual intercourse. A fold of membrane called the *hymen* may sometimes be found at or near the vaginal (VAJ-ih-nal) canal opening.

#### THE GREATER VESTIBULAR GLANDS

Just above and to each side of the vaginal opening are the mucus-producing *greater vestibular* (ves-TIB-u-lar), or *Bartholin's, glands.* These glands open into an area near the vaginal opening known as the *vestibule.*

### The Vulva and the Perineum

The external parts of the female reproductive system form the *vulva* (VUL-vah), which includes two pairs of lips, or *labia* (LA-be-ah); the *clitoris* (KLIT-o-ris), which is a small organ of great sensi-

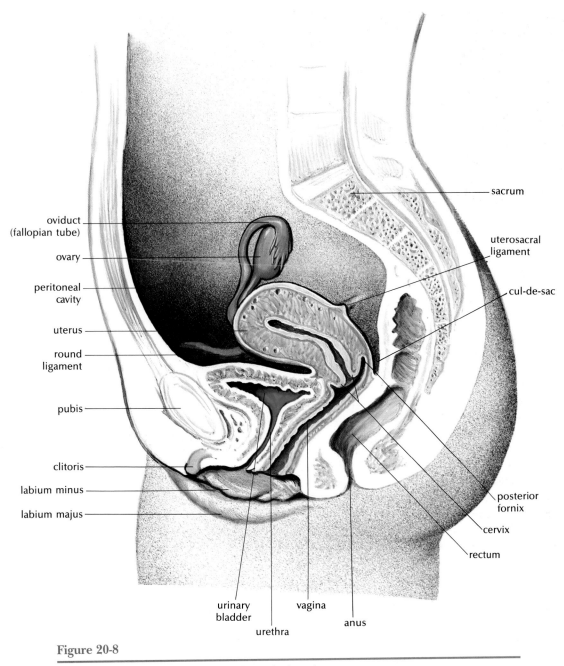

**Figure 20-8**

Female reproductive system, as seen in sagittal section.

tivity; and related structures (Fig. 20-9). Although the entire pelvic floor in both the male and female is properly called the *perineum* (per-ih-NE-um) (see Fig. 7-10), those who care for pregnant women usually refer to the limited area between the vaginal opening and the anus as the perineum.

## The Menstrual Cycle

In the female, as in the male, reproductive function is controlled by hormones from the pituitary gland as regulated by the hypothalamus. Female activity differs, however, in that it is cyclic, that is, shows reg-

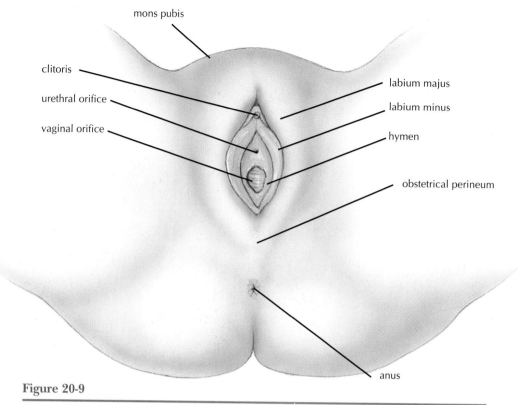

mons pubis

clitoris

urethral orifice

vaginal orifice

labium majus

labium minus

hymen

obstetrical perineum

anus

**Figure 20-9**

The external female genitalia.

ular patterns of increases and decreases in hormone levels. These changes are regulated by hormonal feedback.

The length of the menstrual cycle varies between 22 and 45 days in normal women, but 28 days is taken as an average, with the first day of menstrual flow being considered the first day of the cycle (Fig. 20-10).

At the start of each cycle, under the influence of FSH produced by the pituitary, a follicle begins to develop in the ovary. This follicle produces increasing amounts of *estrogen* as the ovum matures. (*Estrogen* is the term used for a group of related hormones, the most active of which is estradiol.) The estrogen is carried in the bloodstream to the uterus, where it starts preparing the endometrium for a possible pregnancy. This preparation includes thickening of the endometrium and elongation of the glands that produce the uterine secretion. Estrogen in the blood also acts as a feedback messenger to inhibit the release of FSH and stimulate the release of LH from the pituitary (see Fig. 11-2).

About 1 day before ovulation, there is an **LH surge,** a sharp rise in LH. This hormone causes ovulation and transforms the ruptured follicle into the corpus luteum. The corpus luteum produces some estrogen and large amounts of *progesterone.* Under the influence of these hormones, the endometrium continues to thicken, the glands and blood vessels increasing in size. The rising levels of estrogen and progesterone feed back to inhibit the release of FSH and LH from the pituitary (Fig. 20-11). During this time the ovum makes its journey to the uterus by way of the oviduct. If the ovum is not fertilized while passing through the uterine tube, it dies within 2 to 3 days and then disintegrates.

If fertilization does not occur, the corpus luteum degenerates and the levels of estrogen and progesterone decrease. Without the hormones to support growth, the endometrium degenerates. Small hemorrhages appear in this lining, producing the bleeding known as *menstrual flow.* Bits of endometrium break away and accompany the flow of blood. The average duration of this discharge is 2 to 6 days.

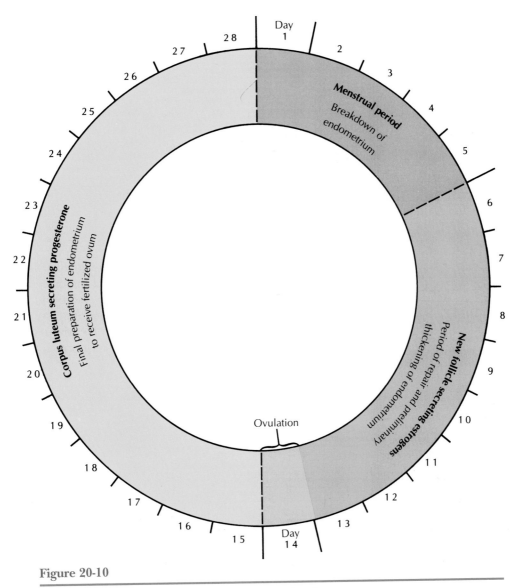

**Figure 20-10**

Summary of events in an average 28-day menstrual cycle.

Before the flow ceases, the endometrium begins to repair itself through the growth of new cells. The low levels of estrogen and progesterone allow the release of FSH from the anterior pituitary. This causes a new ovum to begin to ripen within the ovaries, and the cycle begins anew.

The activity of ovarian hormones as negative feedback messengers is the basis of hormonal methods of contraception (birth control). The hormones act to inhibit the release of FSH and LH from the pituitary, resulting in a menstrual period but no ovulation (Table 20-1).

## ▶ Menopause

*Menopause* (MEN-o-pawz), often called *change of life,* is that period at which menstruation ceases altogether. It ordinarily occurs between the ages of 45 and 55 and is caused by a normal decline in ovarian function. The ovary becomes chiefly scar tissue and no longer produces ova or appreciable amounts of estrogen. Eventually the uterus, oviducts, vagina, and vulva all shrink somewhat in size.

Although menopause is an entirely normal condition, its onset sometimes brings about effects that

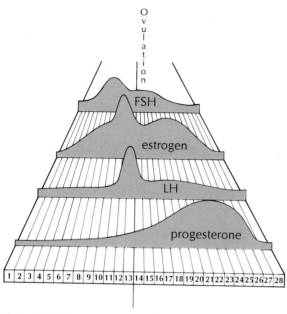

**Figure 20-11**

Hormones in the menstrual cycle. (Redrawn from Djerassi C: Fertility awareness: Jet-age rhythm method? Science 1990;248:1061)

are temporarily disturbing. The decrease in estrogen levels can cause such nervous symptoms as irritability, "hot flashes," and dizzy spells.

Although still a subject of controversy, the use of estrogen replacement therapy is considered to be of overall benefit for menopausal women. Estrogen therapy has proved effective in alleviating hot flashes and the sensitivity of a thinning vaginal mucosa. Some women may require hormone therapy to prevent or halt osteoporosis (bone weakening). It has also been found to reduce the risk of heart attacks, which generally increase among women after menopause. The longer the duration of hormone treatment, the greater is the risk of cancer of the uterus (endometrial cancer). Reducing the dosages and the duration of the therapy lessens this risk, as does giving the estrogen in combination with progesterone (progestin) to prevent overgrowth of the endometrium. Another factor that must be considered involves the risk of blood clots, which is highest among women who smoke. Some studies have also shown an increased risk of breast cancer in the wake of long-term estrogen therapy. A woman who has a family history of cancer of the breast or of the uterus should avoid such treatment.

## ▶ Contraception

**Contraception** is defined as the use of artificial methods to prevent fertilization of the ovum or implantation of the fertilized ovum. Table 20-1 presents a brief description of the main contraceptive methods currently in use, along with some advantages and disadvantages of each. The list is given in rough order of decreasing effectiveness.

The spread of AIDS and other sexually transmitted diseases has stimulated interest in the development of better condoms, including a female condom, as these not only prevent conception but also reduce the chances of infection.

Most other advances involve new methods for administering birth control hormones. Capsules of synthetic progesterone implanted under the skin are effective for 5 years and avoid the side effects of estrogen. This same drug may be given by injection at 3-month intervals.

Although not strictly a contraceptive, the drug **RU 486** is taken after conception to terminate an early pregnancy. It blocks the action of progesterone, causing the uterus to shed its lining and release the fertilized egg. It must be combined with administration of prostaglandins to expel the uterine tissue.

## TABLE 20-1
## Main Methods of Contraception Currently in Use

| Method | Description | Advantages | Disadvantages |
|---|---|---|---|
| **Surgical** | | | |
| Vasectomy/tubal ligation | Cutting and tying of tubes carrying gametes | Nearly 100% effective; involves no chemical or mechanical devices | Not usually reversible; rare surgical complications |
| **Hormonal** | | | |
| Birth control pill | Estrogen and progesterone or progesterone alone taken orally to prevent ovulation | Highly effective; requires no last-minute preparation | Alters physiology; possible serious side effects |
| Birth control shot | Injection of synthetic progesterone every 3 months to prevent ovulation | Highly effective; lasts for 3 to 4 months | Alters physiology; possible side effects include menstrual irregularity, amenorrhea |
| Birth control implants | Devices containing synthetic progesterone implanted under skin to prevent ovulation | Highly effective; lasts for 5 years | Alters physiology; possible side effects include menstrual irregularity, amenorrhea; expensive |
| **Barrier** | | | |
| Male condom | Sheath that fits over erect penis and prevents release of semen | Easily available; does not affect physiology; protects against sexually transmitted disease (STD) | Must be applied just before intercourse; may slip or tear |
| Female condom | Sheath that fits in vagina, held in place with rings | Easily available; protects against STD | More expensive than male condom; must be applied before intercourse |
| Diaphragm (with spermicide) | Rubber cap that fits over cervix and prevents entrance of sperm | Does not affect physiology; some protection against STD | Must be inserted before intercourse; requires fitting by physician |
| **Other** | | | |
| Spermicide | Chemicals used to kill sperm; best when used in combination with a barrier method | Easily available, does not affect physiology; some protection against STD | Local irritation; must be used just before intercourse |
| Fertility awareness | Abstinence during fertile part of cycle as determined by menstrual history, basal body temperature, or quality of cervical mucus | Does not affect physiology; accepted by certain religions | High failure rate; requires careful record keeping |

## SUMMARY

**I. Structure of reproductive tract**
   **A.** Primary organs—gonads
      1. Testes in male; ovaries in female
      2. Germ cells (sperm and eggs)—produced in gonads by meiosis, which reduces chromosome number from 46 to 23
   **B.** Accessory organs—ducts and exocrine glands

**II. Male reproductive system**
   **A.** Structure
      1. Testes
         **a.** Products
            **(1)** Spermatozoa—produced in seminiferous tubules
               **(a)** Head—mainly nucleus
               **(b)** Middle region—has mitochondria
               **(c)** Tail—for locomotion
               **(d)** Acrosome—covers head (aids in penetration of ovum)
            **(2)** Testosterone—produced in interstitial cells (between tubules)
               **(a)** Development of spermatozoa
               **(b)** Production of secondary sex characteristics
      2. Ducts
         **a.** Epididymis—stores spermatozoa until ejaculation
         **b.** Ductus deferens (vas deferens)
            **(1)** Transports spermatozoa
            **(2)** Part of spermatic cord that suspends testis
         **c.** Ejaculatory duct—receives secretions from seminal vesicle and ductus deferens
         **d.** Urethra—carries semen through penis
      3. Penis—copulatory organ
      4. Scrotum—sac that contains testes
   **B.** Semen
      1. Functions
         **a.** Nourishment and transportation of spermatozoa
         **b.** Lubrication of reproductive tract
         **c.** Neutralization of female reproductive tract

      2. Formation
         **a.** Seminal vesicles
         **b.** Prostate
         **c.** Bulbourethral glands
   **C.** Hormonal control of reproduction
      1. Hypothalamus—produces releasing hormone, causing anterior pituitary to release FSH and LH
      2. FSH—stimulates formation of spermatozoa
      3. LH (ICSH)—stimulates production of testosterone
      4. Testosterone—regulates continuous hormone production by acting as negative feedback messenger to hypothalamus

**III. Female reproductive tract**
   **A.** Structure
      1. Ovaries—produce ova and hormones
      2. Oviducts—carry ovum to uterus by means of cilia in lining and by peristalsis
      3. Uterus—site for development of fertilized egg
         **a.** Endometrium—lining of uterus
         **b.** Cervix—narrow region at bottom
         **c.** Broad ligament—supports uterus, ovaries, oviducts
      4. Vagina—canal for copulation, birth, menstrual flow
         **a.** Hymen—membrane that covers opening of vagina
         **b.** Greater vestibular glands—secrete mucus
      5. Vulva—external female genitalia
         **a.** Labia—two liplike pairs of tissue
         **b.** Clitoris—sensitive tissue
         **c.** Perineum—pelvic floor
   **B.** Menstrual cycle
      1. First hormone is FSH from anterior pituitary, which starts ripening of ovum within follicle
      2. Production of estrogen by follicle
         **a.** Development of endometrium for possible pregnancy
         **b.** Feedback to hypothalamus to inhibit FSH and start production of LH
      3. LH—surge occurs 24 hours before ovulation

**a.** Ovulation—caused by LH on day 14 of 28-day cycle

**b.** Conversion of follicle to corpus luteum by LH

**4.** Production of progesterone (and some estrogen) by corpus luteum

    **a.** Continued development of endometrium

    **b.** Feedback to inhibit release of LH

**5.** No fertilization

    **a.** Degeneration of corpus luteum

    **b.** Drop in hormone levels

    **c.** Menstruation

**IV. Menopause**—cessation of menstruation

    **A.** Characteristics

      **1.** Degeneration of reproductive organs

      **2.** Possibility of disturbing side effects

      **3.** Increased risk of osteoporosis and heart disease

    **B.** Therapy—estrogen replacement

**V. Contraception**—use of methods to prevent fertilization or implantation of ovum

## QUESTIONS FOR STUDY AND REVIEW

**1.** In what fundamental respect does reproduction in some single-celled animals such as the ameba differ from that in most animals?

**2.** Name the sex cells of both the male and the female.

**3.** Name all the parts of the male reproductive system and describe the function of each.

**4.** Name the principal parts of the female reproductive system and describe the function of each.

**5.** Describe the process of ovulation.

**6.** Beginning with the first day of the menstrual flow, describe the events of one complete cycle, including the role of the various hormones.

**7.** What is menopause? What causes it? What are some of the changes that take place in the body as a result?

**8.** What are the values of estrogen replacement therapy after menopause? What are some of the dangers of such therapy?

**9.** Define *contraception*. Describe methods of contraception that involve (1) barriers; (2) chemicals; (3) hormones; (4) prevention of implantation.

# CHAPTER 21

# Development and Heredity

## Behavioral Objectives

After careful study of this chapter, you should be able to:

- Describe fertilization and the early development of the fertilized egg
- Describe the structure and function of the placenta
- Briefly describe the four stages of labor
- Cite the advantages of breastfeeding
- Briefly describe the mechanism of gene function
- Explain the difference between dominant and recessive genes
- Describe what is meant by a *carrier* of a genetic trait
- Define *meiosis* and explain the function of meiosis in reproduction
- Explain how sex is determined in humans
- Describe what is meant by the term *sex linked* and list several sex-linked traits
- List several factors that may influence the expression of a gene
- Define *mutation*

Memmler, RL, Cohen, BJ, Wood, DL. *STRUCTURE AND FUNCTION OF THE HUMAN BODY*, 6/e,
© 1996 Lippincott-Raven Publishers

# ◗ Pregnancy

Pregnancy begins with fertilization of an ovum and ends with delivery of the fetus and afterbirth. During this 38-week period of development, known as *gestation* (jes-TA-shun), all body tissues differentiate from a single fertilized egg. Along the way, many changes occur in both the mother and the growing offspring.

## Fertilization and the Start of Pregnancy

When semen is deposited in the vagina, the many spermatozoa immediately wriggle about in all directions, some traveling into the uterus and oviducts (see Fig. 20-6). If an egg cell is present in the oviduct, many spermatozoa cluster around it. Using enzymes they dissolve the coating around the ovum so that eventually one sperm cell can penetrate the cell membrane. The nuclei of the sperm and egg then combine.

The result of this union is a single cell with the full human chromosome number of 46. This new cell, called a *zygote* (ZI-gote), can divide and grow into a new individual. The zygote divides rapidly into two cells and then four cells, and soon a ball of cells is formed. During this time, the ball of cells is traveling toward the uterine cavity, pushed along by the cilia and by the peristalsis of the tube. After reaching the uterus, the little ball of cells burrows into the greatly thickened uterine lining, where it is soon completely covered and implanted.

## Development and Functions of the Placenta

Following *implantation,* a group of cells within the ball becomes an *embryo* (EM-bre-o). The outer cells form projections, called *villi,* that invade the uterine wall and maternal blood channels (venous sinuses). This eventually leads to the formation of the *placenta* (plah-SEN-tah), a flat, circular organ that consists of a spongy network of blood-filled lakes and capillary-containing villi.

The embryo, later called the *fetus* (FE-tus), is connected to the developing placenta by a stalk of tissue that eventually becomes the *umbilical* (um-BIL-ih-kal) *cord.* The cord contains two arteries that carry blood from the fetus to the placenta and one vein that carries blood from the placenta to the fetus

(Fig. 21-1). The placenta serves as the organ of nutrition, respiration, and excretion for the embryo through exchanges that occur between the blood of the embryo and the blood of the mother across the capillaries of the villi.

Another function of the placenta is endocrine in nature. Beginning soon after implantation, some of the embryonic cells produce a hormone called *human chorionic gonadotropin* (ko-re-ON-ik gon-ah-do-TRO-pin) *(HCG).* This hormone stimulates the corpus luteum of the ovary, prolonging its life span (to 11 or 12 weeks) and causing it to secrete increasing amounts of progesterone and estrogen. Progesterone is essential for the maintenance of pregnancy. It promotes endometrial secretion to nourish the embryo, and it decreases the ability of the uterine muscle to contract, thus preventing the embryo from being expelled from the body. During pregnancy progesterone also helps prepare the breasts for the secretion of milk. Estrogen promotes enlargement of the uterus and breasts. By the 11th or 12th week of pregnancy, the corpus luteum is no longer needed; by this time, the placenta itself can secrete adequate amounts of progesterone and estrogen.

## Development of the Embryo

For the first eight weeks of life the developing offspring is referred to as an *embryo* (Fig. 21-2). The beginnings of all body systems are established during this period. The heart and the brain are among the first organs to develop. A primitive nervous system begins to form in the third week. The heart and blood vessels originate during the second week, and the first heartbeat appears during week 4, at the same time that other muscles begin to develop. By the end of the first month, the embryo is about 0.62 cm (0.25 inches) long, with four small swellings at the sides called *limb buds,* which will develop into the four extremities. At this time the heart produces a prominent bulge at the front of the embryo. By the end of the second month, the embryo takes on an appearance that is recognizably human. The study of embryonic development is *embryology* (em-bre-OL-o-je).

## The Fetus

The term *fetus* is used for the developing individual from the beginning of the third month until birth. During this period the organ systems continue to grow and mature. For study, the period of devel-

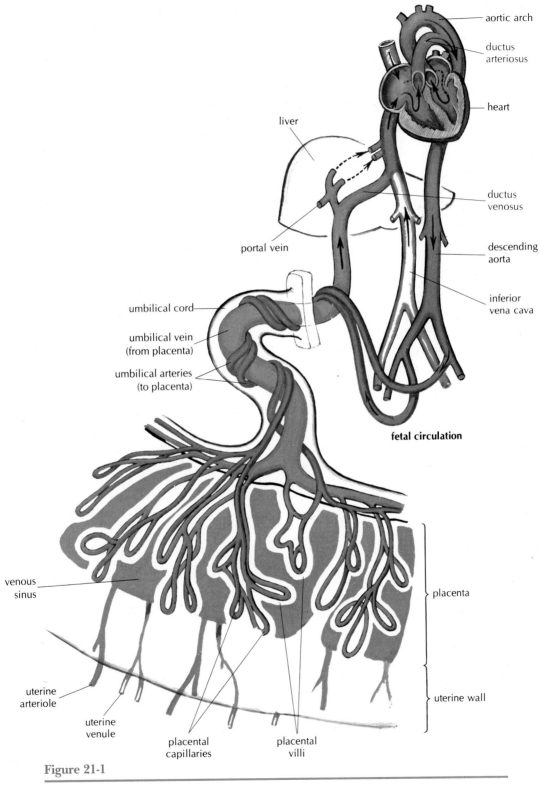

**fetal circulation**

**Figure 21-1**

Fetal circulation and placenta.

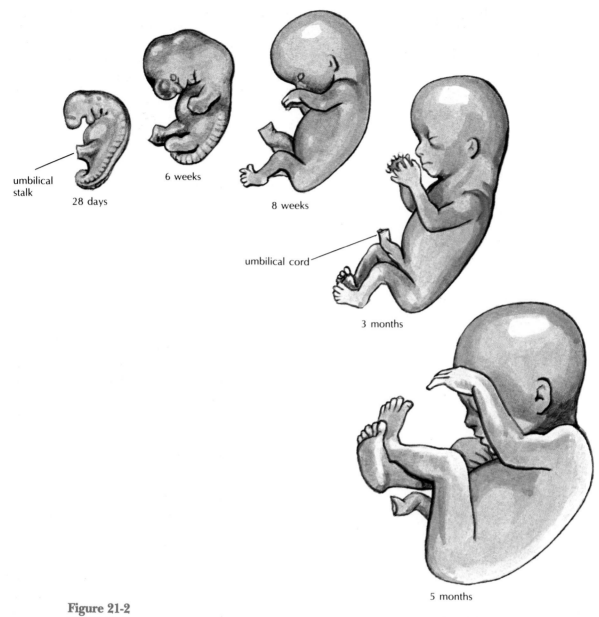

umbilical
stalk

28 days

6 weeks

8 weeks

umbilical cord

3 months

5 months

**Figure 21-2**

Development of an embryo into a fetus.

opment (gestation), may be divided into three equal segments or ***trimesters.*** The most rapid growth occurs during the second trimester (months 4–6). By the end of the fourth month, the fetus is almost 15 cm (6 inches) long, and its external genitalia are sufficiently developed to reveal its sex. By the seventh month, the fetus is usually about 35 cm (14 inches) long and weighs about 1.1 kg (2.4 lbs). At the end of pregnancy, the normal length of the fetus is 45 to 56 cm (18–22.5 inches), and the weight varies from 2.7 to 4.5 kg (6–10 lb).

The ***amniotic*** (am-ne-OT-ik) ***sac,*** which is filled with a clear liquid known as ***amniotic fluid,*** surrounds the fetus and serves as a protective cushion for it (Fig. 21-3). Popularly called the *bag of waters,* the amniotic sac ruptures at birth. During development the skin of the fetus is protected by a layer of cheese-like material called ***vernix caseosa*** (VER-niks KA-se-o-sa).

## The Origin of Tissues

Within the first weeks of development, the cells of the fertilized egg differentiate into three layers that will give rise to all the tissues of the body. These three *germ layers* are the *ectoderm* (outermost), *endoderm* (innermost), and *mesoderm* (middle).

The cells of the ectoderm eventually form all nervous tissue, the epidermis of the skin, hair, nails, and skin glands. The endoderm forms the lining of the respiratory and digestive tracts and some glands. The mesoderm gives rise to all muscle and connective tissue, including bone, cartilage, and blood, and to serous membranes.

Most organs eventually contain cells that originate from all three of these layers. For example, the stomach has nerves (ectoderm), muscle and connective tissue (mesoderm), and an epithelial lining (endoderm). No matter where they appear, all body cells trace their origins to one of these three primitive germinal layers.

### The Mother

The total period of pregnancy, from fertilization of the ovum to birth, is about 266 days, also given as 280 days or 40 weeks from the last menstrual period (LMP). During this time, the mother must supply all the food and oxygen for the fetus and eliminate its waste materials. To support the additional demands of the growing fetus, the mother's metabolism changes markedly, with increased demands being made on several organ systems:

1. The heart pumps more blood to supply the needs of the uterus and its contents.
2. The lungs provide more oxygen to supply the fetus by increasing the rate and depth of respiration.
3. The kidneys excrete nitrogenous wastes from the fetus and from the mother's body.
4. Nutritional needs are increased to provide for the growth of maternal organs (uterus and breasts) and growth of the fetus, as well as for preparation for labor and the secretion of milk.

Nausea and vomiting are common discomforts in early pregnancy. The specific cause is not known, but these symptoms may be due to the great changes in hormone levels that occur at this point. The nausea and vomiting usually last for only a few weeks. Frequency of urination and constipation are often present during the early stages of pregnancy and then usually disappear. They may reappear late in pregnancy as the head of the fetus drops from the abdominal region down into the pelvis, pressing on the rectum and the urinary bladder.

## ▶ Childbirth

The mechanisms that trigger the beginning of uterine contractions for childbirth are still unknown. However, it is recognized that the uterine muscle becomes increasingly sensitive to oxytocin (from the pituitary gland) late in pregnancy. Once labor is started, stimuli from the cervix and vagina produce reflex secretion of this hormone, which in turn increases the uterine contractions.

The process by which the fetus is expelled from the uterus is known as *labor* and *delivery;* it also may be called *parturition* (par-tu-RISH-un). It is divided into four stages:

1. The *first stage* begins with the onset of regular contractions of the uterus. With each contraction, the cervix becomes thinner and the opening larger. Rupture of the amniotic sac may occur at any time, with a gush of fluid from the vagina.
2. The *second stage* begins when the cervix is completely dilated and ends with the delivery of the baby. This stage involves the passage of the fetus, usually head first, through the cervical canal and the vagina to the outside.
3. The *third stage* begins after the child is born and ends with the expulsion of the *afterbirth.* The afterbirth includes the placenta, the membranes of the amniotic sac, and the umbilical cord, except for a small portion remaining attached to the baby's *umbilicus* (um-BIL-ih-kus), or navel.
4. The *fourth stage* begins with the expulsion of the afterbirth and constitutes a period in which bleeding is controlled. Contraction of the uterine muscle acts to close off the blood vessels leading to the placental site.

To prevent the pelvic floor tissues from being torn during childbirth, as often happens, the physician may cut the mother's perineum just before her infant is born and then repair this clean cut imme-

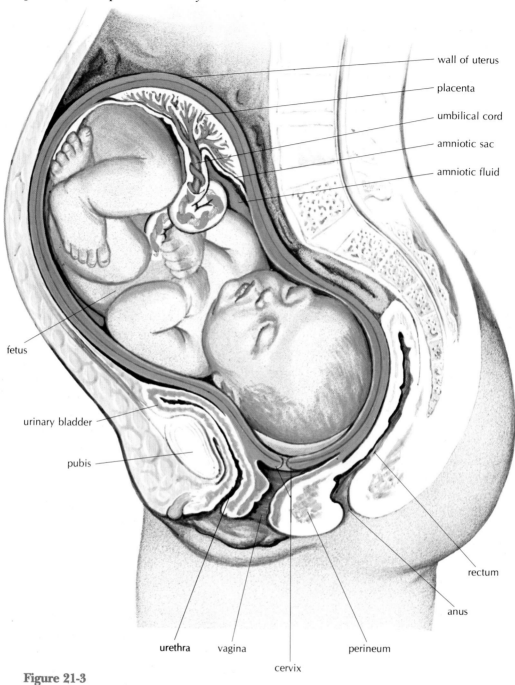

**Figure 21-3**

Midsagittal section of a pregnant uterus with intact fetus.

diately after childbirth; such an operation is called an *episiotomy* (eh-piz-e-OT-o-me). The area between the vagina and the anus that is cut in an episiotomy is referred to as the *surgical* or *obstetrical perineum* (see Fig. 20-9).

## Multiple Births

Until recently, statistics indicated that twins occurred in about 1 of every 80 to 90 births, varying somewhat in different countries. Triplets occurred much less frequently, usually once in several thousand births, while quadruplets occurred very rarely indeed. The birth of quintuplets represented a historic event unless the mother had taken fertility drugs. Now these fertility drugs, usually gonadotropins, are given quite frequently, and the number of multiple births has increased significantly. Multiple fetuses tend to be born prematurely and therefore have a high death rate. However, better care of infants and newer treatments have resulted in more living multiple births than ever.

Twins originate in two different ways and on this basis are divided into two groups:

1. *Fraternal twins* are formed as a result of the fertilization of two different ova by two spermatozoa. Two completely different individuals, as distinct from each other as brothers and sisters of different ages, are produced. Each fetus has its own placenta and surrounding sac.
2. *Identical twins* develop from a single zygote formed from a single ovum fertilized by a single spermatozoon. Obviously, they are always the same sex and carry the same inherited traits. Sometime during the early stages of development the embryonic cells separate into two units. Usually there is a single placenta, although there must be a separate umbilical cord for each individual.

Other multiple births may be fraternal, identical, or combinations of these. The tendency to multiple births seems to be hereditary.

### Live Births and Fetal Deaths

A pregnancy may end before its full term has been completed. The term *live birth* is used if the baby breathes or shows any evidence of life such as heartbeat, pulsation of the umbilical cord, or movement of voluntary muscles. An *immature* or *premature* infant is one born before the organ systems are mature. Infants born before the 37th week of gestation or weighing less than 2500 grams (5.5 lb) are considered *preterm.*

Loss of the fetus is classified according to the duration of the pregnancy:

1. The term *abortion* refers to loss of the embryo or fetus before the 20th week or weight of about 500 grams (1.1 lb). This loss can be either spontaneous or induced.
   a. *Spontaneous abortions* occur naturally with no interference. The most common causes are related to an abnormality of the embryo or fetus. Other causes include abnormality of the mother's reproductive organs, infections, or chronic disorders such as kidney disease or hypertension. *Miscarriage* is the lay term for spontaneous abortions.
   b. *Induced abortions* occur as a result of artificial or mechanical interruption of pregnancy. *Therapeutic abortions* are abortions performed by physicians as a form of treatment for a variety of reasons. With more liberal use of this type of abortion, there has been a dramatic decline in the incidence of death related to illegal abortion.
2. The term *fetal death* refers to loss of the fetus after the eighth week of pregnancy. *Stillbirth* refers to the delivery of an infant that is lifeless.

Immaturity is a leading cause of death in the newborn. After the 20th week of pregnancy, the fetus is considered *viable,* that is, able to live outside the uterus. A fetus expelled before the 28th week or a weight of 1000 grams (2.2 lb) has a less than 50% chance of survival. One born at a point closer to the full 40 weeks stands a much better chance of survival. Increasing numbers of immature infants are being saved because of advances in newborn intensive care.

## ▶ The Mammary Glands and Lactation

The *mammary glands,* or the breasts of the female, are accessories of the reproductive system. They are designed to provide nourishment for the baby after its birth, and the secretion of milk at this time is known as *lactation* (lak-TA-shun).

The mammary glands are constructed in much the same manner as the sweat glands. Each of these glands is divided into a number of lobes composed of glandular tissue and fat, and each lobe is further subdivided. The secretions from the lobes are conveyed through *lactiferous* (lak-TIF-er-us) *ducts,* all of which converge at the nipple (Fig. 21-4).

The mammary glands begin developing during puberty, but they do not become functional until the end of a pregnancy. The hormone *prolactin (PRL),* produced by the anterior lobe of the pituitary, stimulates the secretory cells of the mammary glands. The first of the mammary gland secretions is a thin

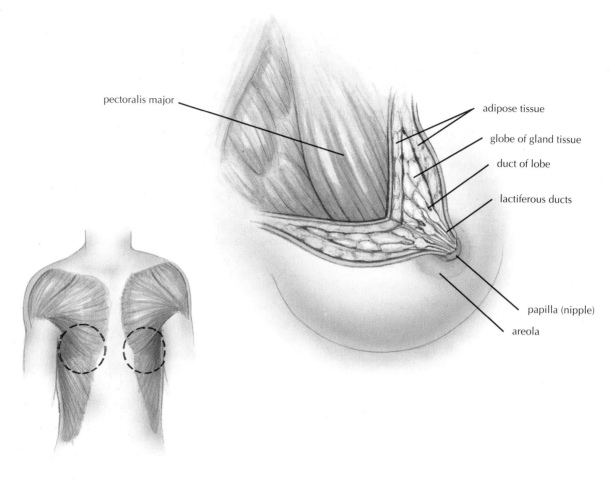

**Figure 21-4**

Section of the breast.

liquid called **colostrum** (ko-LOS-trum). It is nutritious but has a somewhat different composition from milk. Secretion of milk begins within a few days and can continue for several months as long as milk is frequently removed by the suckling baby or by pumping. Stimulation of the breast by the suckling infant causes the release of **oxytocin** from the posterior pituitary gland. This hormone promotes the ejection or *letdown* of milk.

The digestive tract of the newborn baby is not ready for the usual mixed diet of an adult. Mother's milk is more desirable for the young infant than milk from other animals for several reasons, some of which are listed below:

1. Infections that may be transmitted by foods exposed to the outside air are avoided by nursing.
2. Both breast milk and colostrum contain maternal antibodies that help protect the baby against disease-causing organisms.
3. The proportions of various nutrients and other substances in human milk are perfectly suited to the human infant. Substitutes are not exact imitations of human milk. Nutrients are present in more desirable amounts if the mother's diet is well balanced.
4. The psychologic and emotional benefits of nursing are of infinite value to both the mother and the infant.

## ▶ Heredity

Often we are struck by the resemblance of a baby to one or both of its parents, yet rarely do we stop to consider *how* various traits are transmitted from parents to offspring. This subject—heredity—has fascinated humans for thousands of years; the Old Testament contains numerous references to heredity (although, of course, the word was unknown in biblical times). However, it was not until the 19th century that methodic investigation into heredity was begun. At that time an Austrian monk, Gregor Mendel, discovered through his experiments with garden peas that there was a precise pattern in the appearance of differences among parents and their progeny (offspring). Mendel's most important contribution to the understanding of heredity was the demonstration that there are independent units of hereditary influence. Later, these independent units were given the name *genes*.

## Genes and Chromosomes

Genes are actually segments of DNA (deoxyribonucleic acid) contained in the threadlike chromosomes within the nucleus of each cell. Genes act by controlling the manufacture of enzymes, which are necessary for all the chemical reactions that occur within the cell. When body cells divide by the process of mitosis, the DNA that makes up the chromosomes is duplicated and distributed to the daughter cells so that each daughter cell gets exactly the same kind and number of chromosomes as were in the original cell. Each chromosome (aside from the Y chromosome, which determines sex) may carry thousands of genes, and each gene carries the code for a specific trait (characteristic). These traits constitute the physical, biochemical, and physiologic makeup of every cell in the body.

In humans, every cell except the sex cells contains 46 chromosomes. These chromosomes exist in pairs. One member of each pair was received at the time of fertilization from the father of the offspring, and one was received from the mother. These paired chromosomes, except for the pair that determines sex, are alike in size and appearance and carry genes for the same traits. Thus, the genes for each trait exist in pairs.

### DOMINANT AND RECESSIVE GENES

Another of Mendel's discoveries was that genes can be either dominant or recessive. A *dominant* gene is one that expresses its effect in the cell regardless of whether the gene at the same site on the matching chromosome is the same as or different from the dominant gene. The gene need be re-ceived from only one parent to be expressed in the offspring. The effect of a *recessive* gene will not be evident unless the gene at that site in the matching chromosome of the pair is also recessive. Thus, a recessive trait will appear only if the recessive genes for that trait are received from both parents. For example, genes for dark eyes are dominant, whereas genes for light eyes are recessive. Light eyes will appear in the offspring only if genes for light eyes are received from both parents.

Clearly, a recessive gene will not be expressed if it is present in the cell together with a dominant gene. However, it can be passed on to offspring and may thus appear in future generations. An individual who shows no evidence of a trait but has a recessive gene for that trait is described as a *carrier* of the gene.

### DISTRIBUTION OF CHROMOSOMES TO OFFSPRING

The reproductive cells (ova and spermatozoa) are produced by a special process of cell division called *meiosis.* This process divides the chromosome number in half so that each reproductive cell has 23 chromosomes. Moreover, the division occurs in such a way that each cell receives one member of each chromosome pair that was present in the original cell. The separation occurs at random, meaning that either member of the original pair may be included in a given germ cell. Thus, the maternal and paternal sets of chromosomes get mixed up and redistributed at this time, leading to increased variety within the population.

### SEX DETERMINATION

The two chromosomes that determine the sex of the offspring, unlike the other 22 pairs of chromosomes, are not matched in size and appearance. The female X chromosome is larger than most other chromosomes and carries genes for other characteristics in addition to that for sex. The male Y chromosome is smaller than other chromosomes and mainly determines sex. A female has two X chromosomes in each body cell; a male has one X and one Y. By the process of meiosis, each male sperm cell receives either an X or a Y chromosome, whereas every egg cell can have only an X chromosome. If a sperm with an X chromosome fertilizes an ovum, the resulting infant will be female; if a sperm with a Y chromosome fertilizes an ovum, the resulting infant will be male (Fig. 21-5).

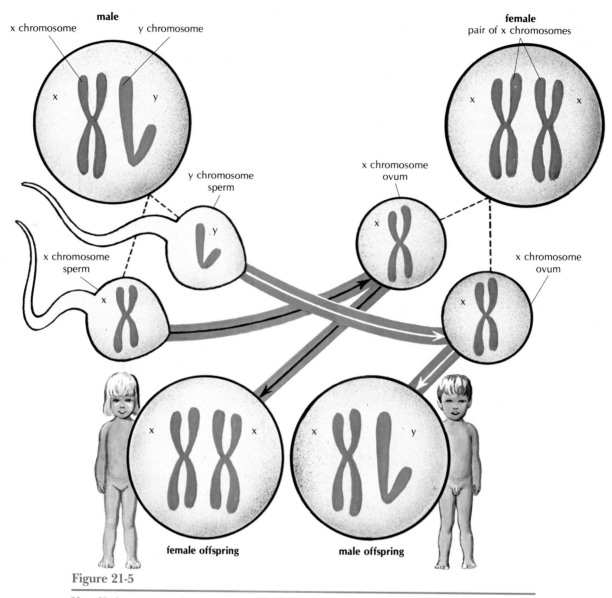

**Figure 21-5**

If an X chromosome from a male unites with an X chromosome from a female, the child is female; if a Y chromosome from a male unites with an X chromosome from a female, the child is male.

## SEX-LINKED TRAITS

Any trait that is carried on a sex chromosome is said to be *sex-linked*. Because the Y chromosome carries few traits aside from sex determination, most sex-linked traits are carried on the X chromosome and are described as *X-linked*. Examples are hemophilia (a hereditary bleeding disorder), certain forms of baldness, and red–green color blindness. Sex-linked traits appear almost exclusively in males. The reason for this is that most of these traits are re-cessive, and if a recessive gene is located on the X chromosome in a male it cannot be masked by a matching dominant gene. Thus, a male who has only one recessive gene for the trait will exhibit the characteristic, whereas a female must have two recessive genes to show the trait.

### Hereditary Traits

Some observable hereditary traits are skin, eye, and hair color and facial features. Also influenced by genetics are less clearly defined traits such as weight, body build, life span, and susceptibility to disease. Some human traits, including the traits involved in many genetic diseases, are determined by a single pair of genes; most, however, are the result of two or more gene pairs acting together in what is termed *multifactorial inheritance.* This accounts for the wide range of variations within populations in such characteristics as coloration, height, and weight, all of which are determined by more than one pair of genes.

### GENE EXPRESSION

The effect or expression of a gene may be influenced by a variety of factors, including the sex of the individual and the presence of other genes. For example, the genes for certain types of baldness and certain types of color blindness may be inherited by either males or females, but the traits appear mostly in males under the effects of male sex hormone.

Environment also plays a part in the expression of genes. One inherits a potential for a given size, for example, but one's actual size is additionally influenced by such factors as nutrition, development, and gen-

eral state of health. The same is true of life span and susceptibility to diseases.

### GENE MUTATION

As a rule, chromosomes replicate exactly during cell division. Occasionally, however, for reasons not yet clearly understood, the genes or chromosomes change. This change may be in a single gene or in the number of chromosomes. Alternatively, it may consist of chromosomal breakage, in which there is loss or rearrangement of gene fragments. Such changes are termed genetic *mutations.* Mutations may occur spontaneously or may be induced by some agent, such as ionizing radiation or chemicals, described as a *mutagenic agent.*

If the mutation occurs in an ovum or a sperm cell involved in reproduction, the altered trait will be inherited by the offspring. The vast majority of harmful mutations never are expressed because the affected fetus dies and is spontaneously aborted. Most remaining mutations are so inconsequential that they have no visible effect. Beneficial mutations, on the other hand, tend to survive and increase as a population evolves.

## SUMMARY

I. **Pregnancy**
 A. Fertilization
  1. Occurs in oviduct
  2. Zygote (fertilized egg)—formed by fusion of egg and sperm nuclei
 B. Early pregnancy
  1. Embryo (developing egg)—travels into uterus and implants
  2. Placenta
   a. Formation—formed by cells around embryo and lining of uterus; connected to embryo by umbilical cord
   b. Functions
    (1) Provision of nourishment to embryo
    (2) Gas exchange with embryo
    (3) Removal of waste from embryo
    (4) Production of chorionic gonadotropin—maintains corpus luteum until 11–12 weeks of pregnancy
  3. Amniotic sac—fluid-filled sac around embryo that serves as protective cushion

  4. Fetus—embryo becomes fetus in third month
II. **Childbirth**
 A. Four stages
  1. Contractions
  2. Delivery of baby
  3. Expulsion of afterbirth
  4. Contraction of uterus
 B. Multiple births—incidence increased because of fertility drugs
 C. Abortion—loss of embryo or fetus before week 20 of pregnancy; may be spontaneous or induced
  1. Fetal death—loss of fetus after 8 weeks of pregnancy
III. **Lactation**
 A. Hormones
  1. Prolactin
  2. Oxytocin
 B. Advantages
  1. Reduces infections
  2. Provides best form of nutrition

3. Transfers antibodies
4. Emotional satisfaction

IV. **Heredity**
  A. Genes
    1. Hereditary units
    2. Segments of DNA in chromosomes within nucleus
    3. Control manufacture of enzymes
    4. Types
      **a.** Dominant gene—always expressed
      **b.** Recessive gene—expressed only if gene received from both parents
        **(1)** Carrier—person with recessive gene that is not apparent but can be passed to offspring
  B. Chromosomes
    1. 46 in each human body cell (except sex cells)
    2. Pairs—matched in size and appearance; one member of each pair from each parent
    3. X and Y chromosomes—determine sex
    4. Meiosis—form of cell division
      **a.** Production of sex cells
      **b.** Halving of chromosome number from 46 to 23
      **c.** Random separation of maternal and paternal chromosomes into different daughter cells
    5. Sex chromosomes—females XX, males XY

      **a.** X chromosome—larger than most others; carries traits in addition to sex determination
      **b.** Sex-linked traits—appear mostly in males (e.g. hemophilia, baldness, red–green color blindness)
  C. Hereditary traits—genes determine physical, biochemical, and physiologic characteristics of every cell
    1. Some traits determined by single gene pairs
    2. Most traits determined by multiple gene pairs (multifactorial inheritance)
      **a.** Produces a range of variations in a population
      **b.** Examples: height, weight, coloration, susceptibility to disease
    3. Factors affecting gene expression
      **a.** Sex
      **b.** Presence of other genes
      **c.** Environment
    4. Mutation
      **a.** Change in genes or chromosomes
      **b.** May be passed to offspring if occurs in germ cells
      **c.** Mutagenic agents—induce mutation
        **(1)** Ionizing radiation
        **(2)** Chemicals

## QUESTIONS FOR STUDY AND REVIEW

1. Distinguish among the following: zygote, embryo, fetus.
2. Name two auxiliary structures that are designed to serve the fetus. What is their origin and what are their functions?
3. Is blood in the umbilical arteries relatively low or high in oxygen? in the umbilical vein?
4. Describe some of the changes that take place in the mother's body during pregnancy.
5. What is the major event of each of the four stages of labor and delivery?
6. What is an episiotomy? Why is it done?
7. Distinguish between fraternal and identical twins.
8. Define *abortion; premature infant; preterm infant.*
9. Name two hormones involved in lactation. Where do they originate and what do they do?
10. List some of the advantages associated with breastfeeding a baby.
11. How many chromosomes are there in a human body cell? in a human sex cell?
12. What process results in the distribution of chromosomes to the offspring?
13. What sex chromosomes are present in a male? in a female?
14. From which parent does a male child receive an X-linked gene?
15. Explain how two normal parents can give birth to a child with a recessive hereditary disease.
16. Explain the great variation in the color of skin, hair, and eyes in humans.
17. Define *mutation;* name two types of mutagenic agents.

# Suggestions for Further Study

Amenta PS: Histology and Human Anatomy, 6th ed. New York, Piccin, 1991

Beck WS: Hematology, 5th ed. Cambridge, MA, MIT Press, 1994

Bishop ML, Duben-Engelkirk JL, Fody EP: Clinical Chemistry, 2nd ed. Philadelphia, JB Lippincott, 1992

Cormack DH: Esssential Histology. Philadelphia, JB Lippincott, 1992

Carroll HJ, Oh MS: Water, Electrolyte and Acid-Base Metabolism, 2nd ed. Philadelphia, JB Lippincott, 1989

Dorland's Illustrated Medical Dictionary, 28th ed. Philadelphia, WB Saunders, 1994

Ganong WF: Review of Medical Physiology, 16th ed. San Mateo, CA, Appleton and Lange, 1993

Guyton AC: Medical Physiology, 8th ed. Philadelphia, WB Saunders, 1990

Gosling JA, Harris PF, Whitmore IV, Humpherson JR, Willan PLT: Atlas of Human Anatomy. Philadelphia, JB Lippincott, 1987

Grant's Atlas of Anatomy, 9th ed. Baltimore, Williams and Wilkins, 1991

Guyton AC: Basic Neuroscience: Anatomy and Physiology, 2nd ed. Philadelphia, WB Saunders, 1991

Hole JW: Essentials of Human Anatomy and Physiology, 5th ed. Boston, WC Brown, 1994

Hollinshead WH, Rosse C: Textbook of Anatomy, 4th ed. Philadelphia, JB Lippincott, 1985

Holman SR: Essentials of Nutrition for the Health Professions. Philadelphia, JB Lippincott, 1987

Melloni J et al: Melloni's Illustrated Review of Human Anatomy. Philadelphia, JB Lippincott, 1988

Pennington J: Bowes and Church's Food Values of Portions Commonly Used, 16th ed. Philadelphia, JB Lippincott, 1993

Ross MH et al: Histology: A Text and Atlas, 3rd ed. Baltimore, Williams and Wilkins, 1994

Rothwell NV: Understanding Genetics, A Molecular Approach. New York, John Wiley and Sons, 1993

Tortora GJ, Grabowski SR: Principles of Anatomy and Physiology, 7th ed. New York, Harper Collins, 1993

Williams PL et al (eds): Gray's Anatomy, 37th ed. New York, Churchill Livingstone, 1989

# Glossary

**Abduction** (ab-DUK-shun) Movement away from the midline; turning outward

**Abortion** (ah-BOR-shun) Loss of an embryo or fetus before the 20th week of pregnancy

**Absorption** (ab-SORP-shun) Transfer of digested food molecules from the digestive tract into the circulation

**Accommodation** (ah-kom-o-DA-shun) Coordinated changes in the eye that enable one to focus on near and far objects

**Acetylcholine** (as-e-til-KO-lene) **(Ach)** Neurotransmitter; released at synapses within the nervous system and at the neuromuscular junction

**Acid** (AH-sid) Substance that can donate a hydrogen ion to another substance

**Acidosis** (as-ih-DO-sis) Condition that results from a decrease in the pH of body fluids

**Acquired immunodeficiency syndrome (AIDS)** Viral disease that attacks the immune system, specifically the T helper lymphocytes with CD4 receptors

**Acrosome** (AK-ro-some) Caplike structure over the head of the sperm cell that helps the sperm to penetrate the ovum

**ACTH** See Adrenocorticotropic hormone

**Actin** (AK-tin) One of the two contractile proteins in muscle cells, the other being myosin

**Action potential** Sudden change in the electric charge on a cell membrane; nerve impulse

**Active transport** Movement of a substance into or out of a cell in a direction opposite that in which it would normally flow by diffusion

**Adduction** (ad-DUK-shun) Movement toward the midline; turning outward

**Adenosine triphosphate** (ah-DEN-o-sene tri-FOS-fate) **(ATP)** Energy-storing compound found in all cells

**ADH** See Antidiuretic hormone

**Adipose** (AD-ih-pose) Referring to a type of connective tissue that stores fat

**Adrenal** (ah-DRE-nal) Endocrine gland located above the kidney; suprarenal gland

**Adrenaline** (ah-DREN-ah-lin) See Epinephrine

**Adrenocorticotropic** (ah-dre-no-kor-tih-ko-TRO-pik) **hormone (ACTH)** Hormone produced by the pituitary that stimulates the adrenal cortex

**Aerobic** (air-O-bik) Requiring oxygen

**Afferent** (AF-fer-ent) Carrying toward a given point, such as a sensory neuron that carries nerve impulses toward the central nervous system

**Agglutination** (ah-glu-tih-NA-shun) Clumping of cells due to an antigen–antibody reaction

**AIDS** See Acquired immunodeficiency syndrome

**Albumin** (al-BU-min) Protein in blood plasma and other body fluids; helps maintain the osmotic pressure of the blood

**Aldosterone** (al-DOS-ter-one) Hormone released by the adrenal cortex that promotes the reabsorption of sodium and water in the kidneys

**Alkalosis** (al-kah-LO-sis) Condition that results from an increase in the pH of body fluids

**Alveolus** (al-VE-o-lus) Small sac or pouch; usually a tiny air sac in the lungs through which gases are exchanged between the outside air and the blood; tooth socket; pl., alveoli

**Amino** (ah-ME-no) **acid** Building block of protein

**Amniotic** (am-ne-OT-ik) **sac** Fluid-filled sac that surrounds and cushions the developing fetus

**Amphiarthrosis** (am-fe-ar-THRO-sis) Slightly movable joint

**Anabolism** (ah-NAB-o-lizm) Metabolic building of simple compounds into more complex substances needed by the body

**Anaerobic** (an-air-O-bik) Not requiring oxygen

**Anastomosis** (ah-nas-to-MO-sis) Communication between two structures such as blood vessels

303

**Anatomy** (ah-NAT-o-me) Study of body structure

**Anemia** (ah-NE-me-ah) Abnormally low level of hemoglobin or red cells in the blood, resulting in inadequate delivery of oxygen to the tissues

**Angiotensin** (an-je-o-TEN-sin) Substance formed in the blood by the action of the enzyme renin from the kidneys. It increases blood pressure by causing constriction of the blood vessels and stimulating the release of aldosterone from the adrenal cortex.

**Anion** (AN-i-on) Negatively charged particle (ion)

**Anorexia** (an-o-REK-se-ah) Loss of appetite. Anorexia nervosa is a psychologic condition in which a person may become seriously, even fatally, weakened from lack of food.

**Anoxia** (ah-NOK-se-ah) See Hypoxia

**ANS** See Autonomic nervous system

**Antagonist** (an-TAG-o-nist) Muscle that has an action opposite that of a given movement; substance that opposes the action of another substance

**Anterior** (an-TE-re-or) Toward the front or belly surface; ventral

**Antibody** (AN-te-bod-e) Substance produced in response to a specific antigen

**Antidiuretic** (an-ti-di-u-RET-ik) **hormone (ADH)** Hormone released from the posterior pituitary gland that increases the reabsorption of water in the kidneys, thus decreasing the volume of urine excreted

**Antigen** (AN-te-jen) **(Ag)** Foreign substance that produces an immune response

**Antiserum** (an-te-SE-rum) Serum containing antibodies; may be given for the purpose of providing passive immunity

**Aorta** (a-OR-tah) Large artery that carries blood out of the left ventricle of the heart

**Aponeurosis** (ap-o-nu-RO-sis) Broad sheet of fibrous connective tissue that attaches muscle to bone or to other muscle

**Appendicular** (ap-en-DIK-u-lar) **skeleton** Part of the skeleton that includes the bones of the upper extremities, lower extremities, shoulder girdle, and hips

**Arachnoid** (ah-RAK-noyd) Middle layer of the meninges

**Areolar** (ah-RE-o-lar) Referring to loose connective tissue

**Arteriole** (ar-TE-re-ole) Vessel between a small artery and a capillary

**Artery** (AR-ter-e) Vessel that carries blood away from the heart

**Atom** (AT-om) Fundamental unit of a chemical element

**ATP** See Adenosine triphosphate

**Atrioventricular** (a-tre-o-ven-TRIK-u-lar) **(AV) node** Part of the conduction system of the heart

**Atrium** (A-tre-um) One of the two upper chambers of the heart; adj., atrial

**Attenuated** (ah-TEN-u-a-ted) Weakened

**Autonomic** (aw-to-NOM-ik) **nervous system (ANS)** The part of the nervous system that controls smooth muscle, cardiac muscle, and glands; motor portion of the visceral or involuntary nervous system

**AV node** See Atrioventricular node

**Axial** (AK-se-al) **skeleton** The part of the skeleton that includes the skull, spinal column, ribs, and sternum

**Axilla** (ak-SIL-ah) Hollow beneath the arm where it joins the body; armpit

**Axon** (AK-son) Fiber of a neuron that conducts impulses away from the cell body

**Basal ganglia** (BA-sal GANG-le-ah) Gray masses in the lower part of the forebrain that aid in muscle coordination

**Base** Substance that can accept a hydrogen ion (H+); substance that donates a hydroxide ion (OH−)

**Basophil** (BA-so-fil) Granular white blood cell that shows large, dark blue cytoplasmic granules when stained with basic stain

**B cell** Agranular white blood cell that produces antibodies in response to an antigen; B lymphocyte

**Bile** Substance produced in the liver that emulsifies fats

**Blood urea nitrogen (BUN)** Amount of nitrogen from urea in the blood; test to evaluate kidney function

**Bradycardia** (brad-e-KAR-de-ah) Heart rate of less than 60 beats per minute

**Brain stem** Portion of the brain that connects the cerebrum with the spinal cord; contains the midbrain, pons, and medulla oblongata

**Bronchiole** (BRONG-ke-ole) Microscopic terminal branch of a bronchus

**Bronchus** (BRONG-kus) Large air passageway in the lung; pl., bronchi (BRONG-ki)

**Buffer** (BUF-er) Substance that prevents sharp changes in the pH of a solution

**BUN** See Blood urea nitrogen

**Bursa** (BER-sah) Small, fluid-filled sac found in an area subject to stress around bones and joints; pl., bursae (BER-se)

**Cancer** (KAN-ser) Tumor that spreads to other tissues; a malignant growth

**Capillary** (CAP-ih-lar-e) Microscopic vessel through which exchanges take place between the blood and the tissues

**Carbohydrate** (kar-bo-HI-drate) Simple sugar or compound made from simple sugars linked together, such as starch or glycogen

**Carbon dioxide** (di-OX-ide) **(CO₂)** The gaseous waste product of cellular metabolism

**Carcinogen** (kar-SIN-o-jen) Cancer-causing substance

**Carrier** Individual who has a gene that is not expressed but that can be passed to offspring

**Cartilage** (KAR-tih-lij) Type of hard connective tissue

**CAT** See Computed tomography

**Catabolism** (kah-TAB-o-lizm) Metabolic breakdown of substances into simpler substances; includes the digestion of food and the oxidation of nutrient molecules for energy

**Cation** (KAT-i-on) Positively charged particle (ion)

**Cecum** (SE-kum) Small pouch at the beginning of the large intestine

**Cell** Basic unit of life

**Cell membrane** Outer covering of a cell; regulates what enters and leaves cell; plasma membrane

**Cellular respiration** Series of reactions by which nutrients are oxidized for energy within the cell

**Central nervous system (CNS)** Part of the nervous system that includes the brain and spinal cord

**Centrifuge** (SEN-trih-fuje) Instrument for separating materials in a mixture based on density

**Centriole** (SEN-tre-ole) Rod-shaped body near the nucleus of a cell; functions in cell division

**Cerebellum** (ser-eh-BEL-um) Small section of the brain located under the cerebral hemispheres; functions in coordination, balance, and muscle tone

**Cerebral** (SER-e-bral) **cortex** The very thin outer layer of gray matter on the surface of the cerebral hemispheres

**Cerebrospinal** (ser-e-bro-SPI-nal) **fluid (CSF)** Fluid that circulates in and around the brain and spinal cord

**Cerebrovascular** (ser-e-bro-VAS-ku-lar) **accident (CVA)** Condition involving bleeding from the brain or obstruction of blood flow to brain tissue, usually as a result of hypertension or atherosclerosis; stroke

**Cerebrum** (SER-e-brum) Largest part of the brain; composed of the cerebral hemispheres

**Cerumen** (seh-RU-men) Earwax; adj., ceruminous (seh-RU-min-us)

**Cervix** (SER-vix) Constricted portion of an organ or part, such as the lower portion of the uterus; neck.; adj., cervical

**Chemoreceptor** (ke-mo-re-SEP-tor) Receptor that detects chemical changes

**Cholesterol** (ko-LES-ter-ol) An organic fat-like compound found in animal fat, bile, blood, myelin, liver, and other parts of the body

**Choroid** (KO-royd) Pigmented middle layer of the eye

**Chromosome** (KRO-mo-some) Dark-staining, threadlike body in the nucleus of a cell; contains genes that determine hereditary traits

**Chyle** (kile) Milky-appearing fluid absorbed into the lymphatic system from the small intestine. It consists of lymph and droplets of digested fat.

**Chyme** (kime) Mixture of partially digested food, water, and digestive juices that forms in the stomach

**Cilia** (SIL-e-ah) Hairs or hairlike processes such as eyelashes or microscopic extensions from the surface of a cell; sing., cilium

**Circumduction** (ser-kum-DUK-shun) Circular movement at a joint

**CNS** See Central nervous system

**Coagulation** (ko-ag-u-LA-shun) Clotting, as of blood

**Cochlea** (KOK-le-ah) Coiled portion of the inner ear that contains the organ of hearing

**Collagen** (KOL-ah-jen) Flexible white protein that gives strength and resilience to connective tissue such as bone and cartilage

**Colloidal** (kol-OYD-al) **suspension** Mixture in which suspended particles do not dissolve but remain distributed in the solvent because of their small size (*e.g.,* cytoplasm); colloid

**Colon** (KO-lon) Main portion of the large intestine

**Complement** (KOM-ple-ment) Group of blood proteins that help antibodies to destroy foreign cells

**Compound** Substance composed of two or more chemical elements

**Computed tomography** (to-MOG-rah-fe) **(CT)** Imaging method in which multiple x-ray views taken from different angles are analyzed by computer to show a cross section of an area; used to detect tumors and other abnormalities; also called *computed axial tomography* (*CAT*)

**Congenital** (con-JEN-ih-tal) Present at birth

**Conjunctiva** (kon-junk-TI-vah) Membrane that lines the eyelid and covers the anterior part of the sclera (white of the eye)

**Contraception** (con-trah-SEP-shun) Prevention of fertilization of an ovum or implantation of a fertilized ovum; birth control

**Cornea** (KOR-ne-ah) Clear portion of the sclera that covers the front of the eye

**Coronary** (KOR-on-ar-e) Referring to the heart or to the arteries supplying blood to the heart

**Corpus callosum** (kal-O-sum) Thick bundle of myelinated nerve cell fibers, deep within the brain, that carries nerve impulses from one cerebral hemisphere to the other

**Corpus luteum** (LU-te-um) Yellow body formed from ovarian follicle after ovulation; produces progesterone

**Cortex** (KOR-tex) Outer layer of an organ such as the brain, kidney, or adrenal gland

**Covalent** (KO-va-lent) **bond** Chemical bond formed by the sharing of electrons between atoms

**CSF** See Cerebrospinal fluid

**CT** See Computed tomography

**Cutaneous** (ku-TA-ne-us) Referring to the skin

**Cytology** (si-TOL-o-je) Study of cells

**Cytoplasm** (SI-to-plazm) Substance that fills the cell and holds the organelles

**Defecation** (def-e-KA-shun) Act of eliminating undigested waste from the digestive tract

**Deglutition** (deg-lu-TISH-un) Act of swallowing

**Dehydration** (de-hi-DRA-shun) Excessive loss of body fluid

**Dendrite** (DEN-drite) Fiber of a neuron that conducts impulses toward the cell body

**Deoxyribonucleic** (de-OK-se-ri-bo-nu-kle-ik) **acid (DNA)** Genetic material of the cell; makes up the chromosomes in the nucleus of the cell

**Dermis** (DER-mis) True skin; deeper part of the skin

**Dextrose** (DEK-strose) Glucose; simple sugar

**Diabetes mellitus** (di-ah-BE-teze mel-LI-tus) Disease in which glucose is not oxidized adequately in the tissues for energy because of insufficient insulin

**Dialysis** (di-AL-ih-sis) Method for separating molecules in solution based on differences in their ability to pass through a semipermeable membrane

**Diaphragm** (DI-ah-fram) Dome-shaped muscle under the lungs that flattens during inhalation; separating membrane or structure

**Diaphysis** (di-AF-ih-sis) Shaft of a long bone

**Diarthrosis** (di-ar-THRO-sis) Freely movable joint; synovial joint

**Diastole** (di-AS-to-le) Relaxation phase of the cardiac cycle; adj., diastolic (di-as-TOL-ik)

**Diencephalon** (di-en-SEF-ah-lon) Region of the brain between the cerebral hemispheres and the brainstem; contains the thalamus, hypothalamus, and pituitary gland

**Diffusion** (dih-FU-zhun) Movement of molecules from a region where they are in higher concentration to a region where they are in lower concentration

**Digestion** (di-JEST-yun) Process of breaking down food into absorbable particles

**Dilation** (di-LA-shun) Widening of a part, such as the pupil of the eye, a blood vessel, or the uterine cervix; dilatation

**Distal** (DIS-tal) Farther from the origin of a structure or from a given reference point

**DNA** See Deoxyribonucleic acid

**Dominant** (DOM-ih-nant) Referring to a gene that is always expressed if present

**Dorsal** (DOR-sal) Toward the back; posterior

**Duct** Tube or vessel

**Ductus deferens** (DEF-er-enz) Tube that carries sperm cells from the testis to the urethra; vas deferens

**Duodenum** (du-o-DE-num) First portion of the small intestine

**Dura mater** (DU-rah MA-ter) Outermost layer of the meninges

**ECG** See Electrocardiograph

**Echocardiograph** (ek-o-KAR-de-o-graf) Instrument to study the heart by means of ultrasound; the record produced is an echocardiogram

**EEG** See Electroencephalograph

**Effector** (ef-FEK-tor) Muscle or gland that responds to a stimulus; effector organ

**Efferent** (EF-fer-ent) Carrying away from a given point, such as a motor neuron that carries nerve impulses away from the central nervous system

**Ejaculation** (e-jak-u-LA-shun) Expulsion of semen through the urethra

**EKG** See Electrocardiograph

**Electrocardiograph** (e-lek-tro-KAR-de-o-graf) **(EKG, ECG)** Instrument to study the electric activity of the heart; record made is an electrocardiogram.

**Electroencephalograph** (e-lek-tro-en-SEF-ah-lo-graf) **(EEG)** Instrument used to study electric activity of the brain; record made is an electroencephalogram

**Electrolyte** (e-LEK-tro-lite) Compound that forms ions in solution; substance that conducts an electric current in solution

**Electron** (e-LEK-tron) Negatively charged particle located in an orbital outside the nucleus of an atom

**Element** (EL-eh-ment) One of the substances from which all matter is made; substance that cannot be decomposed into a simpler substance

**Embryo** (EM-bre-o) Developing offspring during the first 2 months of pregnancy

**Emulsify** (e-MUL-sih-fi) To break up fats into small particles; n., emulsification

**Endocardium** (en-do-KAR-de-um) Membrane that lines the heart chambers and covers the valves

**Endocrine** (EN-do-krin) Referring to a gland that secretes directly into the bloodstream

**Endometrium** (en-do-ME-tre-um) Lining of the uterus

**Endoplasmic reticulum** (en-do-PLAS-mik re-TIK-u-lum) **(ER)** Network of membranes in the cytoplasm of a cell

**Endosteum** (en-DOS-te-um) Thin membrane that lines the marrow cavity of a bone

**Endothelium** (en-do-THE-le-um) Epithelium that lines the heart, blood vessels, and lymphatic vessels

**Enzyme** (EN-zime) Organic catalyst; speeds the rate of a reaction but is not changed in the reaction

**Eosinophil** (e-o-SIN-o-fil) Granular white blood cell that shows beadlike, bright pink cytoplasmic granules when stained with acid stain; acidophil

**Epicardium** (ep-ih-KAR-de-um) Membrane that forms the outermost layer of the heart wall and is continuous with the lining of the pericardium; visceral pericardium

**Epidermis** (ep-ih-DER-mis) Outermost layer of the skin

**Epiglottis** (ep-e-GLOT-is) Leaf-shaped cartilage that covers the larynx during swallowing

**Epimysium** (ep-ih-MIS-e-um) Sheath of fibrous connective tissue that encloses a muscle

**Epinephrine** (ep-ih-NEF-rin) Neurotransmitter and hormone; released from neurons of the sympathetic nervous system and from the adrenal medulla; adrenaline

**Epiphysis** (eh-PIF-ih-sis) End of a long bone

**Episiotomy** (eh-piz-e-OT-o-me) Cutting of the perineum between the vaginal opening and the anus to reduce the tearing of tissue in childbirth

**Epithelium** (ep-ih-THE-le-um) One of the four main types of tissue; forms glands, covers surfaces, and lines cavities; adj., epithelial

**ER** See Endoplasmic reticulum

**Erythrocyte** (eh-RITH-ro-site) Red blood cell

**Erythropoietin** (eh-rith-ro-POY-eh-tin) Hormone released from the kidney that stimulates the production of red blood cells in the bone marrow

**Esophagus** (eh-SOF-ah-gus) Tube that carries food from the throat to the stomach

**Estrogen** (ES-tro-jen) Group of female sex hormones that promotes development of the uterine lining and maintains secondary sex characteristics

**Eustachian** (u-STA-shun) **tube** Tube that connects the middle ear cavity to the throat; auditory tube

**Excretion** (eks-KRE-shun) Removal and elimination of metabolic waste products from the blood

**Exocrine** (EK-so-krin) Referring to a gland that secretes through a duct

**Extracellular** (EK-strah-sel-u-lar) Outside the cell

**Fascia** (FASH-e-ah) Band or sheet of fibrous connective tissue

**Fascicle** (FAS-ih-kl) Small bundle, as of muscle cells or nerve cell fibers

**Feces** (FE-seze) Waste material discharged from the large intestine; excrement; stool

**Feedback** Return of information into a system so that it can be used to regulate that system

**Fertilization** (fer-til-ih-ZA-shun) Union of an ovum and a spermatozoon

**Fetus** (FE-tus) Developing offspring from the third month of pregnancy until birth

**Fibrin** (FI-brin) Blood protein that forms a blood clot

**Filtration** (fil-TRA-shun) Movement of material through a semipermeable membrane under mechanical force

**Fissure** (FISH-ure) Deep groove

**Flagellum** (flah-JEL-lum) Long whiplike extension from a cell used for locomotion; pl., flagella

**Flexion** (FLEK-shun) Bending motion that decreases the angle between bones at a joint

**Follicle** (FOL-lih-kl) Sac or cavity, such as the ovarian follicle or hair follicle

**Follicle-stimulating hormone (FSH)** Hormone produced by the anterior pituitary that stimulates development of ova in the ovary and spermatozoa in the testes

**Fontanelle** (fon-tah-NEL) Area in the infant skull where bone formation has not yet occurred; "soft spot"

**Foramen** (fo-RA-men) Opening or passageway, as into or through a bone; pl., foramina (fo-RAM-in-ah)

**Fossa** (FOS-sah) Hollow or depression, as in a bone; pl., fossae (FOS-se)

**Fovea** (FO-ve-ah) Small pit or cup-shaped depression in a surface; the fovea centralis near the center of the retina is the point of sharpest vision

**FSH** See Follicle-stimulating hormone

**Gamete** (GAM-ete) Reproductive cell; ovum or spermatozoon

**Gamma globulin** (GLOB-u-lin) Protein fraction in the blood plasma that contains antibodies

**Ganglion** (GANG-le-on) Collection of nerve cell bodies located outside the central nervous system.

**Gastrointestinal** (gas-tro-in-TES-tih-nal) **(GI)** Pertaining to the stomach and intestine or the digestive tract as a whole

**Gene** Hereditary unit; portion of the DNA on a chromosome

**Genetic** (jeh-NET-ik) Pertaining to the genes or heredity

**Gestation** (jes-TA-shun) Period of development from conception to birth

**GH** See Growth hormone

**GI** See Gastrointestinal

**Gingiva** (JIN-jih-vah) Tissue around the teeth; gum

**Glomerular** (glo-MER-u-lar) **filtrate** Fluid and dissolved materials that leave the blood and enter the kidney nephron through Bowman's capsule

**Glomerulus** (glo-MER-u-lus) Cluster of capillaries in Bowman's capsule of the nephron

**Glucose** (GLU-kose) Simple sugar; main energy source for the cells; dextrose

**Glycogen** (GLI-ko-jen) Compound built from glucose molecules that is stored for energy in liver and muscles

**Golgi** (GOL-je) **apparatus** System of membranes in the cell that formulates special substances

**Gonad** (GO-nad) Sex gland; ovary or testis

**Gonadotropin** (gon-ah-do-TRO-pin) Hormone that acts on a reproductive gland (ovary or testis; *e.g.,* FSH, LH)

**Gram (g)** Basic unit of weight in the metric system

**Gray matter** Nervous tissue composed of unmyelinated fibers and cell bodies

**Growth hormone (GH)** Hormone produced by anterior pituitary that promotes growth of tissues; somatotropin

**Gustatory** (GUS-tah-to-re) Pertaining to the sense of taste (gustation)

**Gyrus** (JI-rus) Raised area of the cerebral cortex; pl., gyri (JI-ri)

**Hematocrit** (he-MAT-o-krit) **(Hct)** Volume percentage of red blood cells in whole blood; packed cell volume

**Hemoglobin** (he-mo-GLO-bin) **(Hb)** Iron-containing protein in red blood cells that transports oxygen

**Hemolysis** (he-MOL-ih-sis) Rupture of red blood cells; v., hemolyze (HE-mo-lize)

**Hemopoiesis** (he-mo-poy-E-sis) Production of blood cells; hematopoiesis

**Hemostasis** (he-mo-STA-sis) Stoppage of bleeding

**Heparin** (HEP-ah-rin) Substance that prevents blood clotting; anticoagulant

**Heredity** (he-RED-ih-te) Transmission of characteristics from parent to offspring via the genes; the genetic make-up of the individual

**Hilum** (HI-lum) See hilus

**Hilus** (HI-lus) Area where vessels and nerves enter or leave an organ

**Histamine** (HIS-tah-mene) Substance released from tissues during an antigen–antibody reaction

**Histology** (his-TOL-o-je) Study of tissues

**HIV** See Human immunodeficiency virus

**Homeostasis** (ho-me-o-STA-sis) State of balance within the body; maintenance of body conditions within set limits

**Hormone** Secretion of an endocrine gland; chemical messenger that has specific regulatory effects on certain other cells

**Human immunodeficiency virus (HIV)** The virus that causes AIDS

**Hydrolysis** (hi-DROL-ih-sis) Splitting of large molecules by the addition of water, as in digestion

**Hyperglycemia** (hi-per-gli-SE-me-ah) Abnormal increase in the amount of glucose in the blood

**Hypertonic** (hi-per-TON-ik) Describing a solution that is more concentrated than the fluids within a cell

**Hypophysis** (hi-POF-ih-sis) Pituitary gland

**Hypothalamus** (hi-po-THAL-ah-mus) Region of the brain that controls the pituitary and maintains homeostasis

**Hypotonic** (hi-po-TON-ik) Describing a solution that is less concentrated than the fluids within a cell

**ICSH** Interstitial cell-stimulating hormone; see Luteinizing hormone

**Ileum** (IL-e-um) The last portion of the small intestine

**Immunity** (ih-MU-nih-te) Power of an individual to resist or overcome the effects of a particular disease or other harmful agent

**Immunization** (ih-mu-nih-ZA-shun) Use of a vaccine to produce immunity

**Implantation** (im-plan-TA-shun) The embedding of the fertilized egg into the lining of the uterus

**Inferior** (in-FE-re-or) Below or lower

**Inferior vena cava** (VE-nah KA-vah) Large vein that drains the lower part of the body and empties into the right atrium of the heart

**Insertion** (in-SER-shun) End of a muscle attached to a movable part

**Integument** (in-TEG-u-ment) Skin; adj., integumentary

**Intercellular** (in-ter-SEL-u-lar) Between cells

**Interstitial** (in-ter-STISH-al) Between; pertaining to spaces or structures in an organ between active tissues

**Intracellular** (in-trah-SEL-u-lar) Within a cell

**Ion** (I-on) Charged particle formed when an electrolyte goes into solution

**Iris** (I-ris) Circular colored region of the eye around the pupil

**Islets** (I-lets) Groups of cells in the pancreas that produce hormones; islets of Langerhans (LAHNG-er-hanz)

**Isometric** (i-so-MET-rik) **contraction** Muscle contraction in which there is no change in muscle length but an increase in muscle tension, as in pushing against an immovable force

**Isotonic** (i-so-TON-ik) Describing a solution that has the same concentration as the fluid within a cell

**Isotonic contraction** Muscle contraction in which the tone within the muscle remains the same but the muscle shortens to produce movement

**Isotope** (I-so-tope) Form of an element that has the same atomic number as another but a different atomic weight

**Jejunum** (je-JU-num) Second portion of the small intestine

**Joint** Area of junction between two or more bones; articulation

**Juxtaglomerular** (juks-tah-glo-MER-u-lar) **(JG) apparatus** Structure in the kidney composed of cells of the afferent arteriole and distal convoluted tubule that secretes the enzyme renin when blood pressure decreases below a certain level

**Keratin** (KER-ah-tin) Protein that thickens and protects the skin; makes up hair and nails

**Kidney** (KID-ne) An organ of excretion

**Lacrimal** (LAK-rih-mal) Referring to tears or the tear glands

**Lactation** (lak-TA-shun) Secretion of milk

**Lacteal** (LAK-te-al) Capillary of the lymphatic system; drains digested fats from the villi of the small intestine

**Lactic** (LAK-tik) **acid** Organic acid that accumulates in muscle cells functioning without oxygen

**Larynx** (LAR-inks) Structure between the pharynx and trachea that contains the vocal cords; voice box

**Lateral** (LAT-er-al) Farther from the midline; toward the side

**Lens** Biconvex structure of the eye that changes in thickness to accommodate for near and far vision; crystalline lens

**Leukocyte** (LU-ko-site) White blood cell

**Ligament** (LIG-ah-ment) Band of connective tissue that connects a bone to another bone; thickened portion or fold of the peritoneum that supports an organ or attaches it to another organ

**Lipid** (LIP-id) Type of organic compound, one example of which is a fat

**Liter** (LE-ter) **(L)** Basic unit of volume in the metric system.

**Lumen** (LU-men) The central opening of an organ or vessel

**Lung** An organ of respiration

**Luteinizing** (LU-te-in-i-zing) **hormone** Hormone produced by the anterior pituitary that induces ovulation and formation of the corpus luteum in females; in males it stimulates cells in the testes to produce testosterone and is called *interstitial cell-stimulating hormone* (*ICSH*)

**Lymph** (limf) Fluid in the lymphatic system

**Lymphatic duct** Vessel of the lymphatic system

**Lymph node** Mass of lymphoid tissue along the path of a lymphatic vessel that filters lymph and harbors white blood cells active in immunity

**Lymphocyte** (LIM-fo-site) Agranular white blood cell that functions in immunity

**Lysosome** (LI-so-some) Cell organelle that contains digestive enzymes

**Macrophage** (MAK-ro-faj) Large phagocytic cell that develops from a monocyte

**Macula** (MAK-u-lah) Spot; flat, discolored spot on the skin such as a freckle or measles lesion; small yellow spot in the retina of the eye that contains the fovea, the point of sharpest vision. Also called *macule*

**Magnetic resonance imaging (MRI)** A method for studying tissue based on nuclear movement following exposure to radio waves in a powerful magnetic field

**Malnutrition** (mal-nu-TRISH-un) State resulting from lack of food, lack of an essential component of the diet, or faulty use of food in the diet

**Mastication** (mas-tih-KA-shun) Act of chewing

**Medial** (ME-de-al) Nearer the midline of the body

**Mediastinum** (me-de-as-TI-num) Region between the lungs and the organs and vessels it contains

**Medulla** (meh-DUL-lah) Inner region of an organ; marrow

**Medulla oblongata** (ob-long-GAH-tah) Part of the brain stem that connects the brain with the spinal cord

**Megakaryocyte** (meg-ah-KAR-e-o-site) Very large cell that gives rise to blood platelets

**Meiosis** (mi-O-sis) Process of cell division that halves the chromosome number in the formation of the reproductive cells

**Melanin** (MEL-ah-nin) Dark pigment found in skin, hair, parts of the eye, and certain parts of the brain

**Membrane** Thin sheet of tissue

**Mendelian** (men-DE-le-en) **laws** Principles of heredity discovered by an Austrian monk named Gregor Mendel

**Meninges** (men-IN-jeze) Three layers of fibrous membranes that cover the brain and spinal cord

**Menopause** (MEN-o-pawz) Time at which menstruation ceases

**Menses** (MEN-seze) The monthly flow of blood from the female reproductive tract

**Mesentery** (MES-en-ter-e) The membranous peritoneal ligament that attaches the small intestine to the dorsal abdominal wall

**Mesocolon** (mes-o-KO-lon) Peritoneal ligament that attaches the colon to the dorsal abdominal wall

**Metabolic rate** Rate at which energy is released from nutrients in the cells

**Metabolism** (meh-TAB-o-lizm) The physical and chemical processes by which an organism is maintained

**Metastasis** (meh-TAS-tah-sis) Spread of tumor cells; pl., metastases (meh-TAS-tah-seze)

**Meter** (ME-ter) **(m)** Basic unit of length in the metric system

**Microbiology** (mi-kro-bi-OL-o-je) Study of microscopic organisms

**Micrometer** (MI-kro-me-ter) **(μm)** 1/1000th of a millimeter; micron; also an instrument for measuring through a microscope (pronounced mi-KROM-eh-ter)

**Micturition** (mik-tu-RISH-un) Act of urination; voiding of the urinary bladder

**Midbrain** Upper portion of the brainstem

**Mineral** (MIN-er-al) Inorganic substance; in the diet, an element needed in small amounts for health

**Mitochondria** (mi-to-KON-dre-ah) Cell organelles that manufacture ATP with the energy released from the oxidation of nutrients

**Mitosis** (mi-TO-sis) Type of cell division that produces two daughter cells exactly like the parent cell

**Mitral** (MI-tral) **valve** The valve between the left atrium and left ventricle of the heart; bicuspid valve

**Mixture** Blend of two or more substances

**Molecule** (MOL-eh-kule) Particle formed by chemical bonding of two or more atoms; smallest subunit of a compound

**Monocyte** (MON-o-site) Agranular white blood cell active in phagocytosis

**MRI** See Magnetic resonance imaging

**Mucosa** (mu-KO-sah) Lining membrane that produces mucus; mucous membrane

**Mucus** (MU-kus) Thick protective fluid secreted by mucous membranes and glands; adj., mucous

**Murmur** Abnormal heart sound

**Mutation** (mu-TA-shun) Change in a gene or a chromosome

**Myelin** (MI-el-in) Fatty material that covers and insulates the axons of some neurons

**Myocardium** (mi-o-KAR-de-um) Middle layer of the heart wall; heart muscle

**Myoglobin** (MI-o-glo-bin) Compound that stores oxygen in muscle cells

**Myometrium** (mi-o-ME-tre-um) The muscular layer of the uterus

**Myosin** (MI-o-sin) One of the two contractile proteins in muscle cells, the other being actin

**Negative feedback** A self-regulating system in which the result of an action is the control over that action; a method for keeping body conditions within a normal range and maintaining homeostasis

**Nephron** (NEF-ron) Microscopic functional unit of the kidney

**Nerve** Bundle of neuron fibers outside the central nervous system

**Nerve impulse** Electric charge that spreads along the membrane of a neuron; action potential

**Neurilemma** (nu-rih-LEM-mah) Thin sheath that covers certain peripheral axons; aids in regeneration of the axon

**Neuroglia** (nu-ROG-le-ah) Supporting and protective cells of the central nervous system; glial cells

**Neuromuscular junction** Point at which a nerve fiber contacts a muscle cell

**Neuron** (NU-ron) Conducting cell of the nervous system

**Neurotransmitter** (nu-ro-TRANS-mit-er) Chemical released from the ending of an axon that enables a nerve impulse to cross a synaptic junction

**Neutrophil** (NU-tro-fil) Phagocytic granular white blood cell; polymorph; poly; PMN; seg

**Node** Small mass of tissue such as a lymph node; space between cells in the myelin sheath

**Norepinephrine** (nor-epi-ih-NEF-rin) Neurotransmitter similar to epinephrine; noradrenaline

**Nucleotide** (NU-kle-o-tide) Building block of DNA and RNA

**Nucleus** (NU-kle-us) Largest organelle in the cell, containing the DNA, which directs all cell activities; group of neurons in the central nervous system; in chemistry, the central part of an atom

**Olfactory** (ol-FAK-to-re) Pertaining to the sense of smell (olfaction)

**Ophthalmic** (of-THAL-mik) Pertaining to the eye

**Organ** (OR-gan) Body part containing two or more tissues functioning together for specific purposes

**Organelle** (or-gan-EL) Specialized subdivision within a cell

**Organic** (or-GAN-ik) Referring to compounds found in living things and containing carbon, hydrogen, and oxygen

**Organism** (OR-gan-izm) Individual plant or animal; any organized living thing

**Organ of Corti** (KOR-te) Receptor for hearing located in the cochlea of the internal ear

**Origin** (OR-ih-jin) Source; beginning; end of a muscle attached to a nonmoving part

**Osmosis** (os-MO-sis) Movement of water through a semipermeable membrane

**Osmotic** (os-MOT-ik) **pressure** Tendency of a solution to draw water into it; is directly related to the concentration of the solution

**Ossicle** (OS-ih-kl) One of three small bones of the middle ear: malleus, incus, or stapes

**Ossification** (os-ih-fih-KA-shun) Process of bone formation

**Osteoblast** (OS-te-o-blast) Bone-forming cell

**Osteoclast** (OS-te-o-clast) Cell that breaks down bone

**Osteocyte** (OS-te-o-site) Mature bone cell; maintains bone but does not divide

**Osteoporosis** (os-te-o-po-RO-sis) Abnormal loss of bone tissue with tendency to fracture

**Ovary** (O-vah-re) Female reproductive gland

**Ovulation** (ov-u-LA-shun) Release of a mature ovum from a follicle in the ovary

**Ovum** (O-vum) Female reproductive cell or gamete; pl., ova

**Oxidation** (ok-sih-DA-shun) Chemical breakdown of nutrients for energy

**Oxygen** (OK-sih-jen) ($O_2$) The gas needed to completely break down nutrients for energy within the cell

**Oxygen debt** Amount of oxygen needed to reverse the effects produced in muscles functioning without oxygen

**Pacemaker** Sinoatrial (SA) node of the heart; group of cells or artificial device that sets the rate of heart contractions

**Pancreas** (PAN-kre-as) Large, elongated gland behind the stomach; produces digestive enzymes and hormones (*e.g.,* insulin)

**Papilla** (pah-PIL-ah) Small nipple-like projection or elevation

**Parasympathetic nervous system** Craniosacral division of the autonomic nervous system

**Parathyroid** (par-ah-THI-royd) Any of four small glands embedded in the capsule enclosing the thyroid gland; produces hormone that regulates calcium in the blood

**Parietal** (pah-RI-eh-tal) Pertaining to the wall of a space or cavity

**Parturition** (par-tu-RISH-un) Childbirth; labor

**Pelvis** (PEL-vis) Basin-like structure, such as the lower portion of the abdomen or the upper flared portion of the ureter (renal pelvis)

**Penis** (PE-nis) Male organ of urination and sexual intercourse

**Pericardium** (per-ih-KAR-de-um) Fibrous sac lined with serous membrane that encloses the heart

**Perichondrium** (per-ih-KON-dre-um) Layer of connective tissue that covers cartilage

**Perineum** (per-ih-NE-um) Pelvic floor; external region between the anus and genital organs

**Periosteum** (per-e-OS-te-um) Connective tissue membrane covering a bone

**Peripheral** (peh-RIF-er-al) Located away from a center or central structure

**Peripheral nervous system (PNS)** All the nerves and nervous tissue outside the central nervous system

**Peristalsis** (per-ih-STAL-sis) Wavelike movements in the wall of an organ or duct that propel its contents forward

**Peritoneum** (per-ih-to-NE-um) Serous membrane that lines the abdominal cavity and forms outer layer of abdominal organs; forms supporting ligaments for some organs

**pH** Symbol indicating hydrogen ion (H+) concentration; scale that measures the relative acidity and alkalinity (basicity) of a solution

**Phagocytosis** (fag-o-si-TO-sis) Engulfing of large particles through the cell membrane

**Pharynx** (FAR-inks) Throat; passageway between the mouth and esophagus

**Physiology** (fiz-e-OL-o-je) Study of the function of living organisms

**Pia mater** (PI-ah MA-ter) Innermost layer of the meninges

**Pineal** (PIN-e-al) **body** Gland-like organ in the brain that is regulated by light; involved in sleep–wake cycles

**Pinocytosis** (pi-no-si-TO-sis) Intake of small particles and droplets by the cell membrane

**Pituitary** (pih-TU-ih-tar-e) **gland** Endocrine gland located under and controlled by the hypothalamus; releases hormones that control other glands; hypophysis

**Placenta** (plah-SEN-tah) Structure that nourishes and maintains the developing individual during pregnancy

**Plasma** (PLAZ-mah) Liquid portion of the blood

**Plasma cell** Cell that produces antibodies; derived from a B cell

**Platelet** (PLATE-let) Cell fragment that forms a plug to stop bleeding and acts in blood clotting; thrombocyte

**Pleura** (PLU-rah) Serous membrane that lines the chest cavity and covers the lungs

**Plexus** (PLEK-sus) Network of vessels or nerves

**PNS** See Peripheral nervous system

**Pons** (ponz) Area of the brain between the midbrain and medulla; connects the cerebellum with the rest of the central nervous system

**Portal system** Venous system that carries blood to a second capillary bed before it returns to the heart

**Posterior** (pos-TE-re-or) Toward the back; dorsal

**Prime mover** Muscle that performs a given movement; agonist

**Progeny** (PROJ-eh-ne) Offspring

**Progesterone** (pro-JES-ter-one) Hormone produced by the corpus luteum and placenta; maintains the lining of the uterus for pregnancy

**Proprioceptor** (pro-pre-o-SEP-tor) Sensory receptor that aids in judging body position and changes in position; located in muscles, tendons, and joints

**Prostaglandins** (pros-tah-GLAN-dinz) Group of hormones produced by many cells; these hormones have a variety of effects

**Protein** (PRO-tene) Organic compound made of amino acids; contains nitrogen in addition to carbon, hydrogen, and oxygen (some contain sulfur or phosphorus)

**Prothrombin** (pro-THROM-bin) Clotting factor; converted to thrombin during blood clotting

**Proton** (PRO-ton) Positively charged particle in the nucleus of an atom

**Proximal** (PROK-sih-mal) Nearer to point of origin or to a reference point

**Pulse** Wave of increased pressure in the vessels produced by contraction of the heart

**Pupil** (PU-pil) Opening in the center of the eye through which light enters

**Receptor** (re-SEP-tor) Specialized cell or ending of a sensory neuron that can be excited by a stimulus; also a site in the cell membrane to which a special substance (*e.g.*, hormone, antibody) may attach

**Recessive** (re-SES-iv) Referring to a gene that is not expressed if a dominant gene for the same trait is present

**Reflex** (RE-flex) A simple, rapid, automatic response involving few neurons

**Reflex arc** (ark) A pathway through the nervous system from stimulus to response; commonly involves a receptor, sensory neuron, central neuron(s), motor neuron, and effector

**Refraction** (re-FRAK-shun) Bending of light rays as they pass from one medium to another of a different density

**Renin** (RE-nin) Enzyme released from the juxtaglomerular apparatus of the kidneys that indirectly increases blood pressure by activating angiotensin

**Resorption** (re-SORP-shun) Loss of substance, such as that of bone or a tooth

**Respiration** (res-pih-RA-shun) Exchange of oxygen and carbon dioxide between the outside air and body cells

**Reticuloendothelial** (reh-tik-u-lo-en-do-THE-le-al) **system** Protective system consisting of highly phagocytic cells in body fluids and tissues, such as the spleen, lymph nodes, bone marrow, and liver

**Retina** (RET-ih-nah) Innermost layer of the eye; contains light-sensitive cells (rods and cones)

**Retroperitoneal** (ret-ro-per-ih-to-NE-al) Behind the peritoneum, as are the kidneys, pancreas, and abdominal aorta

**Ribonucleic** (RI-bo-nu-kle-ik) **acid (RNA)** Substance needed for protein manufacture in the cell

**Ribosome** (RI-bo-some) Small body in the cytoplasm of a cell that is a site of protein manufacture

**RNA** See Ribonucleic acid

**Rugae** (RU-je) Folds in the lining of an organ such as the stomach or urinary bladder; sing., ruga (RU-gah)

**Saliva** (sah-LI-vah) Secretion of the salivary glands; moistens food and contains an enzyme that digests starch

**SA node** See Sinoatrial node

**Sclera** (SKLE-rah) Outermost layer of the eye; made of tough connective tissue; "white" of the eye

**Scrotum** (SKRO-tum) Sac in which testes are suspended

**Sebum** (SE-bum) Oily secretion that lubricates the skin; produced by sebaceous glands; adj., sebaceous (se-BA-shus)

**Semen** (SE-men) Mixture of sperm cells and secretions from several glands of the male reproductive tract

**Semicircular canal** Bony canal in the internal ear that contains receptors for the sense of dynamic equilibrium; there are three semicircular canals in each ear

**Sensory adaptation** The gradual loss of sensation when sensory receptors are exposed to continuous stimulation

**Septum** (SEP-tum) Dividing wall, as between the chambers of the heart or the nasal cavities

**Serosa** (se-RO-sah) Serous membrane; epithelial membrane that secretes a thin, watery fluid

**Serum** (SE-rum) Liquid portion of blood without clotting factors; thin, watery fluid; adj., serous (SE-rus)

**Sex-linked** Referring to a gene carried on a sex chromosome, usually the X chromosome

**Sinoatrial** (si-no-A-tre-al) **(SA) node** Tissue in the upper wall of the right atrium that sets the rate of heart contractions; pacemaker of the heart

**Sinusoid** (SI-nus-oyd) Enlarged capillary that serves as a blood channel

**Solute** (SOL-ute) Substance that is dissolved in another substance (the solvent)

**Solution** (so-LU-shun) Mixture, the components of which are evenly distributed

**Solvent** (SOL-vent) Substance in which another substance (the solute) is dissolved

**Spermatozoon** (sper-mah-to-ZO-on) Male reproductive cell or gamete; pl., spermatozoa

**Sphincter** (SFINK-ter) Muscular ring that regulates the size of an opening

**Sphygmomanometer** (sfig-mo-mah-NOM-eh-ter) Device used to measure blood pressure

**Spleen** Lymphoid organ in the upper left region of the abdomen

**Stasis** (STA-sis) Stoppage in the normal flow of fluids, such as blood, lymph, urine, or contents of the digestive tract

**Stenosis** (sten-O-sis) Narrowing of a duct or canal

**Sterilization** (ster-ih-li-ZA-shun) Process of killing every living microorganism on or in an object; procedure that makes an individual incapable of reproduction

**Steroid** (STE-royd) Category of lipids that includes the hormones of the sex glands and the adrenal cortex

**Stethoscope** (STETH-o-skope) Instrument for conveying sounds from the patient's body to the examiner's ears

**Stimulus** (STIM-u-lus) Change in the external or internal environment that produces a response

**Subcutaneous** (sub-ku-TA-ne-us) Under the skin

**Sudoriferous** (su-do-RIF-er-us) Producing sweat; referring to the sweat glands

**Sulcus** (SUL-kus) Shallow groove, as between convolutions of the cerebral cortex; pl., sulci (SUL-si)

**Superior** (su-PE-re-or) Above; in a higher position

**Superior vena cava** (VE-nah KA-vah) Large vein that drains the upper part of the body and empties into the right atrium of the heart

**Surfactant** (sur-FAK-tant) Substance in the alveoli that prevents their collapse by reducing surface tension of the fluids within

**Suspension** (sus-PEN-shun) Mixture that will separate unless shaken

**Suture** (SU-chur) Type of joint in which bone surfaces are closely united, as in the skull; stitch used in surgery to bring parts together or to stitch parts together in surgery

**Sympathetic nervous system** Thoracolumbar division of the autonomic nervous system

**Synapse** (SIN-aps) Junction between two neurons or between a neuron and an effector

**Synarthrosis** (sin-ar-THRO-sis) Immovable joint

**Synovial** (sin-O-ve-al) Pertaining to a thick lubricating fluid found in joints, bursae, and tendon sheaths; pertaining to a freely movable (diarthrotic) joint

**System** (SIS-tem) Group of organs functioning together for the same general purposes

**Systemic** (sis-TEM-ik) Referring to a generalized infection or condition

**Systole** (SIS-to-le) Contraction phase of the cardiac cycle; adj., systolic (sis-TOL-ik)

**Tachycardia** (tak-e-KAR-de-ah) Heart rate over 100 beats per minute

**Target tissue** Tissue that is capable of responding to a specific hormone

**T cell** Lymphocyte active in immunity that matures in the thymus gland; destroys foreign cells directly; T lymphocyte

**Tendon** (TEN-don) Cord of fibrous connective tissue that attaches a muscle to a bone

**Testis** (TES-tis) Male reproductive gland; pl., testes (TES-teze)

**Testosterone** (tes-TOS-ter-one) Male sex hormone produced in the testes; promotes the development of sperm cells and maintains secondary sex characteristics

**Tetanus** (TET-an-us) Constant contraction of a muscle; infectious disease caused by a bacterium (*Clostridium tetani);* lockjaw

**Tetany** (TET-an-e) Muscle spasms due to abnormal calcium metabolism, as in parathyroid deficiency

**Thalamus** (THAL-ah-mus) Region of the brain located in the diencephalon; chief relay center for sensory impulses traveling to the cerebral cortex

**Thorax** (THO-raks) Chest; adj., thoracic (tho-RAS-ik)

**Thrombocyte** (THROM-bo-site) Blood platelet; cell fragment that participates in clotting

**Thymus** (THI-mus) Endocrine gland in the upper portion of the chest; stimulates development of T cells

**Thyroid** (THI-royd) Endocrine gland in the neck

**Thyroid-stimulating hormone (TSH)** Hormone produced by the anterior pituitary that stimulates the thyroid gland

**Thyroxine** (thi-ROK-sin) Hormone produced by the thyroid gland; increases metabolic rate and needed for normal growth

**Tissue** Group of similar cells that performs a specialized function

**Tonsil** (TON-sil) Mass of lymphoid tissue in the region of the pharynx

**Tonus** (TO-nus) Also tone. Partially contracted state of muscle

**Toxin** (TOK-sin) Poison

**Toxoid** (TOK-soyd) Altered toxin used to produce active immunity

**Trachea** (TRA-ke-ah) Tube that extends from the larynx to the bronchi; windpipe

**Tract** Bundle of neuron fibers within the central nervous system

**Trait** Characteristic

**Tricuspid** (tri-KUS-pid) **valve** Valve between the right atrium and right ventricle of the heart

**TSH** See Thyroid-stimulating hormone

**Tympanic** (tim-PAN-ik) **membrane** Membrane between the external and middle ear that transmits sound waves to the bones of the middle ear; eardrum

**Umbilical** (um-BIL-ih-kal) **cord** Structure that connects the fetus with the placenta; contains vessels that carry blood between the fetus and placenta

**Umbilicus** (um-BIL-ih-kus) Small scar on the abdomen that marks the former attachment of the umbilical cord to the fetus; navel

**Urea** (u-RE-ah) Nitrogen waste product excreted in the urine; end product of protein metabolism

**Ureter** (U-re-ter) Tube that carries urine from the kidney to the urinary bladder

**Urethra** (u-RE-thrah) Tube that carries urine from the urinary bladder to the outside of the body

**Urinary bladder** Hollow organ that stores urine until it is eliminated

**Urine** (U-rin) Liquid waste excreted by the kidneys

**Uterus** (U-ter-us) Muscular, pear-shaped organ in the female pelvis within which the fetus develops during pregnancy

**Uvea** (U-ve-ah) Middle coat of the eye, including the choroid, iris, and ciliary body; vascular and pigmented structures of the eye

**Uvula** (U-vu-lah) Soft, fleshy, V-shaped mass that hangs from the soft palate

**Vaccine** (vak-SENE) Substance used to produce active immunity; usually a suspension of attenuated or killed pathogens given by inoculation to prevent a specific disease

**Vagina** (vah-JI-nah) Lower part of the birth canal that opens to the outside of the body; female organ of sexual intercourse

**Valve** Structure that prevents fluid from flowing backward, as in the heart, veins, and lymphatic vessels

**Vas deferens** (DEF-er-enz) Tube that carries sperm cells from the testis to the urethra; ductus deferens

**Vasectomy** (vah-SEK-to-me) Surgical removal of part or all of the ductus (vas) deferens; usually done on both sides to produce sterility

**Vasoconstriction** (vas-o-kon-STRIK-shun) Decrease in the diameter of a blood vessel

**Vasodilation** (vas-o-di-LA-shun) Increase in the diameter of a blood vessel

**Vein** (vane) Vessel that carries blood toward the heart

**Vena cava** (VE-nah KA-vah) A large vein that carries blood into the right atrium of the heart; superior vena cava or inferior vena cava

**Venous sinus** (VE-nus SI-nus) Large channel that drains deoxygenated blood

**Ventilation** (ven-tih-LA-shun) Movement of air into and out of the lungs

**Ventral** (VEN-tral) Toward the front or belly surface; anterior

**Ventricle** (VEN-trih-kl) Cavity or chamber; one of the two lower chambers of the heart; one of the four chambers in the brain in which cerebrospinal fluid is produced; adj., ventricular (ven-TRIK-u-lar)

**Venule** (VEN-ule) Very small vein that collects blood from the capillaries

**Vertebra** (VER-teh-brah) A bone of the spinal column; pl., vertebrae (VER-teh-bre)

**Vesicle** (VES-ih-kl) Small sac or blister filled with fluid

**Vestibule** (VES-tih-bule) Part of the internal ear that contains receptors for the sense of equilibrium; any space at the entrance to a canal or organ

**Villi** (VIL-li) Small finger-like projections from the surface of a membrane; projections in the lining of the small intestine through which digested food is absorbed; sing., villus

**Viscera** (VIS-er-ah) Organs in the ventral body cavities, especially the abdominal organs

**Vitamin** (VI-tah-min) Organic compound needed in small amounts for health

**Vitreous** (VIT-re-us) **body** Soft, jellylike substance that fills the eyeball and holds the shape of the eye; vitreous humor

**White matter** Nervous tissue composed of myelinated fibers

**X-ray** Ray or radiation of extremely short wavelength that can penetrate opaque substances and affects photographic plates and fluorescent screens; an image made by means of such rays

**Zygote** (ZI-gote) Fertilized ovum; cell formed by the union of a sperm and an egg

# Biologic Terminology

Biologic terminology is based on an understanding of a relatively few basic elements. These elements—roots, prefixes, and suffixes—form the foundation of almost all biologic terms. A useful way to familiarize yourself with each term is to learn to pronounce it correctly and say it aloud several times. Soon it will become an integral part of your vocabulary.

The foundation of a word is the word root. Examples of word roots are *abdomin-*, referring to the belly region; and *aden-*, pertaining to a gland. A word root is often followed by a vowel to facilitate pronunciation, as in *abdomino-* and *adeno-*. We then refer to it as a "combining form." The hyphen appended to a combining form indicates that it is not a complete word; if the hyphen precedes the combining form, then it commonly appears as the terminal element of the word, as in *-cyte*, meaning "cell."

A prefix is a part of a word that precedes the word root and changes its meaning. For example, the prefix *sub-* in *subcutaneous* means "below." A suffix, or word ending, is a part that follows the word root and adds to or changes its meaning. The suffix *-ase* means "enzyme", as in *lipase*, an enzyme that digests fat.

Many biologic words are compound words; that is, they are made up of more than one root or combining form. Examples of such compound words are *histology* (study of tissues), *erythrocyte* (red blood cell), and many more difficult words, such as *sternoclavicular* (indicating relations to both the sternum and the clavicle).

A general knowledge of language structure and spelling rules is also helpful in mastering biologic terminology. For example, adjectives include words that end in *-al*, as in *sternal* (the noun is *sternum*), and words that end in *-ous*, as in *mucous* (the noun is *mucus*).

The following list includes some of the most commonly used word roots, combining forms, prefixes, and suffixes, as well as examples of their use.

**a-, an-**  absent, deficient, lack of: *amorphous, atrophy, anaerobic*

**ab-**  away from: *abduction, aboral*

**abdomin-, abdomino-**  belly or abdomen: *abdominal, abdominopelvic*

**acou-**  hearing, sound: *acoustic*

**acr-, acro-**  extreme ends of a part, especially of the extremities: *acromion, acrosome*

**ad-**  (sometimes converted to *ac-, af-, ag-, ap-, as-, at-,*) toward, added to, near: *adrenal, accretion, agglomerated, afferent*

**aden-, adeno-,**  gland: *adenectomy, adenoid, adenoma*

**-agogue**  inducing, leading, stimulating: *cholagogue, galactagogue*

**amb-, ambi-,**  both, on two sides: *ambidextrous, ambivalent*

**angio-**  vessel: *angiogram, angiotensin*

**ant-, anti-**  against; to prevent, suppress, or destroy: *antibiotic, anticoagulant, antihistamine*

**ante-**  before, ahead of: *antecedent, antenatal, antepartum*

**antero-**  position ahead of or in front of (*i.e.,* anterior to) another part: *anterolateral, anteroventral*

**arthr-, arthro-**  joint or articulation: *arthritis, arthroscope, arthrosis*

**-ase**  enzyme: *lipase, protease*

**audio-**  sound, hearing: *audiogenic, audiometry, audiovisual*

**aur-**  ear: *aural, auricle*

**aut-, auto-**  self: *autodigestion, autoimmune, autonomic*

**bi-**  two, twice: *bifurcate, bisexual*

**bio-**  life, living organism: *antibiotic, biochemistry, biology*

**blast-, blasto-, -blast**  early stage, immature cell or bud: *blastula, blastophore, erythroblast*

**bleph-, blephar-, blepharo-**  eyelid, eyelash: *blepharoptosis, blepharospasm*

**brachi-**   arm: *brachial, brachiocephalic*
**brady-**   slow: *bradycardia*
**bronch-**   windpipe or other air tubes: *bronchiole, bronchoscope*
**bucc-**   cheek: *buccal*

**cardi-, cardia-, cardio-**   heart: *cardiac, cardiologist, cardiovascular*
**centi-**   relating to 100 (used in naming units of measurements): *centigrade, centimeter*
**cephal-, cephalo-**   head: *cephalic, cephalopelvic*
**cervi-**   neck: *cervical, cervix*
**cheil-, cheilo-,**   lips; brim or edge: *cheilitis, cheilosis*
**cheir-, cheiro-**   (also written **chir-, chiro-**) hand: *cheiromegaly, chiropractic*
**chol-, chole-, cholo-**   bile, gall: *chologogue, cholecyst, cholecystokinin*
**chondr-, chondri-, chondrio-**   cartilage: *chondrocyte, hypochondrium, perichondrium*
**circum-**   around, surrounding: *circumorbital, circumrenal, circumduction*
**-clast**   break: *osteoclast*
**colp-, colpo-**   vagina: *colposcope, colpotomy*
**contra-**   opposed, against: *contraindication, contralateral*
**cost-, costa-, costo**   ribs: *intercostal, costosternal*
**counter-**   against, opposite to: *counteract, counterirritation, countertraction*
**crani-, cranio-**   skull: *craniosacral, cranium*
**crypt-, crypto-**   hidden, concealed: *cryptic, cryptogenic*
**cut-**   skin: *cuticle, subcutaneous*
**cysti-, cysto-**   sac, bladder: *cystic, cystitis, cystoscope*
**cyt-, cyto-, -cyte**   cell: *cytoplasm, syncytium, thrombocyte*

**dactyl-, dactylo-**   digits (usually fingers, but sometimes toes): *dactylitis, polydactyly*
**de-**   remove: *decompose, deoxygenate, detoxify*
**dendr-**   tree: *dendrite*
**dento-, dent-, denti-**   tooth: *dentition, dentin, dentifrice*
**derm-, derma-, dermo-, dermat-, dermato-**   skin: *dermatitis, dermatology, dermis*
**di-, diplo-**   twice, double: *dimorphism, diarthrosis*
**dia-**   through, between, across, apart: *diaphragm, diaphysis*
**dis-**   apart, away from: *dissect, dissolve, distal*
**dors-, dorsi-, dorso-**   back (in the human, this combining form refers to the same regions as postero-): *dorsiflexion, dorsonuchal*

**-ectasis**   expansion, dilation, stretching: *angiectasis, bronchiectasis*
**ecto-**   outside, external: *ectoderm, ectogenous, ectoplasm*
**-emia**   blood: *anemia, glycemia, hyperemia*
**encephal-, encephalo-**   brain: *diencephalon, encephalogram*
**end-, endo-**   within, innermost: *endarterial, endocardium, endothelium*

**enter-, entero-**   intestine: *enteric, mesentery*
**epi-**   on, upon: *epidermis, erythropoiesis*
**eryth-, erythro-**   red: *erythrocyte, erythropoiesis*
**-esthesia**   sensation: *anesthesia, paresthesia*
**eu-**   well, normal, good: *euphoria, eupnea*
**ex-, exo-**   outside, out of, away from: *excretion, exocrine, exophthalmic*
**extra-**   beyond, outside of, in addition to: *extracellular, extrasystole, extravascular*

**fasci-**   fibrous connective tissue layers: *fascia, fascicle*
**-ferent**   to bear, to carry: *afferent, efferent*
**fibr-, fibro-**   threadlike structures, fibers: *fibrillation, fibroblast, myofibril*

**galact-, galacta-, galacto-**   milk: *galactagogue, galactose*
**gastr-, gastro-**   stomach: *gastric, gastrointestinal*
**gen-, geno-**   a relationship to reproduction or sex: *genetic, genotype, regenerate*
**-gen**   an agent that produces or originates: *allergen, fibrinogen, pepsinogen*
**-genic**   produced from, producing: *endogenic, neurogenic*
**genito-**   organs of reproduction: *genitoplasty, genitourinary*
**-geny**   manner of origin, development or production: *ontogeny, progeny*
**glio-, -glia**   gluey material; specifically, the connective tissue of the central nervous system: *gliogenous, neuroglia*
**gloss-, glosso-**   tongue: *glossal, glossopharyngeal*
**gly-, glyco-**   sweet, relating to sugar: *glycogen, glycolytic, glycosuria*
**-gnath, gnatho-**   related to the jaw: *agnathous, prognathic*
**gon-**   seed, knee: *gonad, gonarthritis*
**-gram**   record, that which is recorded: *electrocardiogram, hologram, tomogram*
**-graph**   instrument for recording: *electroencephalograph, sonograph*
**gyn-, gyne-, gyneco-, gyno-**   female, woman: *gynecology, gynandromorph*

**hem-, hema-, hemato-, hemo-**   blood: *hematology, hemoglobin, hemostasis*
**hemi-**   one half: *hemisphere, hemilateral, hemiplegia*
**hepat-, hepato-**   liver: *hepatic, hepatogenous*
**heter-, hetero-**   other, different: *heterogeneous, heterosexual, heterozygous*
**hist-, histo-, histio-**   tissue: *histology, histiocyte*
**homeo-, homo-**   unchanging, the same: *hemeostasis, homosexual*
**hydr-, hydro-**   water: *dehydration, hydrolysis*
**hyper-**   above, over, excessive: *hyperglycemia, hypersecretion, hypertonic*
**hypo-**   deficient, below, beneath: *hypochondrium, hypodermic, hypogastrium*
**hyster-, hystero-**   uterus: *hysterectomy*

**-ia**　state of: *myopia, hypochondria, phobia*

**idio-**　self, one's own, separate, distinct: *idiopathic, idiosyncrasy*

**im-, in-**　in, into, lacking: *implantation, inanimate, infiltration*

**infra-**　below: *infraspinous, infracortical*

**inter-**　between: *intercostal, interstitial*

**intra-**　within a part or structure: *intracranial, intracellular, intraocular*

**iso-**　equal: *isotonic, isometric*

**juxta-**　next to: *juxtaglomerular*

**karyo-**　nucleus: *karyotype, karyoplasm*

**kerat-, kerato-**　cornea of the eye, certain horny tissues: *keratin, keratoplasty*

**lacri-**　tear: *lacrimal*

**lact-, lacto-**　milk: *lactation, lactogenic*

**later-**　side: *lateral*

**leuk-, leuko-**　(also written as *leuc-, leuco-*) white: *leukemia, leukocyte, leukoplakia*

**lig-**　bind: *ligament, ligature*

**-logy, -ology**　study of: *cytology, endocrinology, physiology*

**lyso-, -lysis**　flowing, loosening, dissolution (dissolving of): *hemolysis, paralysis, lysosome*

**macro-**　large, abnormal length: *macrophage, macroblast*. See also **-mega, megalo-**

**mast-, masto-**　breast: *mastectomy, mastitis*

**meg-, mega-, megal-, megalo-**　unusually or excessively large: *megacolon, megaloblast, megakaryocyte*

**men-, meno-**　physiologic uterine bleeding: *menses, menopause*

**mening-, meningo-**　membranes covering the brain and spinal cord: *meninges, meningitis*

**mes-, mesa-, meso-**　middle, midline: *mesencephalon, mesoderm*

**meta-**　change, beyond, after, over, near: *metabolism, metacarpus, metaplasia*

**-meter**　measure: *hemocytometer, sphygmomanometer*

**micro-**　very small: *microscope, microbiology, micrometer*

**mono-**　single: *monocular, monocyte, mononucleosis*

**my-, myo-**　muscle: *myenteron, myocardium, myometrium*

**myel-, myelo-**　marrow (often used in reference to the spinal cord): *myelin, myeloid, myeloblast, myoglobin*

**neo-**　new, strange: *neocortex, neonatal, neoplasm*

**neph-, nephro-**　kidney: *nephrology, nephron*

**neur-, neuro-**　nerve: *neurilemma, neuroglia, neuron*

**noct-, nocti-**　night: *noctambulation, noctiphobia, nocturnal*

**ocul-, oculo-**　eye: *ocular, oculist, oculomotor*

**odont-, odonto-**　tooth, teeth: *odontogenesis, orthodontics*

**-oid**　likeness, resemblance: *lymphoid, myeloid*

**olig-, oligo-**　few, a deficiency: *oligodendrocyte, oligospermia, oliguria*

**onych-, onycho-**　nails: *eponychium, hyponychium*

**oo-, ovi, ovo-**　ovum, egg: *oocyte, oviduct, ovoplasm* (do not confuse with **oophor-**)

**oophor-, oophoro-**　ovary: *oophorectomy, oophoritis, oophoron*. See also **ovar-**

**ophthalm-, ophthalmo-**　eye: *ophthalmia, ophthalmologist, ophthalmoscope*

**orth-, ortho-**　straight, normal: *orthopedics, orthopnea, orthosis*

**oscillo-**　to swing to and fro: *oscilloscope*

**oss-, osseo-, ossi-, oste-, osteo-**　bone, bone tissue: *osseous, ossicle, osteocyte*

**ot-, oto-**　ear: *otolith, otology*

**ovar-, ovario-**　ovary: *ovarian, ovariotomy*. See also **oophor-**

**ox-, -oxia**　pertaining to oxygen: *anoxia, hypoxemia*

**para-**　near, beyond, apart from, beside: *paramedical, parametrium, parathyroid, parasagittal*

**ped-, pedia-**　child, foot: *pedal, pedicel, pediatrician*

**per-**　through, excessively: *percutaneous, perfusion*

**peri-**　around: *pericardium, perichondrium*

**phag-, phago-**　to eat, to ingest: *phage, phagocyte*

**-phagia, -phagy**　eating, swallowing: *aphagia, dysphagia*

**-phasia**　speech, ability to talk: *aphasia, dysphasia*

**-phil, -philic**　to be fond of, to like (have an affinity for): *eosinophil, chromophilic, hydrophilic*

**phot-, photo-**　light: *photoreceptor, photophobia*

**pile-, pili-, pilo-**　hair, resembling hair: *pileous, piliation, pilonidal*

**pleur-, pleuro-**　pleura; serous membrane covering the lung and lining the chest cavity: *pleural, pleurisy*

**-pnea**　air, breathing: *dyspnea, eupnea*

**pneum-, pneuma-, pneumo-, pneumon-, pneumato-**　lung, air: *pneumatic, pneumograph, pneumonia*

**pod-, podo-**　foot: *podiatry, podocyte*

**-poiesis**　making, forming: *erythropoiesis, hematopoiesis*

**poly-**　many: *polychromic, polymer, polymorphic*

**post-**　behind, after, following: *postnatal, postocular, postpartum*

**pre-**　before, ahead of: *premature, premolar, prenatal*

**pro-**　in front of, before: *prosencephalon, prolactin, prothrombin*

**proct-, procto-**　rectum: *proctoscope, proctologist*

**pseud-, pseudo-**　false: *pseudopod, pseudostratified*

**psych-, psycho-**　mind: *psychosomatic, psychotherapy*

**pulmo-, pulmono-**　lung: *pulmonic, pulmonology*

**pyel-, pyelo-**　kidney pelvis: *pyelitis, pyelogram*

**rachi-, rachio-**　spine: *rachidial, rachicentesis*

**radio-** emission of rays or radiation: *radioactive, radiography, radiology*

**re-** again, back: *reabsorption, reaction, regenerate*

**ren-** kidney: *renal, renin*

**retro-** backward, located behind: *retrocecal, retroperitoneal*

**rheo-** flow of matter, or of a current of electricity: *rheology, rheoscope*

**rhin-, rhino-** nose: *rhinencephalon, rhinitis, rhinoplasty*

**-rrhage, -rrhagia** excessive flow: *hemorrhage, menorrhagia*

**salping-, salpingo-** tube: *salpingitis, salpingoscopy*

**scler-, sclero-** hardness; *sclera, sclerosis*

**-scope** instrument used to look into or examine a part: *bronchoscope, cystoscope, endoscope*

**semi-** mild, partial, half: *semiflexion, semilunar, semipermeable*

**-sis** condition or process, usually abnormal: *dermatosis, osteoporosis*

**somat-, somato-** body: *somatic, somatotype*

**sono-** sound: *sonogram, sonography*

**splanchn-, splanchno-** internal organs: *splanchnic, splanchnoptosis*

**sta-, stat-** stop, stand still, remain at rest: *stasis, static*

**sthen-, stheno-, -sthenia, -sthenic** strength: *asthenic, calisthenics*

**stomato-** mouth: *stomatitis*

**sub-** under, below, near, almost: *subclavian, subcutaneous, subluxation*

**super-** over, above, excessive: *superego, supernatant, superficial*

**supra-** location above or over: *supranasal, suprarenal*

**sym-, syn-** with, together: *symbiotic, symphysis, synapse*

**tacho-, tachy-** rapid, fast, swift: *tachycardia, tachypnea*

**tars-, tarso-** eyelid, foot: *metatarsal, tarsitis, tarsoplasty*

**tax-, -taxia, -taxis** order, arrangement: *chemotaxis, taxonomy*

**tens-** stretch, pull: *extension, tensor*

**therm-, thermo-, -thermy** heat: *thermometer, thermotaxis*

**tox-, toxic-, toxico-** poison: *toxemia, toxicology, toxin*

**trache-, tracheo-** windpipe: *trachea, tracheitis, tracheotomy*

**trans-** across, through, beyond: *transorbital, transpiration, transplant, transport*

**tri-** three: *triad, triceps*

**trich-, tricho-** hair: *trichiasis, trichosis*

**-trophic, -trophy** nutrition (nurture): *atrophic, hypertrophy*

**-tropic** turning around, influencing, changing: *adrenotropic, gonadotropic, thyrotropic*

**ultra-** beyond or excessive: *ultrasound, ultraviolent, ultrastructure*

**uni-** one: *unilateral, uniovular, unicellular*

**-uria** urine: *glycosuria, hematuria, pyuria*

**vas-, vaso-** vessel, duct: *vascular, vasectomy, vasodilation*

**viscer-, viscero-** internal organs: *viscera, visceroptosis*

# Appendices

**APPENDIX 1.** Metric Measurements

| Unit | Abbreviation | Metric Equivalent | U.S. Equivalent |
|------|--------------|-------------------|-----------------|
| **Units of length** | | | |
| Kilometer | km | 1000 meters | 0.62 miles; 1.6 km/mile |
| Meter* | m | 100 cm; 1000 mm | 39.4 inches, 1.1 yards |
| Centimeter | cm | $1/100$ m; 0.01 m | 0.39 inches; 2.5 cm/inch |
| Millimeter | mm | $1/1000$ m; 0.001 m | 0.039 inches; 25 mm/inch |
| Micrometer | μm | $1/1000$ mm; 0.001 mm | |
| **Units of weight** | | | |
| Kilogram | kg | 1000 g | 2.2 lb |
| Gram* | g | 1000 mg | 0.035 oz.; 28.5 g/oz |
| Milligram | mg | $1/1000$ g; 0.001 g | |
| Microgram | μg | $1/1000$ mg; 0.001 mg | |
| **Units of volume** | | | |
| Liter* | L | 1000 mL | 1.06 qt |
| Deciliter | dL | $1/10$ L; 0.1 L | |
| Milliliter | mL | $1/1000$ L; 0.001 L | 0.034 oz., 29.4 mL/oz |
| Microliter | μL | $1/1000$ mL; 0.001 mL | |

*Basic unit

## APPENDIX 2. Celsius–Fahrenheit Temperature Conversion Scale

| Celsius to Fahrenheit | Fahrenheit to Celsius |
|---|---|

Use the following formula to convert Celsius readings to Fahrenheit readings:

$°F = 9/5°C + 32$

For example, if the Celsius reading is 37°:

$°F = (9/5 × 37) + 32$
$= 66.6 + 32$
$= 98.6°F$ (normal body temperature)

Use the following formula to convert Fahrenheit readings to Celsius readings:

$°C = 5/9 (°F − 32)$

For example, if the Fahrenheit reading is 68°:

$°C = 5/9 (68 − 32)$
$= 5/9 × 36$
$= 20°C$ (a nice spring day)

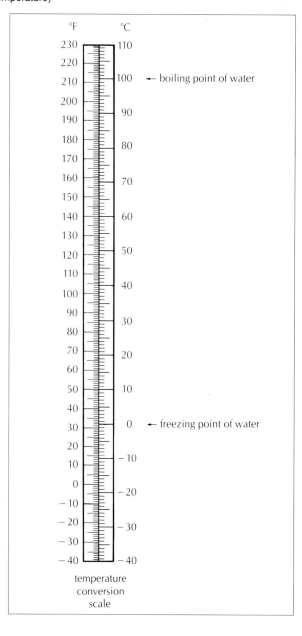

temperature
conversion
scale

# Index

The letter *f* following a page number indicates a figure; *n* denotes a footnote; and *t* indicates tabular material.

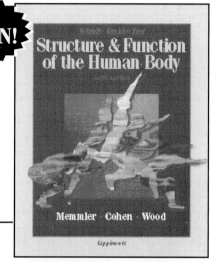